CAMBRIDGE MONOGRAPHS ON PHYSICS

GENERAL EDITORS

M. M. Woolfson, D.SC.

Professor of Theoretical Physics, University of York

J. M. Ziman, D.PHIL., F.R.S.

Professor of Theoretical Physics, University of Bristol

LIQUID STATE PHYSICS – A STATISTICAL MECHANICAL INTRODUCTION

LIQUID STATE PHYSICS–
A STATISTICAL MECHANICAL
INTRODUCTION

CLIVE A. CROXTON

Fellow of Jesus College, Cambridge

CAMBRIDGE UNIVERSITY PRESS

Published by the Syndics of the Cambridge University Press
Bentley House, 200 Euston Road, London NW1 2DB
American Branch: 32 East 57th Street, New York, N.Y.10022

Library of Congress Catalogue Card Number: 72–89803

ISBN: 0 521 20117 9

First published 1974

Printed in Great Britain
at the University Printing House, Cambridge
(Brooke Crutchley, University Printer)

CONTENTS

CHAPTER 3

Numerical solution of the integral equations

CHAPTER 4

The liquid surface

CHAPTER 5

Numerical methods in the theory of liquids

CHAPTER 6

Transport processes

PREFACE

Liquid state physics no longer has the luxury status of an intellectual plaything – a kind of purgatory between gas and solid, a statistical mechanical jungle populated only by the foolhardy and/or academics. Pressing demands are increasingly being made upon the subject by adjacent branches of physics, and many people are being unwittingly drawn into the field from more conventional and immediately rewarding routes in physics. A subject which has not, as yet, satisfactorily accounted for the solid–fluid transition, nor which has yielded more than three further hard sphere virial coefficients in the 75 years since Boltzmann's original calculations, is evidently beset with problems: but there lies the source of the fascination.

This book is meant to be a guide to the uninitiated, and perhaps to broaden the outlook of those already there. It represents a very personal account of my own journey into the subject, and in consequence the content and approach might be considered in some respects somewhat idiosyncratic. I have made some attempt to present the material objectively, although inevitably one's own particular interests and viewpoints should and do assert themselves. Nonetheless, I felt the time was ripe to include separate chapters on the liquid surface and on the machine simulation methods, and to expand the by now routine statistical mechanical developments of the pair distribution associated with the names of Kirkwood, Born and coworkers, Bogolyubov, Percus and Yevick, and others. Of necessity some selection is inevitable, and I have arbitrarily restricted the discussion to the non-critical domain of the so-called 'simple' liquids, including the quantum liquids to a certain extent, although liquid water does receive a brief mention. I have tried throughout the book to stop off and identify the physics, and at all costs to avoid the sterility which so often tends to go hand in hand with the mathematics.

The majority of the material in this book has been presented in the form of graduate lectures at the Cavendish Laboratory, Cambridge. It has also been presented at the Centre for Advanced

Studies in Physics, Delhi, and at the Raman Research Institute, Bangalore, whilst on a British Council Fellowship. I am grateful to all those who provided helpful comments and illuminating discussions on those occasions. I am particularly grateful to John Ziman for his encouragement and kind words in the early, difficult days of the manuscript; to all those at the Press who have been involved in the production of this book; to Norman March, Bob Ferrier and Denis White for their generous support and encouragement at every stage of my career in liquid state physics; to the Raman Research Institute, Bangalore, for their extremely generous hospitality; to the Division of Applied Chemistry, NRC, Ottawa; to Paul Barnes for countless discussions – not all of them about physics; and to Sarah, my wife, to whom this book is dedicated.

C. A. CROXTON

Cavendish Laboratory
October 1973

CHAPTER I

THEORY OF IMPERFECT GASES

1.1 Introduction

All the thermodynamic properties of a system are determinable from a knowledge of the partition function. For an equilibrium one-component quasi-classical system the N-body partition function is given in terms of the phase integral as

$$Z_N = \frac{1}{N!\,h_0^{3N}} \int \exp(-\beta H)\,\mathrm{d}\Omega, \qquad (1.1.1)$$

where the Hamiltonian of the system is

$$H = \sum_{i=1}^{N} \frac{\boldsymbol{p}_i^2}{2m} + \Phi_N(\boldsymbol{q}_1, \boldsymbol{q}_2, \ldots, \boldsymbol{q}_N), \qquad (1.1.2)$$

and provided it is separable into its kinetic and configurational components as (1.1.2) implies, then determination of Z_N reduces to evaluation of the configurational integral Z_Q

$$Z_N = \left(\frac{2\pi mkT}{h_0^2}\right)^{3N/2} Z_Q, \qquad (1.1.3)$$

where
$$Z_Q = \frac{1}{N!}\int \exp[-\beta\Phi_N(\boldsymbol{q}_1, \boldsymbol{q}_2, \ldots, \boldsymbol{q}_N)]\,\mathrm{d}\Omega. \qquad (1.1.4)$$

This is where the problem begins, for what do we write for

$$\Phi_N(\boldsymbol{q}_1, \boldsymbol{q}_2, \ldots, \boldsymbol{q}_N)?$$

The total configurational energy is, of course, dependent upon the total configuration $\{\boldsymbol{q}_1, \boldsymbol{q}_2, \ldots, \boldsymbol{q}_N\}$, and this is quite unknown. A certain simplification is obtained by assuming pairwise additivity of the interatomic potential, such that

$$\Phi_N(\boldsymbol{q}_1, \boldsymbol{q}_2, \ldots, \boldsymbol{q}_N) = \sum_{i>j}^{N}\sum^{N} \Phi(|\boldsymbol{q}_i - \boldsymbol{q}_j|), \qquad (1.1.5)$$

where $\Phi(|\boldsymbol{q}_i - \boldsymbol{q}_j|)$ is the potential developed between the pair ij. However, if we consider the physical situation a little more closely we find we are able to effect a series solution for Z_N, or Z_Q in particular, where the various terms have immediately identifiable physical roles.

1.2 The cluster expansions

The cluster expansion of the classical partition function is an important conceptual tool in the mathematical analysis of dense fluids as we shall see later. Its physical significance and immediate applicability is most apparent in the case of low density imperfect gases, and we introduce it here in terms of the Mayer f-function[1], although there are several variations[2–5] on the cluster expansion method.

We take advantage of the pair potential expression of Φ_N given in (1.1.5). Under these circumstances

$$\exp(-\beta\Phi_N) = \prod_{i>j}^{N}\prod^{N} \exp\left(-\frac{\Phi}{kT}(r_{ij})\right). \qquad (1.2.1)$$

We now define the Mayer f-function as

$$f(r_{ij}) = \exp\left(-\frac{\Phi}{kT}(r_{ij})\right) - 1. \qquad (1.2.2)$$

Notice that f_{ij} is a short range function, of the order of the range of the pair potential and is practically zero unless two or more molecules are close to each other. Further it is a bounded function; that is as $r_{ij} \to 0$, so $f_{ij} \to -1$ even though the repulsive forces between atoms cause $\Phi(r_{ij})$ to become large and positive.

From (1.2.2), the N-body partition function may be written

$$Z_N = \frac{1}{N!}\left(\frac{2\pi mkT}{h_0^2}\right)^{3N/2} \int \dots \int \prod_i \prod_j (1+f_{ij})\, d\mathbf{1}\dots d\mathbf{N}. \qquad (1.2.3)$$

Here we have used $\int \dots \int d\mathbf{1}\dots d\mathbf{N}$ to signify integration over all positions of molecules $1 \dots N$.

Expansion of the integrand gives

$$\prod_{n\geqslant i>j\geqslant 1}\prod (1+f_{ij}) = 1 + \sum_i\sum_j f_{ij} + \sum_i\sum_j\sum_k\sum_l f_{ij}f_{kl} + \dots$$
$$+\Sigma\dots\Sigma\dots\Sigma\, f_{ij}f_{kl}\dots f_{mn}\dots + \qquad (1.2.4)$$

The general product in this series consists of the sum over all connected products of the same order, n. By this we mean the sum over all n-body clusters which are topologically distinct. The expansion (1.2.4) can be split up and reassembled in *irreducible clusters* of the following forms:

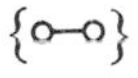 2-body clusters between molecules i, j,

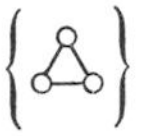 3-body clusters between molecules i, j, k, (1.2.5)

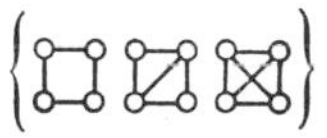 4-body clusters between molecules i, j, k, l

and so on. An irreducible cluster is defined by a mapping or graph in which each point or molecule is connected by an f-bond to at least two other points. For completeness, the exceptional case of two interacting molecules is included as the lowest irreducible cluster. Thus

are not irreducible clusters since the first can be decomposed into two 2-body irreducible clusters, the second into a 3-body and a 2-body irreducible cluster, and the third into two 3-body irreducible clusters.

The number of topologically distinct clusters in each n-body subset increases rapidly with n. Thus there are three 4-body clusters, ten 5-body clusters, and fifty-six 6-body clusters. A recent development by Ree and Hoover[6] enables a drastic reduction to be made in the number of clusters in each subset which need to be evaluated, thereby simplifying the evaluation of the cluster integrals. We shall return to Ree and Hoover's work later.

Each cluster within a given subset is weighted according to the frequency of its occurrence in the expansion (1.2.4). The combinatorial algebra necessary to establish the general expansion (1.2.4) in terms of the correctly weighted n-body subsets (1.2.5) is formidable, but is discussed in several reviews[1,4,7]. Further, the n-body subset must be divided by $n!$ to account for the indistinguishability of topologically identical permutations of the n molecules in the cluster. Cyclic permutation of the labels on the 3-body subset

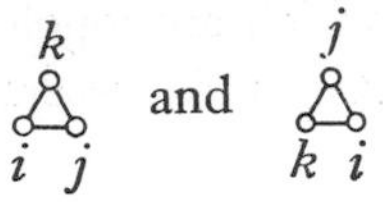

does not produce a topologically distinct diagram. In the evaluation of the quasi-classical partition function we must only sum over physically distinct states. Otherwise we shall be confronted with a discrepancy between the predicted and observed entropies and free energies – the Gibbs paradox.

The partition function in the cluster expansion may then be written, from (1.2.3), (1.2.4):

$$Z_N = \left(\frac{2\pi mkT}{h_0^2}\right)^{3N/2} \int \cdots \int \left\{1 + \frac{1}{2!}\frac{N!}{(N-2)!} \circ\!\!-\!\!\circ + \frac{N!}{3!(N-3)!} \triangle + \frac{N!}{4!(N-4)!}\left[3 \square + 6 \boxslash + \boxtimes\right] \cdots \right\} \mathrm{d}1 \ldots \mathrm{d}\boldsymbol{N}. \tag{1.2.6}$$

The coefficients of the various subsets

$$N!/(N-2)!, N!/(N-3)!, \ldots, N!/(N-n)!$$

represent the number of ways the n-body cluster may be formed from the complete set of N molecules. Notice the absence of the quasi-classical indistinguishability factor $(N!)^{-1}$ from the right hand side of (1.2.6): indistinguishability is being taken care of termwise in the cluster expansion. The weighting factors are given in the 4-body subset. Integration over the expansion yields

$$Z_N = \left(\frac{2\pi mkT}{h_0^2}\right)^{3N/2} \left\{V^N + \frac{N}{2!}(N-1)\,V^{N-1} \bullet\!\!-\!\!\bullet + \frac{N}{3!}(N-1)(N-2)V^{N-2} \blacktriangle + \ldots\right\}, \tag{1.2.7}$$

where the *cluster integrals* are defined as follows:

$$\left.\begin{aligned}
\underset{1\ \ 2}{\bullet\!\!-\!\!\bullet} &\equiv \int \circ\!\!-\!\!\circ \;\mathrm{d}2 = \int f_{12}\,\mathrm{d}2,\\
\underset{1\ \ 2}{\overset{3}{\blacktriangle}} &\equiv \iint \triangle \;\mathrm{d}2\,\mathrm{d}3 = \iint f_{12}f_{23}f_{31}\,\mathrm{d}2\,\mathrm{d}3,\\
\underset{1\ \ 2}{\overset{4\ \ 3}{\boxslash}} &\equiv \iiint \boxslash \;\mathrm{d}2\,\mathrm{d}3\,\mathrm{d}4 = \iiint f_{12}f_{13}f_{14}f_{23}f_{34}\,\mathrm{d}2\,\mathrm{d}3\,\mathrm{d}4.
\end{aligned}\right\} \tag{1.2.8}$$

Then,

$$\ln Z_N = N \ln V + \ln\left\{1 + \frac{N(N-1)}{2!\,V} \bullet\!\!-\!\!\bullet + \frac{N(N-1)(N-2)}{3!\,V^2} \triangle + \dots\right\}$$
$$+\frac{3N}{2}\ln(2\pi mkT/h_0^2). \qquad (1.2.9)$$

The equation of state is determined from the usual relationship

$$P = \frac{1}{\beta}\left(\frac{\partial(\ln Z_N)}{\partial V}\right)_T, \quad \text{where} \quad \beta = (kT)^{-1}. \qquad (1.2.10)$$

Since N is very large we may write N^2 for $N(N-1)$, etc. and for $x \ll 1$, $\ln(1+x) \sim x$, whereupon from (1.2.9), (1.2.10)

$$P = \frac{NkT}{V} - \frac{N^2kT}{2!\,V^2} \bullet\!\!-\!\!\bullet - \frac{2N^3kT}{3!\,V^3} \triangle - \dots$$
$$= \frac{NkT}{V}\left\{1 - \frac{N}{V}\frac{1}{2!} \bullet\!\!-\!\!\bullet - \frac{N^2}{V^2}\frac{2}{3!} \triangle - \dots\right\} \qquad (1.2.11)$$

which is, of course, the *virial expansion*:

$$P = \frac{NkT}{V}\left\{1 + \frac{N}{V}B_2(T) + \frac{N^2}{V^2}B_3(T) + \dots\right\}, \qquad (1.2.12)$$

where $B_2(T), B_3(T), \dots$ are the *virial coefficients*.

We should point out that in applying the approximation

$$\ln(1+x) \sim x$$

to (1.2.9), we require not only that the density N/V is small, but also that the *quantity* N is also small.

Identifying (1.2.11) and (1.2.12) we have the following cluster expansion for P:

$$\frac{P}{\rho kT} = 1 \quad \sum_{l=2} \frac{l-1}{l}\beta_l \rho^{l-1}, \qquad (1.2.13)$$

where $$\beta_l = \frac{1}{(l-1)!}\int \dots \int (\Sigma\Pi f_{ij})_\beta \, d\mathbf{2} \dots d\mathbf{l}, \qquad (1.2.14)$$

the sum of products of f_{ij} being taken over irreducible clusters. The physical significance of the second and subsequent terms in the virial expansion may clearly be attributed to the existence of clusters of 2, 3, 4, ... molecules in simultaneous interaction. This is not to say, of course, that the clusters of molecules remain in permanent association: they will in fact be continuously forming and collisionally disrupting. But as the density of the gas increases the probability of an n-body encounter increases, and the gas departs from idealism.

This approach to the equation of state depends essentially upon ρ being small, that is, below the critical density. As the density increases the convergence deteriorates, and it has now been confirmed that the expansion diverges at high densities and is inapplicable as such at liquid densities(8). The radius of convergence of this series has been investigated by Penrose, and by Lebowitz and Penrose(8).

1.3 The Ursell development

Historically earlier than the Mayer f-function cluster expansion of the partition function, the Ursell expansion of the configurational integral in terms of $\mathcal{U}$-functions again enables the partition function to be written as a series of cluster integrals(2–5). The Ursell approach is, however, more general in that it does not assume additivity of the pair potential. Further this method may be extended to include quantum mechanical systems. The Mayer f-function method appears as a special case of the Ursell development.

Ursell(5) has shown that the N-body Boltzmann factor appearing in the configurational integral may be expressed as a sum of products of $\mathcal{U}$ functions:

$$W(1 \dots N) = \exp(-\Phi_N/kT), \tag{1.3.1}$$

where

$$\left.\begin{aligned} \mathcal{U}(1) &= W(1), \\ \mathcal{U}(12) &= W(12) - W(1)\,W(2), \\ \mathcal{U}(123) &= W(123) - W(12)\,W(3) - W(23)\,W(1) - W(31)\,W(2) \\ &\quad + 2W(1)\,W(2)\,W(3), \\ &\dots \end{aligned}\right\} \tag{1.3.2}$$

It is important to note that the Boltzmann expansion of the n-body $\mathcal{U}$-function *does not* imply permutation of the n molecules within the group. In the case of $\mathscr{U}(123)$ the expansion is expressed in terms of one 3-body cluster, 3 groups of 2-body and 1-body clusters, and one group of 1-body clusters. The coefficients before the various terms are $(-1)^{n-1}(n-1)!$ where n is the number of groups in the term. There are several properties of this expansion which should be noted. Groups of single particle clusters such as $W(1)$, and $W(1)W(2)$, physically represent the Boltzmann factors of isolated non-interacting particles whose potential may be taken as zero, in which case, $W(1) = W(1)W(2) = \dots = 1$. Further the function $\mathscr{U}(1 \dots k)$ is zero for separated configurations; i.e. for all configurations wherein the k molecules are separated into two or more groups at such distances that no interactive coupling exists between the molecules of the several groups. In a dilute gas, for example, where 2-body collisions are frequent but 3-body collisions are rare, the functions $\mathscr{U}(123)$ vanish, as do all $\mathscr{U}$-functions of higher order. The N-body Boltzmann factor $W(1 \dots N)$ would then be expressed entirely in terms of one and two particle $\mathscr{U}$-functions. And if we were only interested in the second virial coefficient involving binary collisions, we should discard all $\mathscr{U}$-functions involving three or more molecules.

The Boltzmann factors for $1, 2, 3, \dots$ particles in terms of the functions may be obtained by inverting (1.3.2) giving

$$\left.\begin{aligned} W(1) &= \mathscr{U}(1) = 1, \\ W(12) &= \mathscr{U}(12) + \mathscr{U}(1)\mathscr{U}(2), \\ W(123) &= \mathscr{U}(123) + \mathscr{U}(12)\mathscr{U}(3) + \mathscr{U}(23)\mathscr{U}(1) + \mathscr{U}(31)\mathscr{U}(2) \\ &\quad + \mathscr{U}(1)\mathscr{U}(2)\mathscr{U}(3), \\ &\quad \dots. \end{aligned}\right\}$$

$$(1.3.3)$$

We can express (1.3.3) diagrammatically, where the bond now represents the $\mathscr{U}$-function, and the unbonded circles represent the isolated particles:

$$\left.\begin{aligned} W(1) &= \circ, \\ W(12) &= \circ\!-\!\circ + \circ \ \circ, \\ W(123) &= \triangle + 3\,\angle\circ + \circ\ \circ\ \circ, \end{aligned}\right\}$$

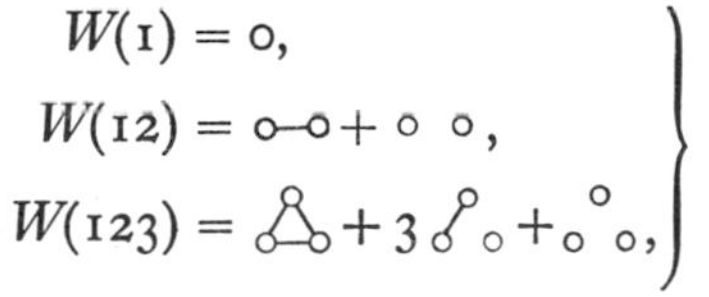

$$(1.3.4)$$

where the factor 3 accounts for the degeneracy of the diagram.

The *distinguishable* N-body Boltzmann factor is then

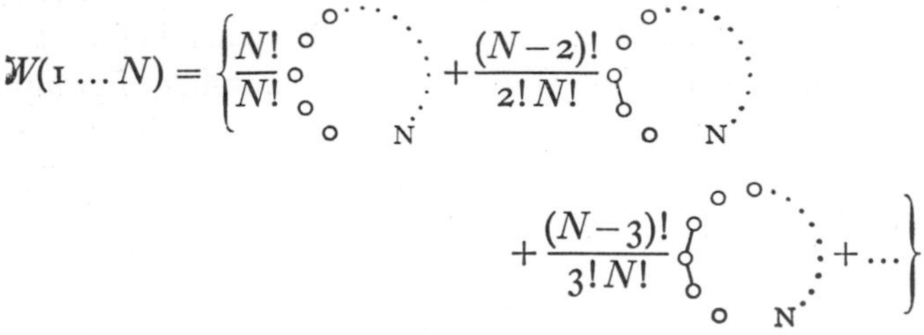

$$W(1 \dots N) = \left\{ \frac{N!}{N!} \overset{\circ}{\underset{\circ}{\circ}} \cdots + \frac{(N-2)!}{2!N!} \overset{\circ}{\underset{\circ}{\circ}} \cdots \right.$$

$$\left. + \frac{(N-3)!}{3!N!} \overset{\circ}{\underset{\circ}{\circ}} \cdots + \dots \right\}. \quad (1.3.5)$$

The number of ways of arranging N molecules n at a time is $N!/(N-n)!$, and the number of arrangements of the n within themselves is $n!$. Since these are topologically identical configurations and in the partition function we must sum only over physically distinct states, we must divide by this figure, giving the factors in (1.3.5). The schematic clusters o—o, o—o—o, ... represent the sum of *all* two-, three-, ... body clusters which develop in the series – we shall return to this point shortly. The unbonded molecules, since they are physically excluded from the interaction, do not contribute to the cluster integrals. The configurational N-body partition function Z_Q is obtained by integrating (1.3.5) over the whole of the volume V available to each of the N molecules, thus:

$$Z_Q = \frac{1}{N!} \int W(1 \dots N) \, d1 \dots dN$$

$$= \left\{ V^N + \frac{N(N-1)}{2!} V^{N-1} \bullet\!\!-\!\!\bullet \right.$$

$$\left. + \frac{N(N-1)(N-2)}{3!} V^{N-2} \bullet\!\!-\!\!\bullet\!\!-\!\!\bullet + \dots \right\}. \quad (1.3.6)$$

The full circles represent integration over the bonded cluster. The V^{N-n} arise from integration over the $(N-n+1)$ isolated centres, relative to molecule 1.

In the course of the expansion (1.3.5), some regrouping has been done to incorporate into the 3-body cluster, for example, such terms as $\triangle$ which do not appear in (1.3.4). These terms have arisen from a special case of the four-body ($ijkl$) cluster for which,

say, $j = k$. As in the case of the f-function analysis, the combinatorial algebra leading to the reclassification in (1.3.5) is considerable.

Writing

$$Z_Q = \frac{V^N}{N!}\left\{1 + \frac{N^2}{2!\,V}\bullet\!\!-\!\!\bullet + \frac{N^3}{3!\,V^2}\bullet\!\!-\!\!\bullet\!\!-\!\!\bullet + \ldots\right\}, \qquad (1.3.7)$$

then the equation of state follows directly:

$$\begin{aligned} P &= kT\left(\frac{\partial(\ln Z_Q)}{\partial V}\right)_T \\ &= \frac{NkT}{V}\left\{1 - \frac{N}{2!\,V}\bullet\!\!-\!\!\bullet - \frac{2N^2}{3!\,V^2}\bullet\!\!-\!\!\bullet\!\!-\!\!\bullet - \ldots\right\}, \end{aligned} \qquad (1.3.8)$$

where the virial coefficients are given by

$$B_2(T) = -\frac{1}{2!}\bullet\!\!-\!\!\bullet$$

$$B_3(T) = -\frac{2}{3!}\bullet\!\!-\!\!\bullet\!\!-\!\!\bullet, \quad \text{etc.} \qquad (1.3.9)$$

It now remains to identify the cluster integrals $\bullet\!\!-\!\!\bullet$, $\bullet\!\!-\!\!\bullet\!\!-\!\!\bullet$, ... over the $\mathscr{U}$-functions. The 3-body cluster integral is, for example, given by

$$\frac{2}{3!}\iint \mathscr{U}'(123)\,\mathrm{d}2\,\mathrm{d}3, \qquad (1.3.10)$$

where integration is relative to 1. The prime signifies the *augmented* $\mathscr{U}$-function since it incorporates terms of the form  arising during the regrouping in (1.3.5). Referring to (1.3.2), and recalling that the single particle Boltzmann factors are unity, it may be quite easily shown that

$$\begin{aligned} \mathscr{U}'(123) = {} & W(123) - W(12)\,W(13) - W(23)\,W(12) \\ & - W(13)\,W(23) + W(12) + W(13) + W(23) - 1. \end{aligned} \qquad (1.3.11)$$

If we define the l-body Ursell function as $\mathscr{U}\{l\}$, then it may be shown, quite generally, that the virial coefficients are directly expressible in terms of these cluster integrals, the jth virial coefficient being given as a combination of the cluster integrals $b_1, b_2, \ldots, b_j$, where

$$b_l = \frac{1}{l!}\int\ldots\int \mathscr{U}(1, 2, \ldots, l)\,\mathrm{d}2\ldots\mathrm{d}l. \qquad (1.3.12)$$

For example, the second virial coefficient $B_2(T)$ is, from (1.3.2),

$$B_2(T) = -\frac{1}{2!}\int \mathcal{U}'(12)\,\mathrm{d}\mathbf{2} = -\frac{1}{2!}\int [W(12)-1]\,\mathrm{d}\mathbf{2}$$

$$= -\frac{1}{2!}2b_2 = -\tfrac{1}{2}\beta_2 \tag{1.3.13}$$

and the third virial coefficient,

$$B_3(T) = -\frac{2}{3!}\iint \mathcal{U}'(123)\,\mathrm{d}\mathbf{2}\,\mathrm{d}\mathbf{3} = -\frac{2}{3!}(3b_3-6b_2^2) = -\tfrac{2}{3}\beta_3 \tag{1.3.14}$$

and it may be shown (the complexity of the combinatorial algebra here becoming apparent),

$$B_4(T) = -\frac{3}{4!}\iiint \mathcal{U}'(1234)\,\mathrm{d}\mathbf{2}\,\mathrm{d}\mathbf{3}\,\mathrm{d}\mathbf{4}$$

$$= -\frac{3}{4!}(4b_4-24b_2b_3+\tfrac{80}{3}b_2^3) = -\tfrac{3}{4}\beta_4. \tag{1.3.15}$$

The virial equation of state may therefore be written

$$\frac{P}{\rho kT} = 1 - \sum_{l=2}\frac{(l-1)}{l}\beta_l\rho^{l-1}, \tag{1.3.16}$$

where
$$\beta_l = \frac{1}{(l-1)!}\int\cdots\int \mathcal{U}'(1\ldots l)\,\mathrm{d}\mathbf{2}\ldots\mathrm{d}\boldsymbol{l} \tag{1.3.17}$$

(compare this with b_l, (1.3.12)). Equation (1.3.16) is seen to be identical to the Mayer f-function virial expression (1.2.14), provided the identification

$$\int\cdots\int \mathcal{U}'(1\ldots l)\,\mathrm{d}\mathbf{2}\ldots\mathrm{d}\boldsymbol{l} = \int\cdots\int(\Sigma\Pi f_{ij})\,\mathrm{d}\mathbf{2}\ldots\mathrm{d}\boldsymbol{l} \tag{1.3.18}$$

is made. Clearly, $(\Sigma\Pi f_{ij})$ represents $\mathcal{U}'(1\ldots l)$ in the approximation of pairwise additivity. Noting that

$$f_{ij} = \exp\left(-\frac{\Phi_{ij}}{kT}\right) - 1 = W(ij)-1, \tag{1.3.19}$$

then the 3-body Mayer expansion is, from the right hand side of (1.3.18),

$$\begin{aligned}(W(12)-1)&(W(23)-1)(W(31)-1)\\ &= W(12)\,W(23)\,W(31) - W(12)\,W(31) - W(23)\,W(12)\\ &\quad - W(23)\,W(31) + W(12) + W(23) + W(31) - 1\end{aligned} \tag{1.3.20}$$

which is indeed seen to be the 3-body augmented $\mathscr{U}'$-function defined in (1.3.11), except that the 3-body function is replaced by the triple product of 2-body functions:

$$W(123) \sim W(12)\,W(23)\,W(31). \qquad (1.3.21)$$

This is a very important and convenient approximation, and has been utilized extensively in liquid state physics where it is generally known as the *superposition approximation*, or *Kirkwood closure*, for reasons which will become apparent in the next chapter on dense fluids. From (1.3.19) it is clear that in terms of the potential function, (1.3.21) implies

$$\Phi(123) = \Phi(12)+\Phi(23)+\Phi(31) \qquad (1.3.22)$$

which is exact for an isolated 3-body interaction to the extent of the assumed additivity of the pair potentials.

It has been pointed out that the Ursell development may be simply extended to include quantum mechanical systems. The virial coefficients in classical statistical mechanics are given in terms of the Boltzmann factors, $W(1\,\ldots\,l)$; in quantum mechanical theory of the equation of state the Boltzmann factor is replaced by its quantum mechanical analogue, the Slater sum, $\mathscr{W}(1\,\ldots\,l)$. Accordingly, we may write for the second virial coefficient (cf. (1.3.13))

$$B_2(T) = -\frac{1}{2!}\int (\mathscr{W}(12)-1)\,d2. \qquad (1.3.23)$$

The Slater sum and the Boltzmann factor are compared below[42]

$$\left.\begin{aligned} \mathscr{W}(1\ldots N) &= N!\left(\frac{h^2}{2\pi mkT}\right)^{3N/2} \sum_\rho \psi_\rho^*(1\ldots N) \\ &\qquad \times \exp(-\mathscr{H}_N/kT)\,\psi_\rho(1\ldots N), \\ W(1\ldots N) &= \exp(-\Phi_N/kT). \end{aligned}\right\} \qquad (1.3.24)$$

$\mathscr{H}_N$ is the quantum mechanical Hamiltonian operator for a system of N particles, and is the expression for the Slater sum. Integration over the momenta accounts for the factor $(h^2/2\pi mkT)^{3N/2}$. The functions $\psi_\rho(1\ldots N)$ are a complete orthonormal set of eigenfunctions which have been properly symmetrized.

Kirkwood[9] has shown that the Slater sum approaches the Boltzmann factor in the high-temperature correspondence limit.

1.4 The cluster expansion of Ree and Hoover

As we have shown in (1.2.12) and (1.3.8), the Mayer–Ursell[5,13] cluster expansions enable us to write the pressure of an imperfect gas in virial form

$$\frac{PV}{NkT} = 1 + B_2\rho + B_3\rho^2 + B_4\rho^3 + \ldots,$$

where the lth virial coefficients B_l is expressed in terms of a number of l-body cluster integrals β_l according to (1.2.13) and (1.3.16)

$$\frac{P}{\rho kT} = 1 - \sum_{l=2} \frac{l-1}{l} \beta_l \rho^{l-1}$$

where in terms of the Mayer f-function (1.2.14)

$$\beta_l = \frac{1}{(l-1)!}\int \ldots \int (\Sigma\Pi f_{ij})_\beta \, d\mathbf{2} \ldots d\boldsymbol{l}$$

whilst in terms of the augmented Ursell $\mathscr{U}$-function (1.3.17)

$$\beta_l = \frac{1}{(l-1)!}\int \ldots \int \mathscr{U}'(1 \ldots l) \, d\mathbf{2} \ldots d\boldsymbol{l}.$$

In both theories the number of topologically distinct clusters, as well as the difficulty in evaluating them, grows rapidly with l. In consequence, only the first few virial coefficients have been evaluated for 'realistic' potentials.

Ree and Hoover[6] introduce the function $\tilde{f}_{ij}$ which is defined by the equation

$$\tilde{f}_{ij} = \exp(-\Phi_{ij}/kT) \tag{1.4.1}$$

and is related to the Mayer f-function by

$$1 = \tilde{f}_{ij} - f_{ij}. \tag{1.4.2}$$

Whenever two points in a cluster are not connected by an f-bond, (1.4.2) is used to introduce $\tilde{f}_{ij} - f_{ij}$ into the diagram. Since the $(\tilde{f}-f)$-bond has value unity it will have no numerical effect, but, as we shall see, it has an outstanding effect on the number of clusters over which we have to sum. When the $(\tilde{f}_{ij}-f_{ij})$-bond is applied to all unconnected points in a cluster and the resulting expression expanded, it is found that the resulting modified clusters can be written in terms of 'straight' f_{ij} bonds, and 'wiggly' $\tilde{f}_{ij}$ bonds. Take, for example, the irreducible cluster expansion of B_4 (cf. (1.2.6))

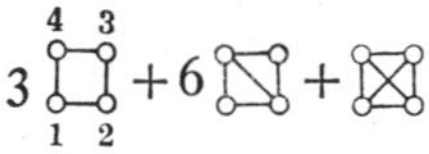

(1.4.3)

and suppose we now incorporate the $(\tilde{f}_{ij}-f_{ij})$ bonds between unconnected points

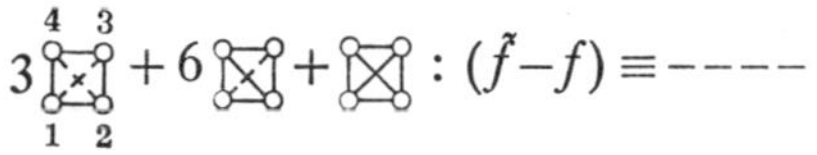

(1.4.4)

The cluster integrals may then be written:

$$3\int f_{12}f_{23}f_{34}f_{14}(\tilde{f}_{13}-f_{13})(\tilde{f}_{24}-f_{24})\,\mathrm{d}1\,\mathrm{d}2\,\mathrm{d}3\,\mathrm{d}4$$
$$+6\int f_{12}f_{23}f_{34}f_{14}f_{13}(\tilde{f}_{24}-f_{24})\,\mathrm{d}1\,\mathrm{d}2\,\mathrm{d}3\,\mathrm{d}4$$
$$+\int f_{12}f_{23}f_{34}f_{14}f_{13}f_{24}\,\mathrm{d}1\,\mathrm{d}2\,\mathrm{d}3\,\mathrm{d}4. \qquad (1.4.5)$$

If we expand the integrands we find there is extensive cancellation, resulting in:

(1.4.6)

The modified cluster expansion for B_5 is

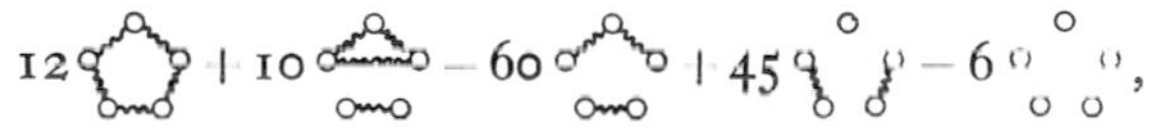

where, for clarity, the f-bonds have been left out: unmodified f-bonds join all the unconnected points. In the f-function formulation, for B_4, B_5 and B_6 there are, respectively 3, 10 and 56 topologically different types of clusters, while in Ree–Hoover $\tilde{f}$-formalism the corresponding modified expressions contain, respectively 2, 5 and 23 topologically different types of modified clusters.

Some of the integrals required for evaluating B_5 and B_6 are zero; for spheres there remain 5 and 22 modified cluster integrals for B_5 and B_6 respectively, for discs the corresponding numbers are 4 and 15. These integrals present formidable geometrical problems in 8-, 10-, 12- and 15-dimensional spaces. Ree and Hoover used a Monte Carlo technique to evaluate the integrals. A trial configuration of the n particles was set up and a check is made to see if any of the modified clusters occurring in B_5 and B_6 correspond to this particular trial configuration. The distances between the particles is of course restricted by the condition that f_{ij} and $\tilde{f}_{ij}$ should be non-zero. If the requirement is violated the configuration is rejected and makes no contribution to the virial coefficient. Furthermore, if the determination of the fifth and sixth virial coefficients is to be of any value, the determination must be accurate which requires the trial configurations to be 'fine-grained' resulting in many hundreds of thousands of trial configurations for each modified cluster.

For the hard sphere gas the first three results (B_2, B_3, B_4) are known exactly[15,16]:

Spheres:
$$\left.\begin{aligned} B_2 \equiv b &= (2\pi/3)\,\sigma^3, \\ B_3/b^2 &= 5/8, \\ B_4/b^3 &= 0.28695, \\ B_5/b^4 &= 0.1103 \pm 0.0003^{(17)}, \\ B_6/b^5 &= 0.0386 \pm 0.0004. \end{aligned}\right\} \tag{1.4.7}$$

For a two-dimensional gas composed of hard discs, the first two coefficients are known exactly[18]:

Discs:
$$\left.\begin{aligned} B_2 \equiv b &= (\pi/2)\,\sigma^2, \\ B_3/b^2 &= 0.78200, \\ B_4/b^3 &= 0.5324 \pm 0.0003,^{(19)} \\ B_5/b^4 &= 0.3338 \pm 0.0005, \\ B_6/b^5 &= 0.1992 \pm 0.0008. \end{aligned}\right\} \tag{1.4.8}$$

Ree and Hoover form a Padé approximant to the virial series. The method of the Padé approximant allows us to estimate the asymtotic behaviour of an infinite series from a knowledge of the first few coefficients. Thus, if we form the Padé approximant $P(3, 3)$ to the virial expansion, we write

$$P(n, m) = \sum_{i=1}^{n} a_i \rho^{i-1} \Big/ \sum_{i=1}^{m} \alpha_i \rho^{i-1}, \tag{1.4.9}$$

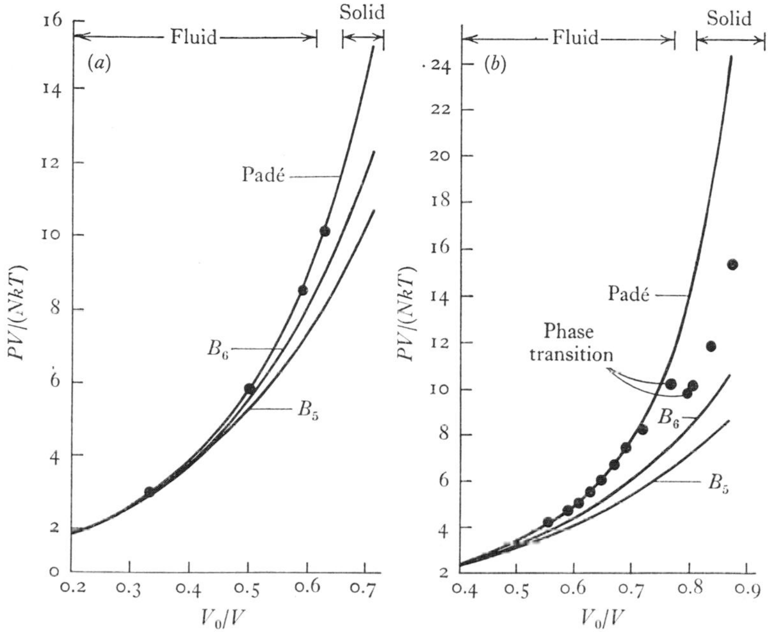

Fig. 1.4.1. (*a*) The hard sphere equation of state and its virial expression. The curve labelled B_5 includes the virial coefficients up to and including B_5. B_6 includes coefficients up to B_6. The other curve is the Padé approximant to the molecular dynamics points (●) of Alder and Wainwright. (*b*) The hard disc equation of state and its virial expression. The two isotherms labelled B_5, B_6 include virial terms up to B_5 and B_6. The Padé approximant to the molecular dynamics points (●) of Alder and Wainwright is also shown. There is no evidence of a van der Waals loop in these series expansions. (Redrawn by permission from Ree and Hoover, *J. Chem. Phys.* **40**, 939 (1964).)

where, in this case, $n = m = 3$. Setting,

$$P(3,3) = \sum_{i=1}^{\infty} B_{i+1}\rho^{i-1} \qquad (1.4.10)$$

with $a_1 \equiv B_2$ and $\alpha_1 \equiv 1$, expansion and comparison of coefficients and substitution of the known values of $B_2 \rightarrow B_6$ yields the equations of state[(6)]:

Spheres:

$$\frac{PV}{NkT} = 1 + \frac{b\rho(1+0.063507b\rho+0.017329b^2\rho^2)}{1-0.561493b\rho+0.081313b^2\rho^2}. \qquad (1.4.11)$$

Discs: $$\frac{PV}{NkT} = 1 + \frac{b\rho(1-0.196703b\rho+0.006519b^2\rho^2)}{1-0.978703b\rho+0.239465b^2\rho^2}. \quad (1.4.12)$$

In fig. 1.4.1 we compare the virial series including B_5, the virial series including B_6, and the Padé approximant to the molecular dynamics results of Alder and Wainwright[20]. The virial series including B_6 agrees closely with the machine curve up to densities of half that of closest packing. The Padé approximant agrees to within 2 per cent over the whole of the fluid branch of the 'experimental' curve. This agreement indicates convergence of the virial series along the entire fluid branch of the hard sphere equation of state.

In the case of discs, the six-term virial series pressure is considerably below ($\sim$ 25 per cent) the molecular dynamics pressure on the fluid side of the two phase region ($V = 1.312V_0$). The next several virial coefficients must presumably be positive to shift the curve onto the 'experimental' isotherm.

Ree and Hoover note that from the Padé hard sphere equation of state (1.4.11) it is possible to determine higher virial coefficients, and they find that B_{20} has the first negative sign, and thereafter approximately every 16th is negative. Whilst from the Padé hard disc equation of state (1.4.12) it is found that all the virial coefficients are positive.

If the Padé approximants (1.4.11) and (1.4.12) are used to estimate B_7 and B_8, we obtain:

Spheres: $B_7/b^6 = 0.0127$ $B_8/b^7 = 0.0040$.

Discs: $B_7/b^6 = 0.115$ $B_8/b^7 = 0.065$.

The inclusion of B_7 in the hard sphere equation of state leads to a slight worsening of the agreement with the machine computations, but this is probably not significant bearing in mind the error associated with the higher order coefficients.

1.5 Quantum mechanical calculation of the second virial coefficient

In the previous sections we have obtained expressions for the virial coefficients on the basis of a cluster expansion of the quasi-classical partition function. It is evident that for a quantum system we must forgo the determinism implicit in the n-body cluster approach although the classical approximation is probably only invalid in the case of helium at sufficiently low temperatures such that the de Broglie wavelength is of the order of the mean atomic separation.

We now show how the second virial coefficient may be calculated in terms of the quantization of the binary interaction of the gas particles. Other than for methodological interest, helium is the only real case: we therefore consider a system having no electronic angular momentum and zero spin. We shall not attempt here to give a full treatment of the problem, but merely indicate the difference in approach between classical and quantum mechanical systems. For a more detailed account from a quantum statistical viewpoint the reader is referred to the literature.

The classical second virial coefficient as developed in the previous sections consists of the volume integral over either the Mayer f-function[1], or the augmented Ursell $\mathscr{U}$-function[5]: in the approximation of additivity of the pair potential, both formalisms reduce to

$$B_2(T) = \int_0^\infty \left\{\exp\left(\frac{\Phi(r)}{kT}\right) - 1\right\} d\mathbf{r} \qquad (1.5.1)$$

and in the classical approximation we assume that the second particle adopts the continuously variable potential $\Phi(r)$ whilst moving in the field of particle 1. This, furthermore, is not restricted to bound states: indeed, at high temperatures such that $kT > \epsilon$, the pair interaction becomes purely collisional. Then the hard core of the pair potential predominates and $B(T) \to 0$ as T increases. Under these conditions the limiting behaviour is described by the perfect gas.

The situation becomes somewhat more complicated for a quantum mechanical system. Even for an *ideal* quantum fluid the exchange degeneracy of the wave functions requires a correction[21]

$$B_2(T)_{\text{exchange}} = \pm\frac{1}{2}\left(\frac{\pi\hbar^2}{mkT}\right)^{\frac{3}{2}}, \qquad (1.5.2)$$

where the $\pm$ sign applies to a boson (fermion) system. We see how the Pauli repulsion between fermions increases the pressure above the ideal value whilst the effect of indistinguishability of bosons leads to a decrease in the pressure below that of an ideal gas.

If we now extend the discussion to an *interacting* system of particles we must recognize that for unbound states there will be a continuum of levels for the two particles (other than quantization imposed by the boundary conditions) whilst in the bound case the energy levels will be discrete, and will develop as eigenvalues of the radial Schrödinger equation. Obviously in evaluating the second virial coefficient a *sum* rather than an integral over the bound states will be necessary. The final quantum mechanical expression of the second virial coefficient incorporating exchange effects is

$$B_2(T)_{\text{quantum}} = \underbrace{\pm\frac{1}{2}\left(\frac{\pi\hbar^2}{mkT}\right)^{\frac{3}{2}}}_{\text{Exchange term}} - 8\sum_l (2l+1)\Big\{\underbrace{\sum_n \exp(-\Phi_n/kT)}_{\text{Bound discrete spectrum}}$$

$$+\underbrace{\frac{1}{\pi}\int_0^\infty \frac{d\delta_l}{dp}\exp(-p^2/mkT)\,dp}_{\text{Collisional continuum}}\Big\}. \qquad (1.5.3)$$

The calculation of $B_2(T)_{\text{quantum}}$ is therefore reduced to the determination of the eigenvalues Φ_n and the partial waveshifts δ_l[22–25].

As we have observed, even for an ideal non-interacting quantum gas there is an apparent 'statistical attraction' in the case of bosons, and a 'statistical repulsion' in the case of fermions. Thus, the first quantum correction to the ideal gas pressure results in an increase (decrease) in the case of an ideal Fermi (Bose) gas. Further, the probability of finding particles close to one another is altered becoming larger in the case of bosons and smaller in the case of fermions. This point has been considered by a number of workers[39,40]. It may be shown that for a two-particle interaction there is an effective potential

$$\Phi_{12}(r_{12}, T) = -kT\ln\left\{1 \pm \exp\left(\frac{-2\pi r_{12}^2}{\lambda^2}\right)\right\}, \qquad (1.5.4)$$

where the upper sign refers to Bose–Einstein, and the lower to Fermi–Dirac statistics. λ is the thermal wavelength. It must be emphasized, however, that Φ_{12} originates solely from the symmetry properties of the wave function. Furthermore it depends upon the temperature and thus cannot be regarded as a true interparticle potential.

Lado[41] has recently obtained a generalization of (1.5.4) incorporating spin effect, when

$$\Phi_{12}(r_{12}, T, \rho_s) = -kT \ln \left\{ 1 \pm \frac{\exp(\pm 2\pi r_{12}^2/\lambda^2)}{(2s+1)} \right\}, \qquad (1.5.5)$$

where s is the spin of the particle and ρ_s is the total density of the system with spin s. In fact this expression is a high-temperature approximation: terms of order λ^3 are neglected. However, Lado shows how the Born–Green hierachy (see chapter 2) may be used to obtain the effective potential at *any* temperature. This potential will then allow the results of wave function symmetrization in quantum fluids to be described in terms of a classical Boltzmann factor.

In the case of a Bose–Einstein gas, Lado[41] chooses units of distance and temperature as follows:

$$a_B = (2.612/\rho_0)^{\frac{1}{3}},$$
$$T_B = h^2/(2\pi m k a_B^2), \qquad (1.5.6)$$

where $\rho_0 = \rho_s/(2s+1)$. T_B is the condensation temperature of a Bose–Einstein ideal gas of density ρ_0. The Born–Green expression was solved iteratively for Φ_{12} for a system of bosons at the reduced temperatures T/T_B = 10, 5, 3, 2, 1.5 and 1.1, and for zero spin, $s = 0$. The pair correlation functions $g_{(2)}(r)$ and computed effective potentials $\beta\Phi_{12}(r)$ are shown in fig. 1.5.1. for these cases. We note that as the temperature is reduced the effective potential becomes weaker and of longer range.

For a Fermi–Dirac gas, convenient units are

$$a_F = (6\pi\rho_0)^{-\frac{1}{3}},$$
$$T_F = \hbar^2/(2mka_F^2),$$

kT_F is the Fermi energy. Lado gives the corresponding distributions and effective potentials for a system of ideal Fermi–Dirac particles of spin $\frac{1}{2}$ at the reduced temperatures T/T_F = 4, 2, 1, 0.5, 0.3 and 0.2. These are shown in fig. 1.5.2. Again a spreading of the effective potential is noticed for decreasing T.

Finally, in table 1.5.1. we give a comparison of the exact values of the internal energy of an ideal quantum gas with the values estimated by using the effective potential[41].

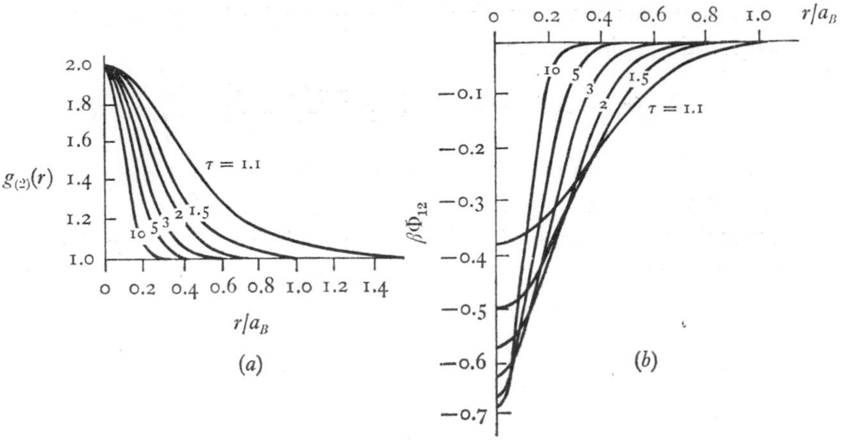

Fig. 1.5.1. (*a*) Pair correlation functions for an ideal gas of spinless bosons. The reduced temperature τ is in units of the Bose–Einstein condensation temperature $\tau = T/T_B$. (*b*) The computed effective potentials, in units of kT, corresponding to the boson pair distributions shown in (*a*). (Redrawn by permission from Lado, *J. Chem. Phys.* **47**, 5369 (1967).)

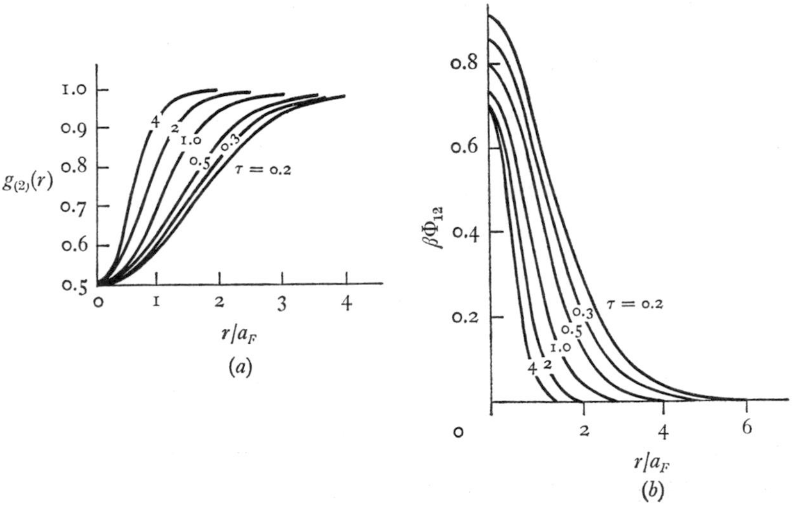

Fig. 1.5.2. (*a*) Pair correlation functions for an ideal gas of spin $\frac{1}{2}$ fermions; $\tau = T/T_F$. (*b*) The computed effective potentials, in units of kT, corresponding to the fermion pair distributions shown in (*a*). (Redrawn by permission from Lado, *J. Chem. Phys.* **47**, 5369 (1967).)

TABLE 1.5.1. *Internal energy of an ideal quantum gas*

	Bosons			Fermions	
	E/NkT_B			E/NkT_F	
T/T_B	Exact	Computed	T/T_F	Exact	Computed
10.0	14.781	14.781	4.0	6.100	6.099
5.0	7.189	7.190	2.0	3.140	3.140
3.0	4.096	4.099	1.0	1.697	1.694
2.0	2.502	2.510	0.5	1.022	1.013
1.5	1.669	1.689	0.3	0.7872	0.7696
1.1	0.959	1.022	0.2	0.6915	0.6634

1.6 The virial coefficients

For a classical system the second virial coefficient is given by (1.5.1) which allows an immediate evaluation of the first-order correction to the ideal gas equation of state from a knowledge of the pair potential $\Phi(r)$. We see from (1.5.1) that at high temperatures the exponent is nearly unity and $B_2(T) \to 0$. Conversely, at low temperatures the region of the pair potential 'sampled' in the collision is negative, and $B_2(T) \to -\infty$. It is evident that there must be a temperature (the Boyle temperature) at which the contributions of the attractive and repulsive interactions are equal. This state is indistinguishable from that of the ideal gas, and $B_2(T)$ is zero. Assuming a Lennard-Jones interaction we may determine the second virial coefficients for a number of gases in reduced coordinates[26, 27]. From fig. 1.6.1 we see that there is good agreement for the classical curves, and the quantum mechanical curve is in excellent agreement with (1.5.3)[26]. The Lennard-Jones parameters may, of course, be determined directly from a measurement of $B_2(T)$[29, 30].

As we have seen in § 1.4, calculations for the first six virial coefficients have been made for hard spheres and discs, and an estimate of B_7 and B_8. For the Lennard-Jones 12–6 potential, however, the results are less substantial[31]. We show the Lennard-Jones virial coefficients B_2[27], B_3[27, 32, 33], B_4[33–36] and B_5[37] in fig. 1.6.2. It is seen that all coefficients up to B_5 become large and negative

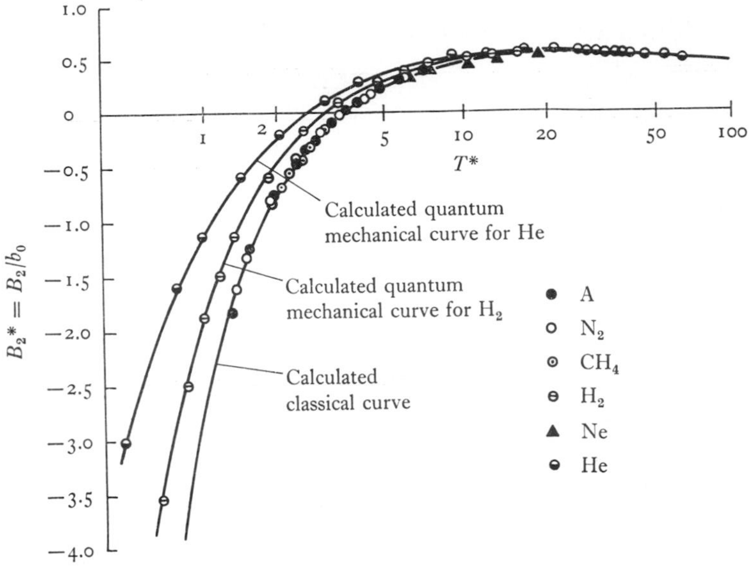

Fig. 1.6.1. Comparison of the reduced second virial coefficient for several Lennard-Jones systems. Quantum deviations are seen to be important at low temperatures for H_2 and He: the classical limit is regained at high temperatures. (Redrawn by permission from Hirschfelder, Curtis and Bird, *Molecular Theory of Gases and Liquids* (Wiley).)

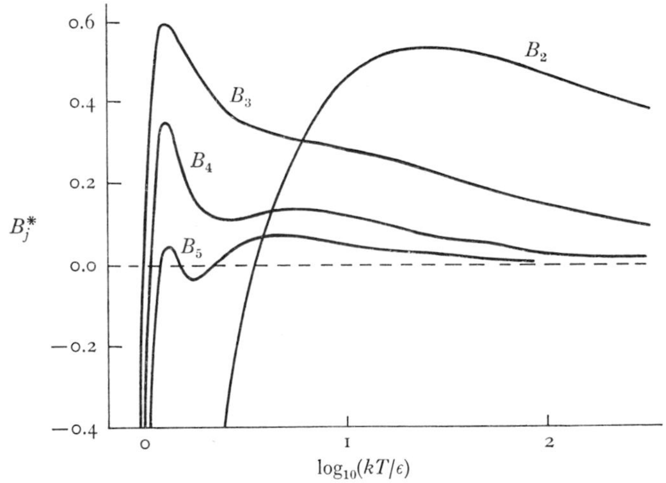

Fig. 1.6.2. The reduced second, third, fourth and fifth virial coefficients as determined for a Lennard-Jones system. (Redrawn by permission from Temperley, Rowlinson and Rushbrooke, *Physics of Simple Liquids* (North-Holland Publ. Co.).)

at low temperatures, which raises doubts about the convergence of the virial expansion at low temperatures, even at low vapour densities. However, assuming that the virial expansion using the above coefficients does converge at the critical density, we may calculate the critical constants from

$$\left(\frac{\partial P}{\partial V}\right)_{T_c} = 0, \quad \left(\frac{\partial^2 P}{\partial V^2}\right)_{T_c} = 0. \qquad (1.6.1)$$

These derivatives may be determined for a virial expansion truncated at the third or higher coefficient. Barker's results are shown in table 1.6.1 and are seen to be converging on a steady value as higher coefficients are included. However, there is no guarantee

TABLE 1.6.1

Coefficient	kT_c/ϵ	b/V_c	$(PV/kT)_c$
B_3	1.445	0.773	0.333
B_4	1.300	0.561	0.352
B_5	1.291	0.547	0.354

$b = \frac{2}{3}\pi\sigma^3$

that these are the *correct* critical ratios. Green and Sengers[38] have shown that the critical point is highly singular – indeed this is seen to be the case in the Yang–Lee theory of condensation[10–14] – and so cannot be reproduced exactly by solving (1.6.1) at the critical point.

Ram and Singh[43] have very recently calculated the quantum corrections to the third and fourth virial coefficients for a Lennard-Jones (6–12) and Axilrod–Teller triple-dipole dispersion potential. It is found that the two-body quantum correction to B_3 is always positive, whilst the three-body correction is negative: this is ascribed to the triplet potential energy being negative for most configurations. A considerable improvement over the classical curve for B_3 is obtained at low temperatures.

CHAPTER 2

EQUILIBRIUM THEORY OF DENSE FLUIDS: THE CORRELATION FUNCTIONS

2.1 Introduction

We saw in the previous chapter how in the limit of low densities we were able to obtain a convergent expansion for the configurational partition function, and thereby determine all the thermodynamic functions of the system including the virial equation of state. However, what we did not have was a detailed knowledge, or indeed *any* knowledge, of the *structure* of the system.

Clearly, at liquid densities the series expansions will fail to converge(1), and anyway the computational labour in evaluating the multi-dimensional high-order cluster integrals which would develop in the highly connected fluid would be overwhelming, even in the economical Ree-Hoover formalism.

So, we are forced to adopt a new mathematical approach – the formalism of *molecular distribution functions*, or *correlation functions*. Instead of trying to evaluate the N-body configurational integral directly, the theory describes the probability of configurational groupings of two, three, and more particles. Further, it may be shown that we can still obtain the same amount of information concerning the system as is obtained from the study of the statistical integral itself. Moreover, in this way we obtain direct information on the molecular structure of the system being studied. Of course the method of correlation functions applies equally well to gases and solids, but we would not generally adopt that approach by choice since other characteristics of these phases suggest a more direct route to the partition function. The formalism is adopted, however, in the case of amorphous solids.

We shall find that the one- and two-body distribution functions, generally denoted by $g_{(1)}$ and $g_{(2)}$ respectively, will be of central importance in the equilibrium theory of liquids. Indeed, the fundamental problem of liquid state physics is to deduce $g_{(2)}$ solely from a knowledge of the pair potential $\Phi(r)$ and it is this problem with which we shall be concerned throughout this chapter. The radial distribution function (RDF), as $g_{(2)}(r)$ is frequently known, arises quite naturally in the theory of scattering of radiation by liquids, developed by Zernike and Prins in 1927[2]. Further, $g_{(2)}(r)$ will be seen to appear in virtually every thermodynamic expression describing the liquid phase. So, our apparent obsession with the pair correlation function is not without reason.

2.2 Generic and specific distributions

From the general (generic) molecular distribution function $f_{(N)}(\boldsymbol{p}_{(N)}, \boldsymbol{q}_{(N)}, t)$, we define

$${}_{s}f_{(N)}(\boldsymbol{p}_{(N)}, \boldsymbol{q}_{(N)}, t)\,\Delta\boldsymbol{p}_{(N)}\Delta\boldsymbol{q}_{(N)} \equiv \text{probability of finding molecule 1 in phase volume } \Delta\boldsymbol{p}_1\Delta\boldsymbol{q}_1, \text{ molecule 2 in phase volume } \Delta\boldsymbol{p}_2\Delta\boldsymbol{q}_2, \ldots \text{ all at time } t.$$

This distribution is termed the *specific distribution*, and is to be distinguished from the *generic* distribution which takes account of the indistinguishability of the particles. Thus,

$${}_{g}f_{(N)}(\boldsymbol{p}_{(N)}, \boldsymbol{q}_{(N)}, t) = N!\,{}_{s}f_{(N)}(\boldsymbol{p}_{(N)}, \boldsymbol{q}_{(N)}, t), \qquad (2.2.1)$$

where the subscripts g, s, refer to generic and specific, respectively. Of course, there will be many more acceptable generic (indistinguishable, non-specific) distributions than specific: $N!$ times as many, in fact.

It follows directly from (2.2.1) that for a subset of h particles

$${}_{g}f_{(h)}(\boldsymbol{p}_{(h)}, \boldsymbol{q}_{(h)}, t) = \frac{N!}{(N-h)!}\,{}_{s}f(\boldsymbol{p}_{(h)}, \boldsymbol{q}_{(h)}, t). \qquad (2.2.2)$$

We may also arrive at the generic $f_{(h)}$ distribution from the N-body distribution quite simply as;

$${}_{g}f_{(h)} = \frac{1}{(N-h)!}\int \cdots \int {}_{g}f_{(N)} \prod_{i=h+1}^{N} \mathrm{d}\boldsymbol{p}_i\,\mathrm{d}\boldsymbol{q}_i. \qquad (2.2.3)$$

Further, distribution functions of consecutive order are related by the identity

$$_{g}f_{(h)} = \frac{1}{(N-h)!}\iint {_{g}f_{(h+1)}}\,\mathrm{d}\boldsymbol{p}_{h+1}\,\mathrm{d}\boldsymbol{q}_{h+1} \tag{2.2.4}$$

as may be shown from (2.2.3). Henceforth it will be assumed that we are dealing with generic distributions since the quasi-classical Gibbs function incorporates indistinguishability. The subscript g will therefore be dropped. In this chapter on equilibrium properties of the liquid we shall assume the integral over momenta to have been performed yielding the usual factor $(2\pi mkT/h_0^2)^{3N/2}$ for a classical system. The formal single, two-body and three-body distribution functions are then given by (neglecting the momentum factor above)

$$\left.\begin{aligned} \rho g_{(1)}(1) &= \frac{\rho^N}{(N-1)!}\int\cdots\int g_{(N)}(1\ldots N)\,\mathrm{d}\mathbf{2}\ldots\mathrm{d}\boldsymbol{N},\\ \rho^2 g_{(2)}(12) &= \frac{\rho^N}{(N-2)!}\int\cdots\int g_{(N)}(1\ldots N)\,\mathrm{d}\mathbf{3}\ldots\mathrm{d}\boldsymbol{N},\\ \rho^3 g_{(3)}(123) &= \frac{\rho^N}{(N-3)!}\int\cdots\int g_{(N)}(1\ldots N)\,\mathrm{d}\mathbf{4}\ldots\mathrm{d}\boldsymbol{N},\end{aligned}\right\} \tag{2.2.5}$$

where $\rho^h g_{(h)}$ represents the h-body distribution function, ρ is the system number density. Alternatively, the recurrence relation (2.2.4) may be used to write

$$g_{(2)}(12) = \frac{\rho}{(N-2)!}\int g_{(3)}(123)\,\mathrm{d}\mathbf{3}. \tag{2.2.6}$$

$g_{(h)}$ is dimensionless and represents the relative probability of finding the subset of h particles in the configuration $\{\boldsymbol{r}_1, \boldsymbol{r}_2, \ldots, \boldsymbol{r}_h\}$, *relative* that is to the probability of the uniform uncorrelated distribution. The relative probability may exceed one, and indeed does so at r_0. At r_0 the probability of finding a particle is greater than the random or uncorrelated distribution given by $g_{(2)}(r) = 1.00$. Of course, at large separations the two particles become uncorrelated, and we have the boundary condition

$$g_{(2)}(\boldsymbol{r}) \to 1.00 \quad \text{as} \quad \boldsymbol{r} \to \infty. \tag{2.2.7}$$

In fact, quite generally,

$$g_{(h)}(\boldsymbol{r}) \to 1.00 \quad \text{as} \quad \boldsymbol{r} \to \infty. \tag{2.2.8}$$

2.3 A formal relation between $g_{(2)}(r)$ and the pair potential

We have seen that the phase partition function can, for certain forms of the Hamiltonian, be separated into the configurational and momentum components:

$$Z = Z_P Z_Q, \tag{2.3.1}$$

where

$$\left.\begin{aligned} Z_P &= \frac{1}{h^{3N}}\int\cdots\int \exp\left\{-\sum_{i=1}^{N}\frac{p_i^2}{2mkT}\right\}\mathrm{d}\boldsymbol{p}_1\ldots\mathrm{d}\boldsymbol{p}_N,\\ Z_Q &= \frac{1}{N!}\int\cdots\int \exp\left\{-\frac{\Phi_N(1\ldots N)}{kT}\right\}\mathrm{d}\boldsymbol{q}_1\ldots\mathrm{d}\boldsymbol{q}_N.\end{aligned}\right\} \tag{2.3.2}$$

The two components are quite independent, and under conditions of equilibrium the momentum integral may be integrated to give

$$Z_P = \left(\frac{2\pi mkT}{h^2}\right)^{3N/2}, \tag{2.3.3}$$

regardless of the atomic structure of the system. We may relate the configurational projection of the generic distribution function to the total potential as follows:

$$\rho^N g_{(N)} = \frac{1}{N!Z_Q}\exp\left\{\frac{\Phi_N(1\ldots N)}{kT}\right\} \tag{2.3.4}$$

and the pair distribution, for example, becomes

$$\rho^2 g_{(2)}(\boldsymbol{r}) = \frac{1}{(N-2)!}\frac{1}{Z_Q}\int\cdots\int \exp\{-\beta \sum_{1\leqslant i<j\leqslant N}\Phi(\boldsymbol{r}_{ij})\}\,\mathrm{d}3\ldots\mathrm{d}N, \tag{2.3.5}$$

where we have assumed additivity of the pair potential. So, we have a formal relation between the pair distribution and the pair potential, but clearly, as before, the expression is as intractible as ever, and even contains the configurational partition function, Z_Q.

We note one other ominous feature. Performing the momentum integration in (2.2.4), we obtain the following recurrence relation between consecutive correlation functions

$$g_{(h)}(\boldsymbol{r}^h) = \frac{\rho}{(N-h)!}\int g_{(h+1)}(\boldsymbol{r}^{h+1})\,\mathrm{d}(\boldsymbol{r}^{h+1}). \tag{2.3.6}$$

That is, $g_{(2)}$ can be determined if we have an explicit knowledge of the triplet correlation $g_{(3)}$. To determine $g_{(3)}$ we need to know $g_{(4)}$, and so on. This will prove to be another difficulty which has somehow to be overcome, and we shall be discussing approximations for the termination or 'closure' of the linked hierachy of equations later in this chapter.

Since,

$$\ln Z = \frac{3N}{2}\ln\left(\frac{2\pi mkT}{h^2}\right) + \ln Z_Q \tag{2.3.7}$$

and from the statistical thermodynamic relationship

$$U = G - T\left(\frac{\partial G}{\partial T}\right)_V = kT^2\frac{\partial}{\partial T}[\ln Z(V,T)]. \tag{2.3.8}$$

The total internal energy U may be written as

$$U = \tfrac{3}{2}NkT + kT^2\left(\frac{\partial Z_Q/\partial T}{Z_Q}\right)$$

$$= \tfrac{3}{2}NkT + \frac{1}{Z_Q}\frac{1}{N!}\int\dots\int \Phi_N \exp\left(-\frac{\Phi_N}{kT}\right)\mathrm{d}\mathbf{1}\dots\mathrm{d}\boldsymbol{N}. \tag{2.3.9}$$

From (2.3.5) this may be written

$$U = \tfrac{3}{2}NkT + \frac{\rho^2}{2}\int g_{(2)}(\boldsymbol{r}_{12})\,\Phi(\boldsymbol{r}_{12})\,\mathrm{d}\boldsymbol{r}_{12}, \tag{2.3.10}$$

where pairwise additivity of the total potential has been assumed. If the molecules are spherical so that the equilibrium system exhibits no directional properties then we may use the scalar expression

$$\frac{U}{V} = \tfrac{3}{2}\rho kT + \frac{4\pi\rho^2}{2}\int g_{(2)}(r)\,\Phi(r)\,r^2\,\mathrm{d}r. \tag{2.3.11}$$

The first term represents the kinetic contribution to the internal energy and is, in fact, the equipartition value for an equilibrium 3-dimensional assembly of N particles. The second term represents the configurational (potential) contribution to the internal energy and for real systems this term is negative. U is a function of both temperature and density; it appears explicitly in the kinetic term and implicitly through the temperature and density dependence of $g_{(2)}(r)$ in the potential term.

The equation of state, or pressure equation is not so easy to determine since it involves the volume derivative[(3)]:

$$P = -\left(\frac{\partial G}{\partial V}\right)_T = kT\frac{\partial}{\partial V}[\ln Z_Q(V, T)]. \qquad (2.3.12)$$

In order to make Φ_N an explicit function of V we make a change in the limits of the integration according to

$$\boldsymbol{r} = \boldsymbol{r}'V^{\frac{1}{3}}. \qquad (2.3.13)$$

This enables us to write r in terms of V and then perform the volume in differentiation.

$$Z_Q = \frac{V^N}{N!}\int_0^1 \dots \int_0^1 \exp\left(-\frac{\Phi(1 \dots N)}{kT}\right) \mathrm{d}\mathbf{1} \dots \mathrm{d}\boldsymbol{N}, \qquad (2.3.14)$$

$$\ln Z_Q = \ln V^N + \ln\left\{\frac{1}{N!}\int \dots \int \exp\left(-\frac{\Phi_N}{kT}\right) \mathrm{d}\mathbf{1} \dots \mathrm{d}\boldsymbol{N}\right\}. \qquad (2.3.15)$$

The first term V^N gives the kinetic contribution to the pressure as NKT/V, from (2.3.12). The second term involves the differential $\partial\Phi_N/\partial V$, and assuming pairwise additivity, we have

$$\frac{\partial \Phi_N}{\partial V} = \sum_{1 \leqslant i < j \leqslant N} \frac{\partial \Phi(r_{ij})}{\partial r_{ij}} \frac{r_{ij}}{3V}. \qquad (2.3.16)$$

The transformation (2.3.13) is now reversed, and by an analogous procedure to that for the calculation of U, we obtain

$$P = \frac{NkT}{V} - \frac{1}{Z_Q}\int \dots \int \exp\left(-\frac{\Phi_N}{kT}(1 \dots N)\right) \frac{\partial \Phi_N}{\partial V} \mathrm{d}\mathbf{1} \dots \mathrm{d}\boldsymbol{N}$$

$$= \frac{NkT}{V} - \frac{\rho}{6}\iint g_{(2)}(r_{12})\, r_{12} \frac{\partial \Phi}{\partial r}(r_{12})\, \mathrm{d}\mathbf{1}\, \mathrm{d}\mathbf{2} \qquad (2.3.17)$$

for spherical particles,

$$= \rho kT - \frac{4\pi\rho^2}{6}\int_0^\infty g_{(2)}(r) \frac{\partial \Phi}{\partial r}(r)\, r^3\, \mathrm{d}r. \qquad (2.3.18)$$

Notice that (2.3.18) for the pressure is effectively quadratic in the density. For a given system pressure and temperature there are evidently two real values of the density. These, of course, are the densities of the coexistent liquid and vapour phases. We need to point out, however, that $g_{(2)}(r)$ will differ in the two phases, so (2.3.18) cannot be regarded quite as a simple quadratic for ρ. At vapour densities the integral in the pressure equation vanishes as the potential interactions diminish, and the expression reduces to the kinetic term in the limit $\rho \to 0$. Equation (2.3.18) expresses the virial theorem of Clausius[(4)] (see Yvon[(5)] (1935)).

The expressions (2.3.11) and (2.3.18) for the internal energy and the equation of state are true strictly only for the inert gases. However, a further class of molecules, namely those with spherical potential energies in which the molecular rotations and vibrations are independent of their environment.

Other expressions for the thermodynamic functions of the fluid in terms of the pair distribution follow. The temperature derivative of (2.3.11) at constant volume yields the specific heat:

$$\frac{C_V}{V} = \tfrac{3}{2}Nk + \frac{4\pi\rho^2}{2}\int_0^\infty \left(\frac{\partial g_{(2)}(r)}{\partial T}\right)_V \Phi(r)\, r^2\, dr. \qquad (2.3.19)$$

It is assumed here that $\Phi(r)$ is independent of temperature so that only the temperature dependence of $g_{(2)}(r)$ appears in (2.3.19). This certainly is not the case for the liquid metals. A precise knowledge of the variation of $g_{(2)}(r)$ with T at constant density is required before the above expression is useful. The first term may be identified as the specific heat to be associated with the kinetic modes of motion, whilst the second term represents the configuration contribution. For an N-body 3-dimensional harmonic oscillator $C_V = 3Nk$. For liquid argon at the triple point $C_V = 2.32Nk$, for liquid sodium $C_V = 3.4k$. The corresponding values for the solid are $2.89Nk$ and $3.1Nk$, respectively. The solid phase is evidently well represented as an assembly of harmonic oscillators. The liquid phase, however, does not have a specific heat close to either model.

We might mention here that formally we could just as easily express the thermodynamic functions in terms of the triplet distribution function $g_{(3)}(123)$. The three-body distribution however, unlike the two-body function, is not directly accessible to experiment, although further information on this function has recently become available (§2.6).

2.4 Equations for the pair correlation function

In the previous section we saw that there was no particular difficulty in establishing a formal relation between the pair distribution function and the pair potential. Furthermore, the radial distribution function (RDF), as the integral of $g_{(2)}(\boldsymbol{r})$ over angle is known, allows us to express the thermodynamic functions in terms of the experimentally accessible pair distribution function. We are not much closer, however, to an adequate theory of $g_{(2)}(r)$ than we were to the evaluation of the partition function, although it is clearly a somewhat less formidable task than the determination of $g_{(N)}(\boldsymbol{r}_1, \ldots, \boldsymbol{r}_N)$. In the case of dilute gases we were able to determine the thermodynamic functions as differentials of the partition function expressed as a series expansion in density. Such an approach is prohibited at liquid densities and we are obliged to find new approaches to the problem.

The theory of the pair distribution function is still under development, and Widom[77] gives some interesting inequalities satisfied by the pair distribution. Nevertheless, two distinct approaches may be discerned. Historically the earlier, the developments of Born–Green–Yvon, Kirkwood and Bogolyubov will be discussed. This class of equations represents the mathematical standpoint and approaches the problem in terms of the coupled hierachy of equations,

$$g_{(h)}(\boldsymbol{r}^h) = \frac{\rho}{(N-h)!}\int g_{(h+1)}(\boldsymbol{r}^{h+1})\,\mathrm{d}(\boldsymbol{r}^{h+1}).$$

Clearly, if we wish to make an explicit determination of one of the lower order distribution functions we must somehow terminate the series of equations: if we are trying to determine $g_{(2)}(r)$ its expression in terms of $g_{(3)}(\boldsymbol{r}^3)$ is of little use. We shall therefore be involved with the 'closure' of the hierachy, and its physical consequences will be examined. We have, in fact, already encountered the 'Kirkwood closure' or 'superposition approximation' as it is also known, in the Ursell $\mathscr{U}$-function expansion for the partition function of a non-ideal gas (§ 1.3).

The second, and more recent, approach is based on the physical concept of 'direct' and 'indirect' correlation. In fact the first discussion of the direct and indirect correlation functions was given by Ornstein and Zernike in 1914 in their treatment of critical fluctuations. The correlation between two particles is considered to be composed of a *direct* effect due to the direct interaction between the two centres, and an *indirect* effect – the first particle affecting the second via its coupling to a third. The object of this latter approach is to express the direct correlation in terms of the pair potential, and as we shall see, various models have been proposed relating the direct correlation to the pair potential.

It is seen that in both treatments the two-body distribution function can only be expressed in the context of the three body distribution. The Born–Green–Yvon, Kirkwood, Bogolyubov approach really differs only in emphasis from the direct–indirect correlation approach – the former being mathematical and the latter physical. Both require some form of closure. There is no difficulty in demonstrating that the direct–indirect correlation approach can be extended to include higher-order correlations, but here we take the indirect correlation to involve only a third body. We therefore neglect higher-order routes involving the correlation of four or more particles.

2.5 Differential equations for the pair distribution: the Born–Green–Yvon equation (BGY)

We shall now consider the various integro-differential equations which have been developed to relate $g_{(2)}(\boldsymbol{r})$ to the pair potential $\Phi(r)$. We first consider the mathematical approach to the problem in terms of the truncation of the hierachy of the equations (2.3.6).

In this form a theory relating $g_{(2)}$ and $\Phi(r)$ was first given by Yvon[5] in 1935; it was subsequently formulated independently by both Bogolyubov[6] (1946) and Born and Green[7] (1946).

A simple physical derivation of the BGY equation may be given as follows[8]. Consider a triplet of particles at the points **1**, **2** and **3**. We may write the pair distribution in terms of the potential of mean force as

$$g_{(2)}(12) = \exp\left(-\frac{\Psi(12)}{kT}\right). \tag{2.5.1}$$

However, the net force on particle 1 will be

$$-g_{(2)}(12)\,\nabla\Psi(12) = -g_{(2)}(12)\,\nabla_1\Phi(12) + \int [-\nabla_1\Phi(13)]\,\rho g_{(3)}(123)\,\mathrm{d}3, \quad (2.5.2)$$

where the first term represents the force on 1 due to particle 2, and the second term is the force due to particles other than the one at 2, since $[\rho g_{(3)}(123)]\,\mathrm{d}3$ represents the probability of finding an atom in d3 at 3, knowing that there are atoms at **1** and **2**. Combining (2.5.1) and (2.5.2) we obtain the BGY equation:

$$-kT\nabla_1 \ln g_{(2)}(12) = \nabla_1\Phi(12) + \rho\int \nabla\Phi(13)\frac{g_{(3)}(123)}{g_{(2)}(12)}\,\mathrm{d}3 \quad (2.5.3)$$

and, generally,

$$-kT\nabla_1 g_{(h)}(\boldsymbol{r}^h) = g_{(h)}(\boldsymbol{r}^h)\sum_{i=2}^{N}\nabla_1\Phi(1,i) + \rho\int \nabla_1\Phi(1,h+1)\,g_{(h+1)}(\boldsymbol{r}^{h+1})\,\mathrm{d}(\boldsymbol{h}+\mathbf{1}). \quad (2.5.3a)$$

An alternative derivation in terms of the hierachical relationship between successive orders of distribution function is now given. In its essentials it follows the initial development of Born, Green and Yvon.

The h-body distribution is, by definition

$$\rho^h g_{(h)}(\boldsymbol{r}^h) = \frac{1}{(N-h)!}\frac{1}{Z_Q}\int\cdots\int \exp(-\beta\Phi_N)\,\mathrm{d}(\boldsymbol{h}+\mathbf{1})\ldots\mathrm{d}\boldsymbol{N}. \quad (2.5.4)$$

This is, of course, the h-body generic projection of the phase distribution $f_{(N)}$. We are particularly interested in the pair distribution $g_{(2)}(\boldsymbol{r})$, and so we write more specifically

$$\rho^2 g_{(2)}(\boldsymbol{r}) = \frac{1}{(N-2)!}\frac{1}{Z_Q}\int\cdots\int \exp(-\beta\Phi_N)\,\mathrm{d}3\ldots\mathrm{d}\boldsymbol{N}. \quad (2.5.5)$$

$\mathrm{d}3\ldots\mathrm{d}\boldsymbol{N}$ represent the volume elements $\mathrm{d}\boldsymbol{V}_3\ldots\mathrm{d}\boldsymbol{V}_N$. As such, (2.5.5) is just as intractable as the initial equations for the configurational integrals. However, we follow BGY and investigate what change in $g_{(2)}(\boldsymbol{r})$ occurs if we vary the position of atom **1**:

$$\rho^2\frac{\partial g_{(2)}(\boldsymbol{r})}{\partial \boldsymbol{r}_1} = \frac{1}{(N-2)!}\frac{1}{Z_Q}\int\cdots\int\left(-\beta\frac{\partial\Phi_N}{\partial\boldsymbol{r}_1}\right)\exp(-\beta\Phi_N)\,\mathrm{d}3\ldots\mathrm{d}\boldsymbol{N}. \quad (2.5.6)$$

Making the usual assumption of pairwise additivity of the total potential, we may write the first few terms as follows:

$$\Phi_N = \Phi_{12}+\Phi_{13}+\ldots+\Phi_{1N}+\Phi_{23}+\Phi_{24}+\ldots+\ldots \qquad (2.5.7)$$

whereupon,
$$\frac{\partial\Phi_N}{\partial \boldsymbol{r}_1} = \frac{\partial\Phi_{12}}{\partial \boldsymbol{r}_1}+\sum_{j=3}^{N}\frac{\partial\Phi_{1j}}{\partial \boldsymbol{r}_1}. \qquad (2.5.8)$$

Inserting this in (2.5.6) we obtain

$$\rho^2\frac{\partial g_{(2)}(\boldsymbol{r})}{\partial \boldsymbol{r}_1} = \frac{-\beta}{(N-2)!\,Z_Q}\int\cdots\int\frac{\partial\Phi_{12}}{\partial \boldsymbol{r}_1}\exp(-\beta\Phi_N)\,\mathrm{d}\mathbf{3}\ldots\mathrm{d}\boldsymbol{N}$$
$$-\frac{\beta}{(N-2)!\,Z_Q}\int\cdots\int\sum_{j=3}^{N}\frac{\partial\Phi_{1j}}{\partial \boldsymbol{r}_1}\exp(-\beta\Phi_N)\,\mathrm{d}\mathbf{3}\ldots\mathrm{d}\boldsymbol{N}. \qquad (2.5.9)$$

On comparing the first term on the right of (2.5.9) with (2.5.5) we see that it is just $-\beta\rho(\partial\Phi_{12}/\partial \boldsymbol{r}_1)g_{(2)}(\boldsymbol{r})$. The second term is the sum of $(N-2)$ terms, i.e. for $j = 3$ to N. The first of these is

$$-\frac{\beta}{(N-2)!\,Z_Q}\int\cdots\int\frac{\partial\Phi_{13}}{\partial \boldsymbol{r}_1}\exp(-\beta\Phi_N)\,\mathrm{d}\mathbf{3}\ldots\mathrm{d}\boldsymbol{N} \qquad (2.5.10)$$

which may be written

$$-\frac{\beta}{(N-2)!\,Z_Q}\int_3\frac{\partial\Phi_{13}}{\partial \boldsymbol{r}_1}\left[\int\cdots\int\exp(-\beta\Phi_N)\,\mathrm{d}\mathbf{4}\ldots\mathrm{d}\boldsymbol{N}\right]\mathrm{d}\mathbf{3}. \qquad (2.5.11)$$

Setting $h = 3$ in the general expression (2.5.4) we see that (2.5.11) simplifies to

$$-\frac{\beta\rho^3}{(N-2)}\int\frac{\partial\Phi_{13}}{\partial \boldsymbol{r}_1}g_{(3)}(123)\,\mathrm{d}\mathbf{3}. \qquad (2.5.12)$$

As remarked above, there are $(N-2)$ similar integrals, which factor just cancels that in the denominator. Equation (2.5.9) finally reduces to the Born–Green–Yvon expression:

$$-kT\nabla_1 g_{(2)}(12) = g_{(2)}(12)\nabla_1\Phi(12)+\rho\int\nabla_1\Phi(13)g_{(3)}(123)\,\mathrm{d}\mathbf{3} \qquad (2.5.13)$$

which is seen to be identical to (2.5.3) obtained by a physical derivation.

It should be pointed out that thus far the BGY relation is exact in as far as the assumption of a pairwise decomposable total potential is valid. As it stands, however, the BGY equation given in (2.5.3) and (2.5.13) represents little more than a formal relationship between $g_{(2)}(12)$ and $g_{(3)}(123)$ and brings us no nearer a tractable expression for the pair distribution. This was a problem we anticipated in §2.3 (equation (2.3.6)). We could of course set up an analogous expression relating $g_{(3)}$ to $g_{(4)}$, and indeed this represents the approach of Cole and Fisher, to which we shall return later. A determination of $g_{(3)}$ in terms of $g_{(4)}$, however, merely postpones rather than overcomes the problem of breaking the linked chain of equations, and not until we terminate the hierachy and effect some form of 'closure' on the equations can we hope to have anything other than a formal solution to our problem.

Kirkwood (1935)[9] was the first to suggest a closure for $g_{(3)}(123)$, and this he did in terms of the pair distribution $g_{(2)}$. Kirkwood's 'closure', or superposition approximation as it is also known, provides a means of expressing the BGY equation entirely in terms of the pair distribution. The superposition approximation as developed by Kirkwood may be written

$$g_{(3)}(123) = g_{(2)}(12)\,g_{(2)}(23)\,g_{(2)}(31) \qquad (2.5.14)$$

which amounts to saying that the triplet distribution is equal to the product of the individual pair distributions. Physically this means that in the triplet of particles 1, 2, and 3, atoms 1 and 2 correlate as if they were independent of the presence of atom 3. Similarly for the other atoms taken pairwise. Clearly this is not the case in actuality, and only becomes valid in the limit of either the three atoms being greatly separated, as in the case of a dilute gas, or when the third atom is at a great distance from the other two. For, as $\boldsymbol{r}_{ij} \rightarrow \infty$, so $g_{(2)}(\boldsymbol{r}_{ij}) \rightarrow 1.00$ and

$$g_{(3)} \rightarrow g_{(2)} \rightarrow 1.00 \quad \text{as} \quad \boldsymbol{r}_{ij} \rightarrow \infty. \qquad (2.5.15)$$

The second situation, when the third atom is distant from the other two, also represents the limiting validity of the superposition approximation:

$$g_{(3)}(123) \rightarrow g_{(2)}(12) \quad \text{as} \quad g_{(2)}(13),\, g_{(2)}(23) \rightarrow 1.00. \qquad (2.5.16)$$

The assertion of the superposition approximation is that the pair correlations remain unaffected by the presence of a third atom even when the three are in strong mutual interaction, as in the case of fluids at liquid densities.

In terms of the potential of mean force we may write

$$g_{(3)}(123) = \exp\left(-\frac{\Psi(123)}{kT}\right), \qquad (2.5.17)$$

which, in terms of the superposition approximation we could write

$$\Psi(123) = \Phi(12)+\Phi(23)+\Phi(31). \qquad (2.5.18)$$

The exact expression would, of course, be

$$\Psi(123) = \Phi(12)+\Phi(23)+\Phi(31)+\Phi(123) \qquad (2.5.19)$$

(cf. § 1.3), where $\Phi(123)$ represents a correction to the sum of pair potentials originating the modification of the direct two-body interactions by the presence of the third. The Kirkwood approximation consists of neglecting the triplet term $\Phi(123)$. We may demonstrate the inconsistency of the approximation as follows. From the recurrence relationship (2.3.6) we have

$$g_{(2)}(12) = \frac{\rho}{(N-2)!}\int g_{(3)}(123)\,\mathrm{d}3 \qquad (2.5.20)$$

which, in the superposition approximation may be written

$$\frac{(N-2)!}{\rho} = \int g_{(2)}(23)\,g_{(2)}(31)\,\mathrm{d}3 \qquad (2.5.21)$$

which is clearly incorrect since the right hand side of (2.5.21) is a function of the separation 12, whilst the left hand side is apparently a constant. Thus the superposition approximation has obvious shortcomings. However, this is not to say that it is not of use for calculating the thermodynamic functions of both liquids and gases. It may be shown[10] that the superposition approximation is a simple mathematical expression representing the sum of a class of diagrams (to all orders of density) in a cluster expression. Furthermore, it enables us to write the BGY equation in closed form:

$$-kT\nabla_1 \ln g_{(2)}(12) = \nabla_1\Phi(12)+\rho\int \nabla\Phi(13)\,g_{(2)}(23)\,g_{(2)}(31)\,\mathrm{d}3. \qquad (2.5.22)$$

This equation is solved subject to the boundary condition

$$g_{(2)}(12) \to 1; \quad r_{12} \to \infty. \tag{2.5.23}$$

Thus we have a non-linear integro-differential equation relating the pair distribution to the pair potential $\Phi(r)$. It can be written in a number of forms[11,12], but none of these is particularly transparent, and need not concern us here.

Of the theories with which we shall be concerned in this book, the BGY theory is certainly the least successful when judged by the quality of its prediction of the thermodynamic functions[13]. However, it is unique in one respect in that the BGY equation in the superposition approximation ceases to yield convergent or physically acceptable solutions at densities, whilst high, are still appreciably less than those appropriate to a solid phase. It was this feature which, solved for hard spheres by Kirkwood, Maun and Alder[14], led Alder and Wainwright[15] and Wood and Jacobson[10] to perform molecular dynamic simulations to find quite clear evidence of a phase transition at about the density the BGY equation ceased to be integrable. Levesque[13] has found similar instabilities in solving the BGY equation for somewhat more realistic potentials. However, as to whether this is an accidental feature of the equation or whether it does describe the onset of solidification remains an open question.

One further point which will be of interest later in this chapter: Rushbrooke has shown that the BGY equation in the superposition approximation may be readily written in the form

$$\frac{\Psi(12)}{kT} = \frac{\Phi(12)}{kT} - \rho \int E(13)\,[g_{(2)}(23) - 1]\,\mathrm{d}3, \tag{2.5.24}$$

where
$$E(r) = \int_r^\infty \frac{\nabla\Phi(t)}{kT} g_{(2)}(t)\,\mathrm{d}t \tag{2.5.25}$$

and $\Psi(12)$ is the potential of mean force. Since $g_{(2)}(t) \to 1$ for sufficiently large t, we see from (2.5.25) that $E(r) \to -\Phi(r)/kT$ for large r. The form of (2.5.24) will be seen to bear strong similarities with a second class of equations for the pair distribution function (§§ 2.12, 2.13).

2.6 A test of the superposition approximation

It is pertinent at this point to ask just how satisfactory, and over what range of density, is the superposition approximation an adequate representation of the triplet distribution function $g_{(3)}(123)$. From the *a priori* considerations of the last section we concluded that the pair approximation to the triplet function had only limiting validity in the case of dilute gases when one or all of the three particles were widely separated. As we shall now see, the rather surprising conclusion on a test of the adequacy of the approximation is that Kirkwood's closure appears satisfactory even at liquid densities.

Whilst the triplet distribution $g_{(3)}(123)$ is experimentally relatively inaccessible the adequacy of the superposition approximation may be investigated by direct comparison of the function $g_{(3)}(123)$ computed by molecular dynamics with the corresponding computed superposition product. If the Kirkwood product of pair correlation functions, i.e. $g_{(2)}(12)g_{(2)}(23)g_{(2)}(31)$ were exact, the equilateral ratio

$$[g_{(3)}(x, x, x)]^{\frac{1}{3}}/g_{(2)}(x) \tag{2.6.1}$$

would be everywhere unity. Alder[17] has determined the above ratio for a rigid sphere fluid at $\rho_0/\rho = 1.60$ (ρ_0 = close packed density). From the data plotted in fig. 2.6.1 we see that the superposition approximation is remarkably accurate over most of the range of intermolecular separation. Alder concludes that the ratio has the value unity to better than 10 per cent.

Rahman[18] has investigated the adequacy of the approximation for a dense Lennard-Jones fluid. The spread about unity is somewhat greater than the hard sphere case, but is nevertheless surprisingly satisfactory and seems to show that what was formerly thought to be purely a low density approximation may be quite adequate at liquid densities (fig. 2.6.1).

The satisfactory nature of the superposition approximation at high densities may presumably be attributed to extensive cancellation amongst the higher order terms in the cluster expansion of the triplet distribution. We shall find further examples of density expansions, valid at low densities, yet apparently satisfactory at

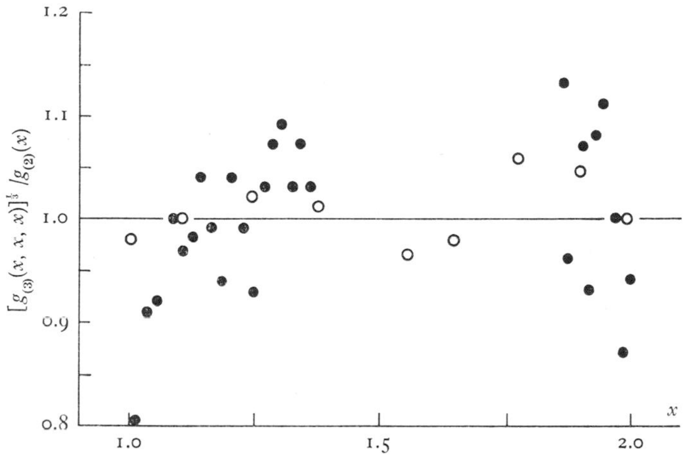

Fig. 2.6.1. A test of the adequacy of the Kirkwood superposition approximation for a hard sphere fluid $\rho_0/\rho = 1.60$, (open circles). The equilateral ratio $[g_{(3)}(x, x, x)]^{\frac{1}{3}}/g_{(2)}(x)$ would be unity for all x if the superposition approximation were exact. It is seen that there is a long range discrepancy which is not systematic, although the actual fluctuation about unity is not large. A test of the adequacy of the Kirkwood superposition approximation for a Lennard-Jones fluid (full circles). The scatter about the Kirkwood estimate is quite considerable. (Redrawn by permission from Rice and Gray, *The Statistical Mechanics of Simple Liquids* (Wiley).)

liquid densities. This fortuitous situation is to be generally attributed to cancellation amongst the higher order clusters and allows us to apply the graphical techniques over a wider range of densities than we might otherwise have anticipated.

Egelstaff, Page and Heard[18] have recently shown that the isothermal pressure derivative of the pair distribution function is rather simply related to the triplet distribution as follows:

$$\rho kT\left(\frac{\partial \overline{\rho_{(2)}(12)}}{\partial P}\right)_T = \int_3 [\overline{\rho_{(3)}(123)} - \overline{\rho\rho_{(2)}(12)}]\, d3 + 2\overline{\rho_{(2)}(12)}. \quad (2.6.2)$$

The triplet distribution function may be decomposed in terms of the superposition product and a supplement:

$$g_{(3)}(123) = g_{(2)}(12) f_{(2)}(23) g_{(2)}(31) - H(123) \quad (2.6.3)$$

$$\simeq g_{(2)}(12) g_{(2)}(23) g_{(2)}(31) - \gamma(12)\gamma(23)\gamma(31), \quad (2.6.4)$$

where the approximation has been made that H is a symmetrical function of γ. Clearly, if the superposition approximation adequately

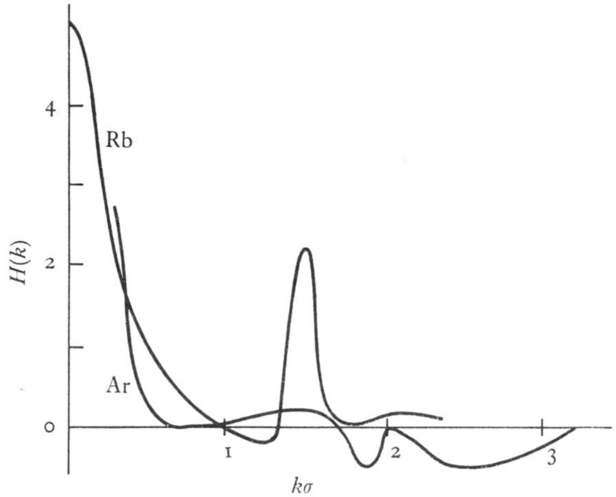

Fig. 2.6.2. The triplet supplement $H(k)$ to the superposition product for Rb and Ar. From the small k behaviour of these functions it is evident that the real-space supplement must have a long range form, even in the vicinity of the critical point for argon (143 °K, 52.5 atm). (Redrawn by permission from Egelstaff *et al.*, *J. Phys. Chem.* **4**, 1453 (1971).)

represented the triplet interaction, then $H = 0$. From an experimental determination of $(\partial \rho_{(2)}(12)/\partial P)_T$, or rather $(\partial S(k)/\partial P)_T$, we may determine $H(k)$. We might anticipate that as we approach the critical point $H(k) \to 0$ for all k. The experimental curve is shown in fig. 2.6.2. for rubidium (333 °K, 420 atm) near the triple point, and argon near the critical point (143 °K, 52.5 atm). We see that for argon, although $H(k) \sim 0$ for $k > 0.6\,\text{Å}^{-1}$, suggesting that the superposition approximation is adequate, it does appear that $H(123)$ has a long range form even at the critical point indicated by the small k behaviour of $H(k)$. In the case of rubidium it is clear that the superposition approximation is quite inadequate, again having a long range tail. The authors fit the trial function

$$\gamma(r) \sim \frac{\alpha}{r} \exp\left(-\frac{r}{\eta}\right) \tag{2.6.5}$$

to the Fourier transform of $H(k)_{(\text{Rb})}$, and conclude that the triplet correlation has a range of about 7 Å.

The pressure derivative $(\partial S(k)/\partial P)_T$ calculated on the basis of the BGY theory would, of course, yield $H = 0$. However, the derivative may also be estimated in the PY and HNC approximations when the adequacy of the supplements H_{PY} and H_{HNC} may be ascertained by comparison with the experimental values. The use of an experimentally determined $\gamma(r)$ in (2.6.4) would presumably allow a great improvement to be made over the superposition approximation, and since the expressions for physical quantities involving the triplet correlation function are integrals over one or more variables, it is possible that the approximation (2.6.4) will suffice for the calculation of these quantities.

Raveché and Mountain[79] have made a real-space investigation of three-body correlations in simple dense liquids by appeal to the isothermal *density* derivative of the pair probability density. They obtain

$$(\partial g_{(2)}(12)/\partial \rho)_T = \beta(\rho\chi_T)^{-1} g_{(2)}(12) \int [g_{(2)}(23) - 1] \times [g_{(2)}(13) - 1]\, d3 + g_{(2)}(12) \times \int g_{(2)}(23)\, g_{(2)}(31)\, [G_{(3)}(123) - 1]\, d3, \tag{2.6.6}$$

where
$$G_{(3)}(123) = \frac{g_{(3)}(123)}{g_{(2)}(12)\, g_{(2)}(23)\, g_{(2)}(31)}. \tag{2.6.7}$$

Clearly, if the superposition approximation were exact, then

$$G_{(3)}(123) \equiv 1.$$

Defining
$$\left.\begin{aligned} g'_{(2)}(12) &= \rho\beta^{-1}\chi_T[\partial \ln g_{(2)}(12)/\partial\rho]_T, \\ C(12) &= \int [g_{(2)}(23) - 1][g_{(2)}(13) - 1]\, d3, \\ F(12|G_{(3)}) &= \int g_{(2)}(23)\, g_{(2)}(13)\, [G_{(3)}(123) - 1]\, d3, \end{aligned}\right\} \tag{2.6.8}$$

(2.6.6) becomes
$$g'_{(2)}(12) = C(12) + F(12|G_{(3)}) \tag{2.6.9}$$

and in the superposition approximation $F \equiv 0$. In fig. 2.6.3 we show the various components of equation (2.6.9) computed from Michels' liquid argon data ($T = 143\,^{\circ}\text{K}, \rho = 1.48 \times 10^{22}$ atoms cm^{-3}). We see that the triplet term $F(12|G_{(3)})$ differs significantly from zero, and is long range oscillatory, in agreement with the data of Egelstaff *et al.* The function $F(12)$ is quite simply related to the $H(123)$ of (2.6.3) above.

An expression for $g'_{(2)}$ may be obtained on the basis of the PY

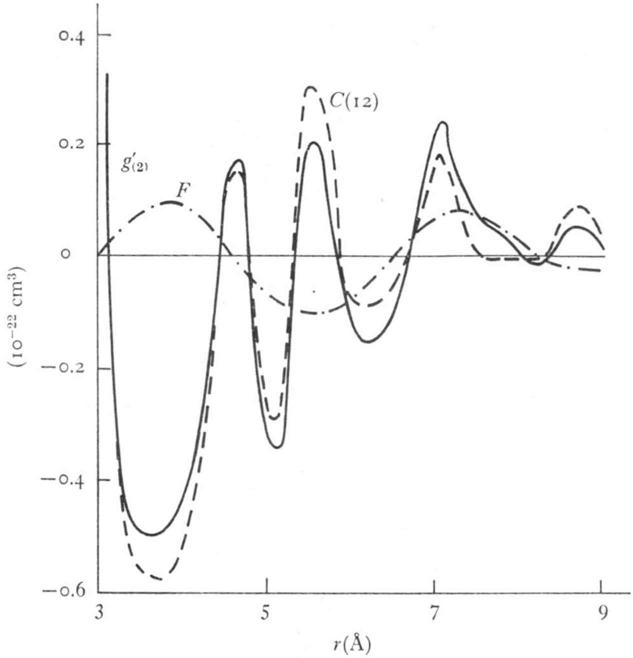

Fig. 2.6.3. The components of (2.6.6), defined in (2.6.8). The triplet term $F(12|G_{(3)})$ would be identically zero if the superposition approximation were exact. Instead F executes a long range oscillatory behaviour, indicative of the inadequacy of the superposition approximation in liquid argon ($T = 143$ °K, $\rho = 1.48 \times 10^{22}$ atoms cm^{-3}). (Redrawn by permission from Raveché and Mountain, *J. Chem. Phys.* **53**, 3101 (1970).)

approximation, and analytic expressions for C and F determined. The components for hard spheres are shown in fig. 2.6.3. where the long range triplet function is seen to develop very strong oscillations which is entirely in accord with Egelstaff, Page and Heard's results for liquid rubidium. The k space form of their triplet function $H(k)$ is seen to be a long range oscillatory function of wavelength $\sim (2\pi/1.5)$ Å which agrees very satisfactorily with the real-space determination. We may also compare the F-function with the distribution of points obtained by Rahman in fig. 2.6.1. The triplet function is again seen from fig. 2.6.4 to be long range oscillatory. The C-function is almost exactly out of phase with F resulting in a small $g'_{(2)}(12)$. A qualitative assessment for $g'_{(2)}(12)$ may be obtained from fig. 2.6.4 for hard spheres.

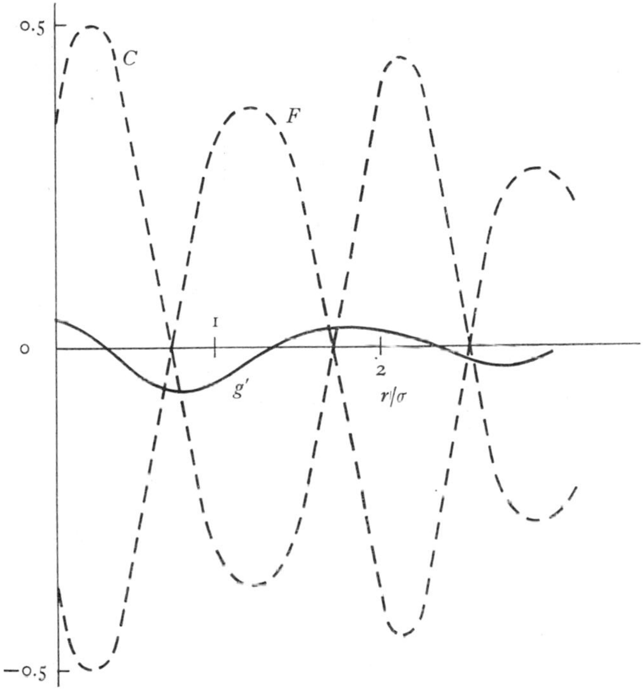

Fig. 2.6.4. The hard sphere derivative term $g'(r_{12})$, convolution term $C(r_{12})$ and three-body functional $F(r_{12}|g_{(3)})$ for $\rho\sigma = 0.883$. (Redrawn by permission from Ravechó and Mountain, *J. Chem. Phys.* **53**, 3101 (1970).)

2.7 The Rice–Lekner (RL) modification of the BGY equation

The qualitative correctness of the BGY equation, in as far as it ceases to be integrable beyond a certain limiting density, suggests that we might profitably investigate possible improvement of the Kirkwood superposition approximation. Bearing in mind the surprisingly satisfactory nature of the pair approximation to the triplet distribution (§2.6) at liquid densities it is to be anticipated that improvement of the superposition approximation will lead to a significant improvement of the radial distribution function.

Rowlinson[20] has calculated the first-order correction δ_4 to the Kirkwood superposition approximation analytically for hard spheres; it is negative and of large magnitude in the contact configuration. However, from fig. 2.6.1 we see that there must be extensive cancellation amongst the higher terms; some method of approximating the entire series is required. Salpeter[21] and Meeron[22] have shown that the exact expression for the triplet distribution function is

$$g_{(3)}(123) = g_{(2)}(12)g_{(2)}(23)g_{(2)}(31)\exp\left\{\sum_{n=1}^{\infty}\rho^n\delta_{n+3}(123)\right\}, \quad (2.7.1)$$

where the coefficients $\delta_{n+3}(123)$ give the contribution to correlations between the three fixed points and n field points. These coefficients are called 'simple 123 irreducible clusters' by Salpeter, and are defined below. Clearly we cannot hope to evaluate more than the first few terms of the infinite series in (2.7.1), but we can, by forming the Padé approximant to the series[23], estimate the asymptotic value of the series from the first few terms. In order to form the simplest Padé approximant we must determine the two coefficients, δ_4 and δ_5.

From the point of view of the potential of mean force, we see that the RL modification of the superposition approximation amounts to

$$\Psi(123) = \Phi(12)+\Phi(23)+\Phi(31)-kT\left\{\sum_{n=1}^{\infty}\rho^n\delta_{n+3}(123)\right\} \quad (2.7.2)$$

(cf. (2.5.19)).

The coefficient $\delta_4(123)$ is given as

$$\delta_4(123) =$$ 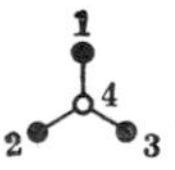 $$\equiv \int f_{14}f_{24}f_{34}\,\mathrm{d}4 \quad (2.7.3)$$

and represents the interaction of the triplet with one field point. The f-functions are defined by the usual Mayer relation:

$$f_{ij} = \exp\{-\Phi(ij)/kT\}-1. \quad (2.7.4)$$

For a hard sphere interaction f_{ij} has the form $f = -1$ $(r < 1.0)$, $f = 0$ $(r \geqslant 1.0)$: it is clear from (2.7.3) that δ_4 is everywhere negative for hard spheres of unit diameter.

For the integrand in (2.7.3) to be non-zero none of the f-bonds must be zero, i.e. particle 4 must be simultaneously in the range of 1, 2 and 3. This may be conveniently dealt with by treating particle 4 as a *point*, and enlarging the spheres to unit *radius*.

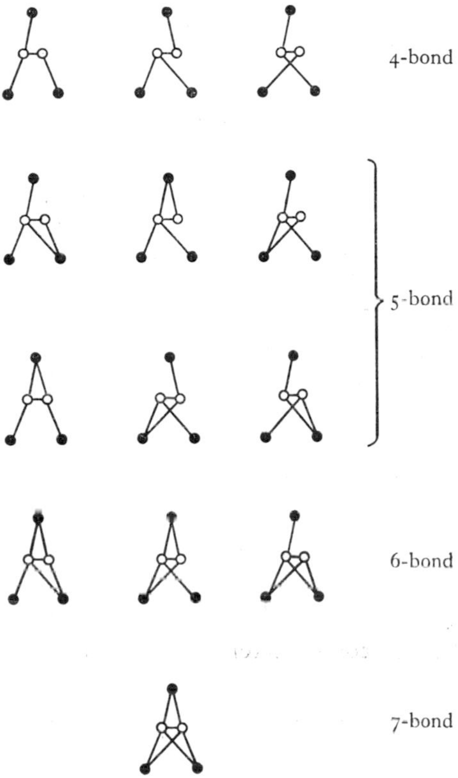

Fig. 2.7.1. Components of the coefficient δ_5. (Redrawn by permission from Rice and Lekner, *J. Chem. Phys.* 42, 3559 (1965).)

Geometrical considerations show that δ_4 amounts to the common volume of overlap of the spheres 1, 2 and 3.

For the evaluation of δ_5 we have to consider the interaction of particles 1, 2 and 3 with *two* field points, and this gives rise to the thirteen cluster diagrams[24] shown in fig. 2.7.1. They are arranged according to the number of bonds in the cluster, and for hard spheres the contribution of the cluster will be negative (positive) according as the number of bonds is odd (even). Clearly there will be extensive cancellation amongst the thirteen clusters, and to avoid unnecessary integration the diagrams in each column are combined into one integral.

Immediate simplification of many of the integrals is possible by integrating over one of the field points. For example

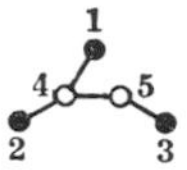

$$\equiv \int f_{14} f_{24} \,\mathrm{d}4 \int f_{35} f_{45} \,\mathrm{d}5 = \int f_{14} f_{24} p_{34} \,\mathrm{d}4, \qquad (2.7.5)$$

where

$$p_{34} = \int f_{35} f_{45} \,\mathrm{d}5 \equiv$$

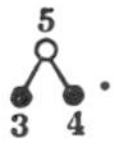

For hard spheres p_{34} is simply the common volume of the spheres 3 and 4 at the separation. Similarly,

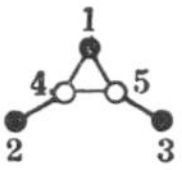

$$\equiv \int f_{14} f_{24} \,\mathrm{d}4 \int f_{15} f_{35} f_{45} \,\mathrm{d}5 = \int f_{14} f_{24} f_{134} \,\mathrm{d}4, \qquad (2.7.6)$$

where $t_{134} \equiv \delta_4(134)$, which is known. Thus, the first column may be quite simply shown to reduce to

$$\int f_{14} f_{24} (p_{34} + t_{134}) \, e_{34} \,\mathrm{d}4, \qquad (2.7.7)$$

where

$$e_{ij} = 1 + f_{ij} = \exp\{-\Phi(ij)/kT\}. \qquad (2.7.8)$$

Substitution of the Salpeter series in the BGY equation (2.5.13) yields

$$-kT\nabla_1 \ln g_{(2)}(12) = \nabla_1 \Phi(12) + \rho \int \nabla_1 \Phi(13) g_{(2)}(13) g_{(2)}(23) \,\mathrm{d}3$$
$$+\rho \int \nabla_1 \Phi(13) g_{(2)}(13) g_{(2)}(23)$$
$$\times \left\{ \exp\left[\sum_{n=1}^{\infty} \rho^n \delta_{n+3}(123) \right] - 1 \right\} \mathrm{d}3 \qquad (2.7.9)$$

which is seen to be of BGY form in the superposition approximation, plus a triplet term.

The asymptotic representation of the series is, in terms of the simplest Padé approximant,

$$\sum_{n=1}^{\infty} \rho^n \delta_{n+3}(123)\,(\text{Padé}) \sim \frac{\rho\delta_4(123)}{1 - \rho\delta_5(123)/\delta_4(123)}. \qquad (2.7.10)$$

The shortcomings of this approximation are immediately apparent: the first term of the series dominates, and the cancellation amongst the higher order terms is simulated by a 'scaling down' of the first term. The approximant would, however, be expected to be a much better approximation to the Salpeter series than $\rho\delta_4+\rho^2\delta_5$, i.e. the 'straight' first two terms of the series. The expectations are confirmed by the numerical calculations reported in the next chapter.

The RL modification to the BGY equation in the Kirkwood superposition approximation has been extended to include Lennard-Jones (6–12) fluids by Rice and Young[25]. The results are in substantial agreement with the molecular dynamics computations of Wood and Parker, and there can be no doubt that this approach represents a considerable improvement on (2.5.22).

We are now perhaps in a slightly better position to understand the surprisingly satisfactory nature of the Kirkwood superposition approximation. We have anticipated that extensive cancellation amongst the higher-order indirect effects on the triplet correlation must occur. This is particularly evident from the Salpeter series where the coefficients δ_m for hard spheres alternate in sign, being $+ve$ $(-ve)$ according as m is odd (even). This we understand from the nature of the hard sphere f-bond in the cluster expansion. The cancellation is, however, rather more complete in the case of the rigid sphere fluid than in a Lennard-Jones system where the coefficients δ_m are not exclusively positive or negative over their entire range. This accounts for the slightly less satisfactory nature of the superposition approximation in the LJ case (cf. fig. 2.6.1).

2.8 The Kirkwood equation (K)

In the derivation of the BGY equation the effect of an elementary displacement of one particle in the group upon the correlation of that group was investigated. The effect was to alter the 'coupling' of the first particle with the other members of the group. An alternative, and more explicit method of doing this is to introduce a coupling parameter, ξ, whose value varies continuously between 0 and 1. The total potential of the system may then be written,

$$\Phi_N = \xi \sum_{j=2}^{N} \Phi(1,j) + \sum_{j=2}^{N} \sum_{k=j+1}^{N} \Phi(j,k). \qquad (2.8.1)$$

The actual situation is, of course, represented by $\xi = 1$. $\xi = 0$ corresponds to complete decoupling of the first particle from the set. By this formal device we may simulate the elementary displacement in the BGY derivation. When the distribution functions are written in terms of the potential (2.8.1), they are accordingly functions of the parameter ξ. Variation of the coupling follows by differentiation with respect to ξ (not $\boldsymbol{r}_1$, as in the BGY derivation). Remembering $g_{(2)}(r, \xi)$ approaches unity as $\xi = 0$ and also as $r \to \infty$, we obtain

$$kT\frac{\partial}{\partial\xi}\ln g_{(2)}(12,\xi) = -\Phi(12)g_{(2)}(12,\xi) - \rho\int\Phi(13)\left\{\frac{g_{(3)}(123)}{g_{(2)}(12,\xi)} - 1\right\}g_{(2)}(12,\xi)\,d3. \quad (2.8.2)$$

Application of the superposition approximation to $g_{(3)}(123)$, and integration with respect to ξ gives

$$kT\ln g_{(2)}(12,\xi) = -\xi\Phi(12) - \rho\int_0^\xi\int\Phi(13)g_{(2)}(13,\xi)\{g_{(2)}(23)-1\}\,d3\,d\xi \quad (2.8.3)$$

which was first derived by Kirkwood and Boggs[26]. It is entirely equivalent to the BGY equation[78] up to the introduction of the superposition approximation, for if we write the BGY equation (2.5.22) in terms of the coupling parameter,

$$kT\nabla_1\ln g_{(2)}(12,\xi) = -\xi\nabla_1\Phi(12) - \rho\xi\int\nabla_1\Phi(13)g_{(2)}(13,\xi)g_{(2)}(23,\xi)\,d3 \quad (2.8.4)$$

and write the space gradient ∇_1 of the Kirkwood equation:

$$kT\nabla_1\ln g_{(2)}(12,\xi) = -\xi\nabla_1\Phi(12) - \rho\int_0^\xi\int\nabla_1[\Phi(13)g_{(2)}(13,\xi)]g_{(2)}(23)\,d3\,d\xi \quad (2.8.5)$$

comparison of the two equations shows that from the gradient in (2.8.5)

$$\nabla_1[\Phi(13)g_{(2)}(13,\xi)] = g_{(2)}(13,\xi)\nabla_1\Phi(13) + \Phi(13)\nabla_1 g_{(2)}(13,\xi) \quad (2.8.6)$$

we obtain the extra non-vanishing term $\Phi(13)\nabla_1 g_{(2)}(13,\xi)$ in the Kirkwood case. The two equations are therefore different because of this term. Numerical solutions of the Kirkwood equation will be considered in chapter 3.

2.9 The equation of Cole[27]

Whilst the Kirkwood closure itself seems to be quite adequate even at liquid densities, numerical solution of the BGY equation in the superposition approximation is quantitatively poor in comparison with the other first-order theories, although it does have the qualitative possibility of exhibiting a solid–fluid phase transition. The question of applying closure in the subspace of molecular *quadruplets* rather than triplets arises as a natural extension of the BGY equation in the spirit of the superposition approximation.

To do this we take the second gradient of (2.5.13):

$$\begin{aligned} -kT\nabla_1^2 g_{(2)}(12) &= [\nabla_1^2\Phi(12)+\nabla_1\Phi(12)\,\nabla_1 g_{(2)}(12)] \\ &\quad +\int [\nabla_1^2\Phi(13)\,g_{(3)}(123)+\nabla_1\Phi(13)\,\nabla_1 g_{(3)}(123)]\,\mathrm{d}3. \end{aligned} \tag{2.9.1}$$

Equation (2.9.1) is a tensor equation for $g_{(2)}(12)$ in terms of $g_{(2)}(123)$ and $\nabla_1 g_{(3)}(123)$. From the general relations (2.5.3 *a*) we have

$$\begin{aligned} \nabla_1 g_{(3)}(123) &= [\nabla_1\Phi(12)+\nabla_1\Phi(13)]\,g_{(3)}(123) \\ &\quad +\int \nabla_1\Phi(14)\,g_{(4)}(1234)\,\mathrm{d}4. \end{aligned} \tag{2.9.2}$$

Substitution of (2.9.2) in (2.9.1) yields

$$\begin{aligned} -(kT)^2\nabla_1^2 g_{(2)}(12) &= g_{(2)}(12)\,[(\nabla_1\Phi(12))^2+kT\nabla_1^2\Phi(12)] \\ &\quad +2kT\nabla_1 g_{(2)}(12)\,\nabla_1\Phi(12) \\ &\quad +\rho\int [kT\nabla_1^2\Phi(13)-(\nabla_1\Phi(13))^2] \\ &\quad \times g_{(3)}(123)\,\mathrm{d}3-\rho^2\iint \nabla_1\Phi(13) \\ &\quad \times \nabla_1\Phi(14)\,g_{(4)}(1234)\,\mathrm{d}3\,\mathrm{d}4. \end{aligned} \tag{2.9.3}$$

For the triplet distribution $g_{(3)}(123)$ we shall assume the Kirkwood form (2.5.14); the problem is now transferred to the closure of the quadruplet distribution $g_{(4)}(1234)$.

The superposition approximation appropriate to a group of four molecules may be selected in several ways. Cole sets

$$g_{(4)}(1234) \equiv g_{(3)}(123)\,g_{(2)}(14)\,g_{(2)}(24)\,g_{(2)}(34). \tag{2.9.4}$$

This formulation obviously has the effect of singling out molecule 4. Using (2.9.4), (2.9.3) may now be written in terms of $g_{(2)}$ only:

$$
\begin{aligned}
-(kT)^2 \frac{\nabla_1^2 g_{(2)}(12)}{g_{(2)}(12)} = {} & [(\nabla_1 \Phi(12))^2 + kT \nabla_1^2 \Phi(12)] \\
& + 2kT \nabla_1 \ln g_{(2)}(12) \nabla_1 \Phi(12) \\
& + \rho \int [kT \nabla_1^2 \Phi(13) - (\nabla_1 \Phi(13))^2] \\
& \times g_{(2)}(13) g_{(2)}(23)\, d3 - \rho^2 \int\int \nabla_1 \Phi(13) \\
& \times \nabla_1 \Phi(14) g_{(2)}(13) g_{(2)}(23) g_{(2)}(14) \\
& \times g_{(2)}(24) g_{(2)}(34)\, d3\, d4. \qquad (2.9.5)
\end{aligned}
$$

This tensor equation is incomparably more difficult to solve than the vector BGY equation. No numerical solutions have, in fact, been obtained, although (2.9.5) has been used by Cole to estimate first-order corrections to the 3-body superposition approximation in a density expansion for dilute gases:

$$g_{(3)}(123) = g_{(2)}(12) g_{(2)}(23) g_{(2)}(31) (1 + \alpha \rho k_{(3)}(123) + \ldots). \quad (2.9.6)$$

$k_{(3)}(123)$ is a known simple function of the set of three particles, and α is a numerical factor which is determined by this procedure.

2.10 The equation of Fisher[(28)]

Fisher proposes a similar approach to a second-order BGY equation to that of Cole, but accounts *exactly* for the three particle correlation function $g_{(3)}(123)$, and assumes a slightly different closure for the four-particle distribution $g_{(4)}(1234)$, thus:

$$g_{(3)}(123) = g_{(2)}(12) g_{(2)}(23) g_{(2)}(31)\, T_{(3)}(123), \qquad (2.10.1)$$

$$g_{(4)}(1234) = g_{(3)}(123) g_{(3)}(234) g_{(3)}(341) g_{(3)}(142). \qquad (2.10.2)$$

Here we have introduced the *indirect* triplet correlation function $T_{(3)}(123)$ according to which correlation amongst the set 123 occurs via a 4th, 5th, ... particle (cf. (2.7.1)). Equation (2.10.1) is of course exact.

Again, from (2.5.3 *a*) we may deduce two coupled vector equations for $h = 2, 3$:

$$-kT\nabla_1 \ln g_{(2)}(12) = \nabla_1\Phi(12) + \rho\int \nabla_1\Phi(13)\, g_{(2)}(13)\, g_{(2)}(23)\, T_{(3)}(123)\, d3 \qquad (2.10.3)$$

which amounts to the exact BGY equation, although in the unknowns $g_{(2)}$ and $T_{(3)}$. Also, we have the equation

$$-kT\nabla_1 \ln T_{(3)}(123) = \rho\int \nabla_1\Phi(14)\, g_{(2)}(14) \times [g_{(2)}(24)\, g_{(2)}(34)\, T_{(3)}(124)\, T_{(3)}(234)\, T_{(3)}(134) - g_{(2)}(24)\, T_{(3)}(124) - g_{(2)}(34)\, T_{(3)}(134)]\, d4. \qquad (2.10.4)$$

These latter two equations are to be solved 'simultaneously' for $g_{(2)}$ and $T_{(3)}$ subject to the conditions

$$\left.\begin{aligned} g_{(2)}(12) &\to 1 \\ T_{(3)}(123) &\to 1 \end{aligned}\right\} \quad \text{as any one of the molecules goes off to infinity.}$$

These latter conditions ensure correct normalization and physical significance of the correlation functions.

Ree, Lee and Ree[29] have taken the second and third equations in the BGY hierachy for $h = 2$ and $h = 3$, and used as closure the approximation

$$g_{(4)}(1234) = \frac{g_{(3)}(123)\, g_{(3)}(234)\, g_{(3)}(341)\, g_{(3)}(421)}{g_{(2)}(12)\, g_{(2)}(13)\, g_{(2)}(14)\, g_{(2)}(23)\, g_{(2)}(24)\, g_{(2)}(34)} \qquad (2.10.5)$$

as suggested several years ago by Fisher and Kopeliovich[30]. They have confined attention to the hard sphere fluid, and published work is still at the level of successive virial coefficients. Nevertheless, this genuinely second-order BGY theory appears to be comparable in accuracy to the direct correlation theories to be reported later in this chapter.

2.11 Abe's series expansion of the BGY equation

Abe[31] has obtained a density expansion of the BGY equation in the superposition approximation by successive resubstitution. Thus, in the BGY equation

$$-kT\nabla_1 g_{(2)}(12) = g_{(2)}(12)\nabla_1\Phi(12) + \rho\int g_{(2)}(12)g_{(2)}(23)g_{(2)}(13)\nabla_1\Phi(13)\,\mathrm{d}\mathbf{3} \qquad (2.11.1)$$

we may substitute for $g_{(2)}(13)\nabla_1\Phi(13)$ from

$$-kT\nabla_1 g_{(2)}(13) = g_{(2)}(13)\nabla_1\Phi(13) + \rho\int g_{(3)}(134)\nabla_1\Phi(14)\,\mathrm{d}\mathbf{4} \qquad (2.11.2)$$

giving

$$-kT\nabla_1 g_{(2)}(12) = g_{(2)}(12)\nabla_1\Phi(12) + \rho\int g_{(2)}(12)g_{(2)}(23) \times\left[-\nabla_1 g_{(2)}(13)kT - \rho\int g_{(2)}(13)g_{(2)}(34) \times g_{(2)}(14)\nabla_1\Phi(14)\,\mathrm{d}\mathbf{4}\right]\mathrm{d}\mathbf{3}. \qquad (2.11.3)$$

Repeating this process for $g_{(2)}(14)\nabla_1\Phi(14)$ yields

$$-kT\nabla_1 g_{(2)}(12) = g_{(2)}(12)\nabla_1\Phi(12) - kT\rho\int g_{(2)}(12)g_{(2)}(23) \times\nabla_1 g_{(2)}(13)\,\mathrm{d}\mathbf{3} + kT\rho^2\iint g_{(2)}(12)g_{(2)}(23) \times g_{(2)}(13)g_{(2)}(34)\left[-\nabla_1 g_{(2)}(14)kT - \rho\int g_{(2)}(14)g_{(2)}(45)g_{(2)}(15)\nabla_1\Phi(15)\,\mathrm{d}\mathbf{5}\right]\mathrm{d}\mathbf{4}\,\mathrm{d}\mathbf{3} + \ldots. \qquad (2.11.4)$$

The general expression of this series is,

$$-kT\nabla_1 \ln g_{(2)}(12) = \nabla_1\Phi(12) + kT\sum_{n=1}^{\infty}(-1)^n\rho^n \times\int_3\cdots\int_{n+2}\{\nabla_1 g_{(2)}(1,n+2)\}g_{(2)}(n+1,n+2) \times\prod_{i=3}^{n+1}\{g_{(2)}(1,i)g_{(2)}(i-1,i)\}\,\mathrm{d}\mathbf{3}\ldots\mathrm{d}(\boldsymbol{n}+\mathbf{2}). \qquad (2.11.5)$$

Each term in the series can be represented by a cluster diagram of a simple closed loop and a complete set of cross links radiating from particle 1.

Abe[32] also establishes a formal solution of the indirect triplet correlation $T_{(3)}(123)$ defined by

$$g_{(3)}(123) = g_{(2)}(12)g_{(2)}(23)g_{(2)}(31)\,T_{(3)}(123)$$

in terms of the infinite power series in the density. In practice, however, this solution is only applicable to gaseous systems at low density.

A useful approximate closed form to the infinite series (2.11.5) has been obtained by Abe who sets the factors $g(1,i)$ equal to unity. The resulting solution is

$$-\frac{\Phi(r)}{kT} = \ln g_{(2)}(r) - \frac{1}{(2\pi)^3\rho}\int\frac{[1-S(k)]^2}{S(k)}\,e^{i\boldsymbol{k}\cdot\boldsymbol{r}}\,d\boldsymbol{k} \qquad (2.11.6)$$

which may be very easily shown to be

$$-\frac{\Phi(r)}{kT} = \ln g_{(2)}(r) - [h(r) - c(r)]. \qquad (2.11.7)$$

This, interestingly enough, is the HNC approximation (equation (2.14.15)). Davison and Lee[87] have established a linear correction to the Abe series (2.11.5) by taking account of the $g(1,i)$ factors. This results in a series extension to (2.11.6) which is truncated at the first supplementary term.

Richardson[33] sought to determine the optimum form of $T_{(3)}(123)$ based on a variational minimization of the free energy of the system. In this approach $T_{(3)}$ turned out to be a constant independent of the coordinates (although it was a function of temperature and density). Clearly this is incorrect if the normalization condition cannot be met, and if its asymptotic form differs from unity.

2.12 Direct and indirect correlation: the Ornstein–Zernike equation

Thus far we have discussed integro-differential relations for the lower order distribution functions in terms of the hierachical relation which exists between adjacent orders of distribution. We found in every case that the h-body distribution can only be expressed in terms of the $(h+1)$-body distribution. Consequently some form of closure device was necessary, and in the majority of cases the Kirkwood superposition approximation was employed. The particular form of the superposition approximation was chosen on the grounds of mathematical expediency, and a subsequent analysis showed that the closure was not altogether physically satisfactory although it had to be admitted that the superposition approximation to the triplet distribution in terms of the pair distribution was surprisingly accurate even at liquid densities. As we shall see in chapter 3, the integro-differential equations in the closure approximation unfortunately appear to amplify any shortcomings in the closure device. Nevertheless, the BGY and K equations do retain the possibility of demonstrating the onset of long range order, a feature possessed by no other class of equations yet developed.

We now move on to discuss a second class of equations which have a somewhat more physical basis. We must first introduce the concepts of *total*, *direct* and *indirect correlation.* The total correlation between two atoms may be considered to be comprised of two components: evidently there will be direct correlation between the two particles, but there will also be correlation on the relative position of 1 and 2 imposed via a *third* particle. The total correlation may therefore be written as the sum of the direct effect plus the indirect effect averaged over all possible positions of the third representative molecule, *subject to its remaining directly correlated with* 1. This is most clearly shown in terms of the Ornstein–Zernike (OZ) integral equation[34]

$$h(12) = c(12)+\rho\int c(13)\,h(23)\,\mathrm{d}3 \qquad (2.12.1)$$

which may be regarded as a definition of the direct correlation function, $c(r)$. The total correlation is defined as

$$h(r) = g_{(2)}(r)-1 \qquad (2.12.2)$$

and so (2.12.1) may be regarded as an integral equation for the pair distribution $g_{(2)}(r)$ in terms of the unknown function $c(r)$. Clearly

(2.12.1) tells us nothing new: it is merely a defining relation, and the problem is transferred to the determination of the direct correlation function. However, several prescriptions for $c(r)$ have been proposed on various grounds and furthermore, $c(r)$ is experimentally more accessible than $g_{(2)}(r)$, (3.6.1). Again, the correlation functions $h(r)$ and $c(r)$ can be defined for pairs, triplets, and higher groups of molecules, but for simplicity only the most important of them – the pair functions – are described here. We consequently drop the subscript (2) in the total correlation approach.

Notice how the three-body dependence of the total correlation $h(r)$ is implicitly included in the integral of the OZ defining relation (2.12.1). The indirect correlation represented in (2.12.1) by the integral expression may in fact be represented by a series expansion of the direct function. Thus, if we repeatedly replace $h(r)$ by $[c(r)+\text{integral}]$ within the integral of (2.12.1) we obtain

$$h(12) = c(12)+\rho\int c(13)\,c(23)\,\mathrm{d}3 + \rho^2\iint c(13)\,c(34)\,c(42)\,\mathrm{d}3\,\mathrm{d}4+\ldots \qquad (2.12.3)$$

The total correlation $h(12)$ is decomposed into a direct correlation of 1 and 2, through $c(12)$, and indirectly through all possible chains of direct correlation within the fluid.

If we now recall Rushbrooke's alternative expression of the BGY equation[35], (2.5.24)

$$\frac{\Psi(12)}{kT} = \frac{\Phi(12)}{kT} - \rho\int E(13)\,[g_{(2)}(23)-1]\,\mathrm{d}3 \qquad (2.12.4)$$

and since, in Boltzmann form

$$g_{(2)}(12) = \exp\left(-\frac{\Psi(12)}{kT}\right) \qquad (2.12.5)$$

then Rushbrooke's equation asymptotically ($r_{12}\rightarrow\infty$) reduces to

$$h(12) \sim -\frac{\Phi(12)}{kT} + \rho\int E(13)\,h(23)\,\mathrm{d}3 \qquad (2.12.6)$$

which bears a strong resemblance to the OZ relation. It enables us to draw some tentative conclusions concerning the long range form

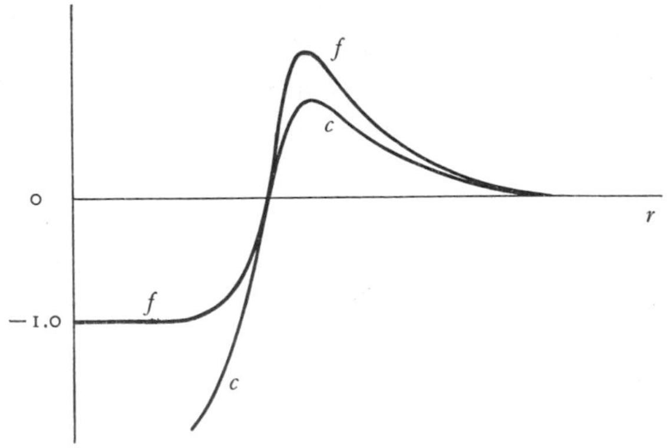

Fig. 2.12.1. A comparison of the direct correlation function $c(r)$, and the Mayer f-function for liquid argon at 148 °K, $\rho = 1$ g cm^{-3}, $kT/\epsilon = 1.23$.

of $c(12)$, for (2.12.6) implies, by direct comparison with the OZ equation, that the direct correlation function varies as the long range form of the pair potential (but for a factor $-1/kT$). This circumstance has enabled Johnson, Hutchinson and March[36] to arrive at some interesting conclusions concerning the asymptotic form of the pair potential. We shall return to this point in chapter 3.

The indirect correlation $\rho \int c(13)\, h(23)\, d3$ has been deduced from the experimental measurement of $h(r)$ and $c(r)$ together with the OZ relation equation (2.12.1)). At low densities the OZ relation might be expected to reduce to

$$h(12) \to c(12), \tag{2.12.7}$$

that is,

$$g(12) \to 1 + c(12). \tag{2.12.8}$$

And again, at low densities we might expect

$$\Psi(12) \to \Phi(12), \tag{2.12.9}$$

whereupon, from (2.12.8), (2.12.9), we are able to write in Boltzmann form:

$$c(12) \to \exp\left\{-\frac{\Phi(12)}{kT}\right\} - 1. \tag{2.12.10}$$

That is, the direct correlation would appear to reduce to the Mayer f-function in the limit of low densities. At $r \sim \sigma$, the atomic diameter, the direct correlation will presumably predominate over the indirect, and indeed we see that it is the direct correlation between atoms 1 and 2 which is responsible for the first peak in the total correlation. This again is strictly valid only in the limit of low densities. Nevertheless, the indentification of the direct correlation function with the Mayer f-function provides the first analytic estimate for $c(r)$. We show in fig. 2.12.1 a comparison of $f(r)$ and $c(r)$ for liquid argon at 148 °K and $kT/\epsilon = 1.23$. At these densities the Mayer f-function is seen to yield an over estimate of the direct correlation.

Two important approximate forms of $c(r)$ have been proposed, and these will be considered in turn in a moment. It will not surprise us that they may be considered as modifications of the simple relation (2.12.10). Before passing on to consider the two approximations some comment is appropriate. The original routes to the hypernetted chain (HNC) and Percus–Yevick (PY) approximations to $c(r)$ were based on quite different lines of reasoning to the approach given here. We first give an apparently *ad hoc* treatment of the HNC and PY approximations in terms of the potential of mean force. We then go on to consider the diagrammatic origins of the two theories when we shall see the physical basis of the approximation.

2.13 The HNC and PY approximations: the $\Psi(r)$ approach

In Boltzmann form, the radial distribution function $g_{(2)}(12)$ may be written in terms of the potential of mean force, thus

$$\ln g_{(2)}(12) = -\frac{\Psi(12)}{kT}. \qquad (2.13.1)$$

The potential of mean force may be split up into two components, much as in the OZ relation:

$$\Psi(12) = \Phi(12) + W(12). \qquad (2.13.2)$$

Here $\Phi(12)$ is the 'direct' pair potential, and $W(12)$ is an unknown effective supplement to the pair potential. $W(12)$ represents the

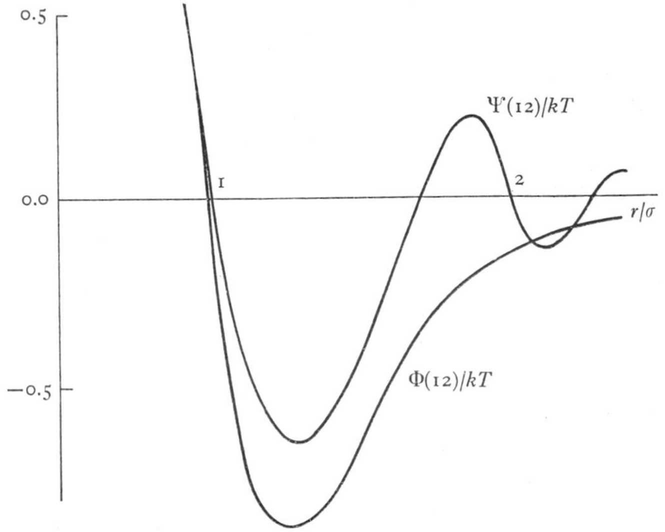

Fig. 2.13.1. A comparison of the pair potential for liquid argon $\Phi(r)$, and the potential of mean force, $\Psi(r)$, in units of kT under the same conditions as in fig. 2.12.1.

mean effect of a third particle averaged over all possible positions. We see from Rushbrooke's form of the BGY equation, (2.5.24)

$$\Psi(12) = \Phi(12) - \rho \int E(13)\, h(23)\, d3,$$

where
$$E(r) = \int_r^\infty \nabla\Phi(t)\, g_{(2)}(t)\, dt,$$

that
$$W(12) = -\rho \int E(13)\, h(23)\, d3. \qquad (2.13.3)$$

The pair potential and the potential of mean force determined from (2.13.1) are shown in fig. (2.13.1) for the data of Mikolaj and Pings[37].

The expression for the direct correlation in the HNC approximation is given as[38]:

$$c_{\text{HNC}}(12) = h(12) - \ln g_{(2)}(12) - \frac{\Phi(12)}{kT}. \qquad (2.13.4)$$

This expression, when inserted in the OZ relation provides a closed equation for the total correlation and radial distribution function. The term, 'hypernetted chain' does not seem particularly appropriate, but the original cluster expansion developed for $c(12)$ in the HNC approximation accounts for the name of the theory. From (2.13.1) and (2.13.2) it is evident that the 'potential tail'

$$W_{\mathrm{HNC}}(12) = kT[c(12)-h(12)]. \tag{2.13.5}$$

The potential of mean force in the HNC approximation is then

$$\frac{\Psi_{\mathrm{HNC}}(12)}{kT} = \frac{\Phi(12)}{kT} - [h(12)-c(12)]. \tag{2.13.6}$$

The component functions have been obtained from the experimental data of Mikolaj and Pings[37].

Similarly, in the Percus–Yevick approximation[39]

$$c_{\mathrm{PY}}(12) = g_{(2)}(12)\left[1-\exp\left(\frac{\Phi(12)}{kT}\right)\right], \tag{2.13.7}$$

i.e.
$$\ln g_{(2)}(12) = -\frac{\Phi(12)}{kT} + \ln[g_{(2)}(12)-c(12)], \tag{2.13.8}$$

and, from (2.13.1) and (2.13.2)

$$\frac{\Psi_{\mathrm{PY}}(12)}{kT} = \frac{\Phi(12)}{kT} - \ln[1+h(12)-c(12)]. \tag{2.13.9}$$

There is a serious discrepancy between the observed and the HNC pair distribution function particularly in the vicinity of the first peak. The pressure equation, or the equation of state, (2.3.18), is particularly susceptible to small errors in the amplitude and/or position of the first peak of the RDF: Gaskell[40] has exposed serious defects in the hyperchain theory when used to calculate fluid pressure. Originating from a low density cluster expansion, it may be shown that cancellation amongst the higher order diagrams is not so complete in the HNC as in the PY approximation for $c(r)$. The HNC approximation appears to improve with decreasing density, and indeed the two approximations coincide at low densities and large r when $\ln(1+h-c) \sim h-c$ that is, when $h-c$ is sufficiently small. The nodes occur in the HNC approximation when $\Psi_{\mathrm{HNC}}(12)$ and $h(12) - 0$, i.e. when

$$\frac{\Phi(12)}{kT} = -c(12), \tag{2.13.10}$$

and in the PY approximation when

$$\frac{\Phi(12)}{kT} = \ln[1 - c(12)]. \qquad (2.13.11)$$

Again, these agree only when $c(12) \ll 1.0$. Furthermore, if we take (2.13.6) and (2.13.9) for the potentials of mean force in the HNC and PY approximations, insertion of the same $c(r)$ and $h(r)$ data yields

$$\Psi_{\text{HNC}} \geqslant \Psi_{\text{PY}} \qquad (2.13.12)$$

since $x \geqslant \ln(1+x)$ for all x.

The *range* of the two approximations for $c(r)$ may be shown to be different. The expression for $c_{\text{PY}}(r)$ (equation (2.13.7)) may be written

$$c_{\text{PY}}(r) = g_{(2)}(r) \exp\left(\frac{\Phi(r)}{kT}\right)\left[\exp\left(-\frac{\Phi(r)}{kT}\right) - 1\right], \qquad (2.13.13)$$

where the expression in the square brackets is, of course, the Mayer f-function:

$$c_{\text{PY}}(r) = g_{(2)}(r) f(r) \exp\left(\frac{\Phi(r)}{kT}\right). \qquad (2.13.14)$$

The f-function has the effect of restricting the range of $c(r)$ to that of the pair potential. And, in particular, for hard spheres $c_{\text{PY}}(r)$ would be expected to be zero beyond the hard sphere radius. Thus in the PY approximation when adjacent spheres are not in contact there can only be indirect correlation – but then this is intuitively obvious. There is no such apparent constriction on the range of $c_{\text{HNC}}(r)$, although the direct correlation must nevertheless be of shorter range than that of the total correlation $h(r)$. We observed earlier, (2.12.10), that a first-order approximation to $c(r)$ might be the Mayer f-function: evidently (2.13.14) represents a refinement of that approximation, although both the HNC and the PY expressions reduce to $f(r)$ in the limit $r \to \infty$, $\rho \to 0$.

The hypernetted chain approximation originated from a class of diagrams retained from a cluster expansion of the direct correlation function. We shall consider the diagrammatic approach in the next section. What is important to observe here is that, like the cluster expansions of the partition function discussed in chapter 1, the HNC approximation is essentially restricted to low density applications. Now we have seen an example, the superposition approximation to the triplet function $g_{(3)}(123)$, where *fortuitously* extensive cancellation amongst the higher order diagrams enables the approximation to be applied beyond its 'official' range of validity with reasonable success. It appears that the higher order cancellation in the HNC approximation is not as extensive as we might have hoped. In fact when we examine the cluster expansions in the HNC and PY approximations we shall find that *diagrammatically* the PY is a more drastic approximation than the HNC. It just so happens that the cancellation is more complete in the PY case. It must be said, however, that the Percus–Yevick theory was not developed on the basis of cluster techniques at all, but rather in terms of a collective coordinate approach, and as such is a theory applicable at all densities. The original approach sought to transform the system Hamiltonian in such a way that the potential component of the Hamiltonian is expressed in terms of single particle functions. Density fluctuations are imagined to propagate throughout the liquid and the Hamiltonian is expressed in terms of the Fourier components of the fluctuations. Approximation enters the theory in the selection of the Fourier components for a system with a finite number $(3N-3)$ degrees of freedom (cf. Debye's theory of the specific heat of a solid). Any assessment of the quantitative error introduced by a chosen procedure is about as difficult to perform as it is to estimate the error introduced in using the superposition approximation. The graphical approach to the PY approximation is due essentially to Stell[41]. It is clear that the PY approximation may be applied at all liquid densities. Numerical predictions resulting from the two theories will be considered in chapter 3.

Substitution of the HNC approximation in the OZ equation yields

$$\ln g_{(2)}(12)+\frac{\Phi(12)}{kT} = \rho\int\left[g_{(2)}(13)-1-\ln g_{(2)}(13)-\frac{\Phi(13)}{kT}\right][g_{(2)}(23)-1]\,d3. \tag{2.13.15}$$

Taking the space gradient ∇_1 of both sides, and after some rearrangement we obtain (HNC–OZ):

$$\nabla_1\left\{\ln g_{(2)}(12)+\frac{\Phi(12)}{kT}\right\} = -\rho\int\frac{\nabla_1\Phi(13)}{kT}g_{(2)}(23)g_{(2)}(13)\,d3$$
$$+\rho\int\left[\nabla_1 g_{(2)}(13)-\nabla_1\ln g_{(2)}(13)-\frac{\nabla_1\Phi(13)}{kT}\right]$$
$$\times[g_{(2)}(23)-1]\,d3$$
$$+\rho\int\frac{\nabla_1\Phi(13)}{kT}g_{(2)}(23)g_{(2)}(13)\,d3 \quad (2.13.16)$$

with which we may compare the BGY equation in the superposition approximation:

$$\nabla_1\left\{\ln g_{(2)}(12)+\frac{\Phi(12)}{kT}\right\} = -\rho\int\frac{\nabla_1\Phi(13)}{kT}g_{(2)}(23)g_{(2)}(13)\,d3.$$

This somewhat laboured comparison enables us to isolate the difference between the BGY and HNC–OZ equations for the pair distribution function. The comparison is rather more direct in Rushbrooke's expression of the BGY equation (2.5.24) in the superposition approximation:

$$\nabla_1\left\{\ln g_{(2)}(12)+\frac{\Phi(12)}{kT}\right\} = -\rho\int E(13)\,[g_{(2)}(23)-1]\,d3$$

whereupon, with (2.13.15) we have

$$E(13) \equiv \nabla_1\left[g_{(2)}(13)-\ln g_{(2)}(13)-\frac{\Phi(13)}{kT}\right], \quad (2.13.17)$$

where $$E(r) = \int_r^\infty \frac{\nabla\Phi(t)}{kT}g_{(2)}(t)\,dt.$$

Thus we have reduced the HNC–OZ equation to a form whereby a distinction may be drawn between the two closure devices. However, this is not a particularly transparent approach to the substantiation of the form for $c_{\text{HNC}}(r)$: this is best done by diagram techniques.

Substitution of the PY approximation in the OZ equation gives

$$g_{(2)}(12)\exp\left(\frac{\Phi(12)}{kT}\right)$$
$$= 1+\rho\int g_{(2)}(13)\left[1-\exp\left(\frac{\Phi(13)}{kT}\right)\right][g_{(2)}(23)-1]\,d3. \quad (2.13.18)$$

Taking the space gradient ∇_1 of both sides, and dividing throughout by $g_{(2)}(12)\exp(\Phi(12)/kT)$ gives

$$\left\{\frac{\nabla_1 g_{(2)}(12)}{g_{(2)}(12)}+\frac{\nabla_1\Phi(12)}{kT}\right\} = \rho\exp\left(-\frac{\Phi(12)}{kT}\right)\int\frac{\nabla_1 g_{(2)}(13)}{g_{(2)}(12)}$$

$$\times\left[1-\exp\left(\frac{\Phi(13)}{kT}\right)\right][g_{(2)}(23)-1]\,d\mathbf{3}$$

$$-\rho\exp\left(-\frac{\Phi(12)}{kT}\right)\int\frac{g_{(2)}(13)}{g_{(2)}(12)}$$

$$\times\left[\frac{\nabla_1\Phi(13)}{kT}\exp\left(\frac{\Phi(13)}{kT}\right)\right]$$

$$\times[g_{(2)}(23)-1]\,d\mathbf{3}, \qquad (2.13.19)$$

i.e.

$$\nabla_1\left\{\ln g_{(2)}(12)+\frac{\Phi(12)}{kT}\right\} = \frac{\rho}{g_{(2)}(12)}\exp\left(-\frac{\Phi(12)}{kT}\right)$$

$$\times\int\left\{\nabla_1 g_{(2)}(13)-\exp\left(\frac{\Phi(13)}{kT}\right)\right.$$

$$\left.\times\left[\nabla_1 g_{(2)}(13)+g_{(2)}(13)\frac{\nabla\Phi(13)}{kT}\right]\right\}$$

$$\times[g_{(2)}(23)-1]\,d\mathbf{3}, \qquad (2.13.20)$$

whereupon, from the Rushbrooke formulation of the BGY equation in the superposition approximation we may make the identification

$$E(13)\equiv\exp\left(-\frac{\Phi(12)}{kT}\right)\left\{\exp\left(\frac{\Phi(13)}{kT}\right)\right.$$

$$\left.\times\left[\nabla_1 g_{(2)}(13)+g_{(2)}(13)\frac{\nabla_1\Phi(13)}{kT}\right]-\nabla_1 g_{(2)}(13)\right\}\Big/ g_{(2)}(12)$$

$$(2.13.21)$$

which may be reduced to

$$\frac{1}{g_{(2)}(12)}\exp\left(-\frac{\Phi(12)}{kT}\right)\nabla_1\left\{g_{(2)}(13)\exp\left(\frac{\Phi(13)}{kT}\right)-g_{(2)}(13)\right\}, \qquad (2.13.22)$$

or in terms of the Mayer f-function

$$-\frac{1}{g_{(2)}(12)}\exp\left(-\frac{\Phi(12)}{kT}\right)\nabla_1\left\{g_{(2)}(13)\exp\left(\frac{\Phi(13)}{kT}\right)f(13)\right\}. \qquad (2.13.23)$$

By direct comparison of (2.13.17) and (2.13.23) for the HNC and PY kernels respectively we again see that the range of the E-function in the latter case is restricted to that of the pair interaction since $f(13)$ and $\nabla_1 f(13)$ are of this range. This is not the case for the HNC identification where, in fact, the range of E is of the order of $g_{(2)}(13)$. This should not surprise us – we have already noted the formal similarity of the BGY equation in the Rushbrooke form and the Ornstein–Zernike relation. From the comparison it appeared that we may make the tentative association (not identification) of $E(13)$ and $c(13)$. The direct correlation in the PY approximation has already been shown to be of shorter range than that of the HNC approximation and this evidently reappears in the form of $E(r)$.

The advantage of expressing the pair distribution in terms of short range functions, of the order of the pair potential, is that these functions are predominantly determined by the nature of the pair interaction. This is clearly a more wieldy situation than one involving longer range functions where the form of the tail is imprecisely known. Moreover, the long range functions will inevitably require knowledge of the small-angle form of the experimentally determined scattering function; measurements at small angles are notoriously difficult.

2.14 The HNC and PY approximations: the diagrammatic approach

Rushbrooke and Scoins[42], by Fourier transforming the Ornstein–Zernike relation, find that they can express the direct correlation function in terms of the expansion[92]

$$c(12) = \sum_{n\geqslant 1} \alpha_{n+1}(12)\rho^{n-1}, \tag{2.14.1}$$

where $$\alpha_2(12) = f(12),$$

i.e. $\alpha_2(12)$ is the Mayer f-function. The coefficients $\alpha_{n+1}(12)$ are given by

$$\alpha_{n+1}(12) = \frac{1}{(n-1)!}\int \dots \int \Sigma\Pi f(ij)\, d3 \dots d(N-1) \tag{2.14.2}$$

so that, for example

$$\alpha_3(12) = \int f(12)f(23)f(13)\, d3 \tag{2.14.3}$$

and,

$$\alpha_4(12) = \frac{1}{2}\Bigg[\int\!\!\int \{f(12)f(23)f(34)f(14)f(24)f(13)$$
$$+f(12)f(23)f(34)f(14)f(24)+f(12)f(23)f(34)f(14)f(13)$$
$$+f(12)f(23)f(34)f(24)f(13)+f(12)f(14)f(34)f(24)f(13)$$
$$+f(12)f(23)f(14)f(24)f(13)+f(23)f(34)f(14)f(24)f(13)$$
$$+f(12)f(23)f(34)f(14)+f(12)f(13)f(34)f(24)$$
$$+f(13)f(23)f(24)f(14)\}\,\mathrm{d}3\,\mathrm{d}4\Bigg]. \qquad (2.14.4)$$

Of the ten terms appearing in α_4 the second, third, fourth and fifth are numerically equal, and so are the eighth and ninth; but we have given the expression in full to illustrate clearly how the weighting factors arise. Of more interest, perhaps, is the observation that eight of these ten terms contain the factor $f(12)$ and therefore decay rapidly with distance. The remaining two terms, of course, decay more slowly having roughly twice the effective range of $f(12)$. It is these latter terms which ensure the continuity of $c(r)$ even for discontinuous potentials. It is obviously of no value to give explicitly the 238 terms involved in α_5, but of these 166 contain $f(12)$ as a factor.

The cluster expansion of the direct correlation function (2.14.1) may evidently be expressed diagrammatically:

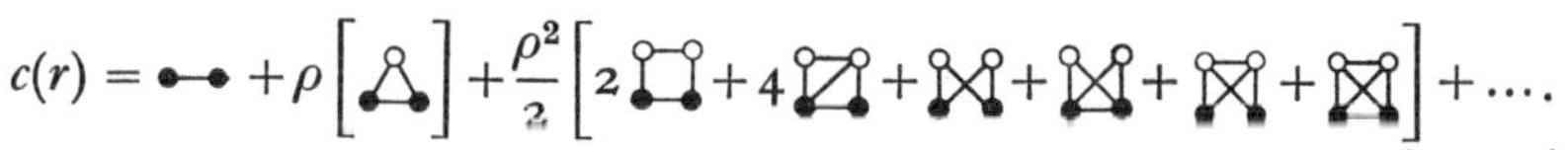

$$c(r) = \bullet\!\!-\!\!\bullet + \rho\left[\,\triangle\,\right] + \frac{\rho^2}{2}\left[2\,\square + 4\,\boxtimes + \boxtimes + \boxtimes + \boxtimes + \boxtimes\right] + \dots. \qquad (2.14.5)$$

Here the topologically identical cluster integrals appearing in (2.14.4) have been combined, giving rise to the weighting factors appearing in the graphical expansion (2.14.5).

In the HNC and PY approximations the first two terms of this expansion are retained, but the third and higher terms are truncated. For example, the third term is

$$\frac{\rho^2}{2}\left[2\,\square + 4\,\boxtimes + \boxtimes + \boxtimes + \boxtimes\right]\text{HNC},$$

$$\frac{\rho^2}{2}\left[2\,\square + 4\,\boxtimes\right] \qquad\qquad \text{PY}. \qquad (2.14.6)$$

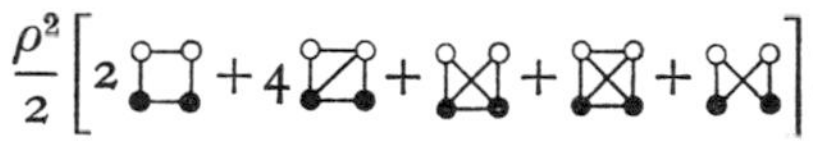

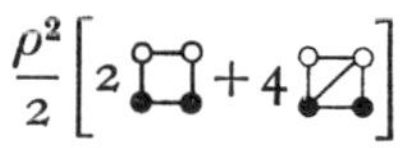

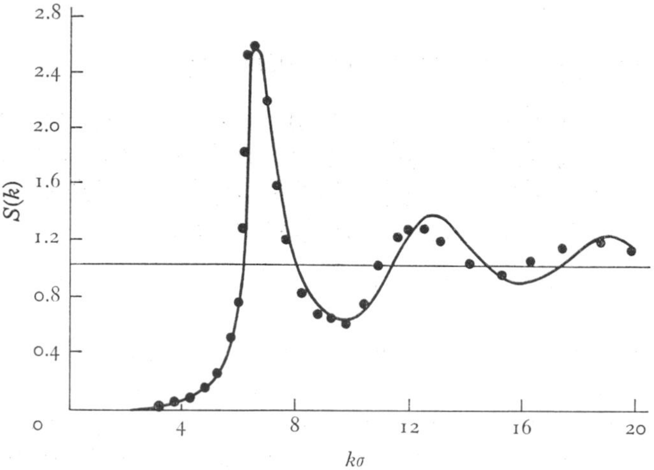

Fig. 2.14.1. Comparison of the structure factor $S(k)$ for liquid rubidium at 40 °C (solid circles) with the analytic hard sphere Percus–Yevick curve obtained by Ashcroft and Lekner. The essentially geometric packing problem is well represented by the hard sphere model, as we see from the quantitative agreement in the vicinity of the principal peak. However, the inadequacy of the detailed form of the model interaction is evidenced by the progressive discrepancy with increasing k. (Redrawn by permission from Ashcroft and Lekner, *Phys. Rev.* **145**, 83 (1966).)

Thus we see that, diagrammatically, the PY is the more drastic approximation, and yet for systems where the repulsive forces dominate the PY approximation is found to be superior! Apart from the extensive cancellation which presumably must occur for the higher order diagrams in (2.14.5), we may understand this apparently paradoxical result by comparing the diagrams dropped. These are

 HNC,

and  PY. (2.14.7)

The numerical value of the contiguous six-bond cluster dropped in the HNC case must be $\sim +1$, whilst the numerical value of the six- and five-bond clusters dropped in the PY case is

$$\sim (+1-1) \sim 0,$$

that is the diagrams dropped tend to cancel one another thereby preserving the cancellation scheme. Of course, for more realistic interactions the cancellation cannot be discussed in such simple terms for then $f(r)$ has a more complicated form. Nevertheless, where repulsive forces dominate we can expect the PY approximation to give a superior representation, and indeed it does. Ashcroft and Lekner[43] have compared the experimental structure factor $S(k)$ for several liquid metals to the PY hard sphere calculation. If the ion–ion interaction is the dominant feature of the pair potential, then a hard sphere approximation may be reasonable. In fig. 2.14.1 we show the comparison for Rb at 40 °C. We see that the principal features of $S(k)$ are reproduced in the PY hard sphere calculation – in particular the first peak which is of importance for resistivity calculations. Quite clearly at liquid densities, $g_{(2)}(r)$ and $S(k)$ are principally determined by geometrical packing considerations. The attractive tail governs the fine structure of the curves. The progressive discrepancy between the experimental and calculated structure factors in fig. 2.14.1 suggests that whilst the periodicity or 'wavelength' of $g_{(2)}(r)$ is quite well described, the assumed form of the repulsive branch of the pair potential is not quite correct. The large k region is dominated by the structure at small separations in real space.

The diagrammatic approach to the HNC and PY approximations consists firstly of identifying various classes of cluster, and then expressing the various correlation functions, $h(r)$, $c(r)$, $g(r)$ and so on in terms of linear combinations of the various classes. This, of course, represents a shorthand method for isolating those cluster integrals which contribute to a given correlation. First, however, we must define the various classes of diagram:

Chains, $C(r)$, are clusters with at least one nodal field point. A *simple chain* is a cluster in which every field point (○) is a node – that is, cutting at a node would cause the diagram to fall into two parts, each having a root or base point (●). Examples of simple chains are given below:

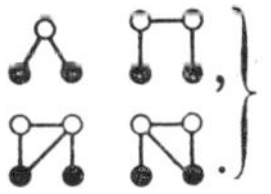

(2.14.8)

A ***netted chain*** is formed from a simple chain by adding *not more than one* field point across each simple chain link. For our purposes we shall not need to distinguish between simple and netted chains – they will all be incorporated into the one class of chains. However, it is important not to confuse netted chains with elementary clusters to be defined below. Note that the chain class has no direct 12-bond: inclusion of such a bond would convert the cluster into a *bundle*.

Bundles, $B(r)$, are clusters containing parallel collections of links between the two root points. Thus there are always at least two *independent* routes from one base point to the other, one of which may be the direct 12-link. Examples of bundles are:

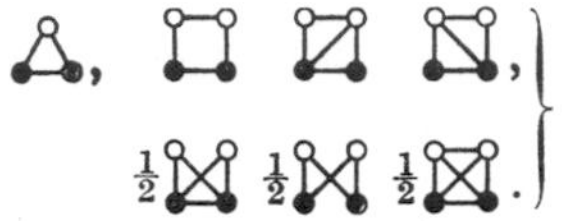

(2.14.9)

Here we have included the symmetry numbers which we shall discuss in a moment.

Elementary clusters, $E(r)$, are those which are neither chains nor bundles. An elementary graph cannot have a direct 12-bond since this would turn it into a bundle. The name is deceptive since the integrations which these clusters represent are usually far from elementary to perform. An example of an elementary diagram is

Each function, $C(r)$, $B(r)$, etc., corresponds to summing over all interaction graphs of the appropriate topological class, where the contribution from any particular graph is given by

$$I/\sigma, \qquad (2.14.10)$$

where I is the integral we obtain by labelling (numbering) the field points and performing the integration over all positions of the field points. σ is the symmetry number of the graph (i.e. the number of automorphisms among the field points which preserve the structure of the graph).

By applying further topological restrictions we find we can express the correlation functions in terms of the classes $C(r)$, $B(r)$ and $E(r)$. Thus, the total correlation function is the sum of all connected graphs, so that we can write generally

$$h(r) = C(r)+B(r)+E(r). \quad (2.14.11)$$

Reference to (2.14.1) and (2.14.5) shows that the direct correlation function does not involve nodes so that we drop the class of chain diagrams:

$$c(r) = B(r)+E(r). \quad (2.14.12)$$

Further we may write

$$\frac{\Phi(r)-\Psi(r)}{kT} = C(r)+E(r) \quad (2.14.13)$$

thereby excluding all direct 12-interaction.

Now, from (2.14.11), (2.14.12), (2.14.13) we have the *exact* relation

$$c(r) = g_{(2)}(r)-1-\ln g_{(2)}(r)-\frac{\Phi(r)}{kT}+E(r). \quad (2.14.14)$$

The HNC approximation consists of setting $E(r) = 0$, thereby dropping the (difficult) class of elementary diagrams. This amounts to assuming a pair correlation function of the form

$$c_{HNC}(r) = h(r)-\ln g_{(2)}(r)-\frac{\Phi(r)}{kT} \quad (2.14.15)$$

which is, of course, the initial relation (2.13.4). Clearly, any improvement in the HNC approximation must take further account of the class of elementary diagrams; a second-order theory designated HNC2 (§2.17) goes some way to include these diagrams.

The Percus–Yevick approximation may be approached graphically as follows. The direct correlation function is given *exactly* as (2.14.12):

$$c(r) = B(r)+E(r).$$

We now set $E(r) = 0$ as in the HNC case. Algebraic manipulation of (2.14.11), (2.14.12), (2.14.13) yields

$$c(r) = h(r)-\ln[1+(y(r)-1)],$$

where

$$y(r) = g_{(2)}(r)\exp\left(\frac{\Phi(r)}{kT}\right).$$

If we now linearize the logarithm, but introduce a new class of diagrams $D(r)$ which preserves the exact relation (2.14.12):

$$c(r) = h(r) - [y(r) - 1] + D(r) \tag{2.14.16}$$

and then set $D(r) = 0$, we obtain

$$c_{\mathrm{PY}}(r) = h(r) - \left[g_{(2)}(r) \exp\left(\frac{\Phi}{kT}\right) - 1\right] \tag{2.14.17}$$

which is exactly the PY approximation to $c(r)$. $D(r)$ involves cluster contributions which can be enumerated. Clearly the PY approximation is more drastic than that of HNC in terms of the classes of diagram dropped, but as we discussed in § 2.13, the PY theory is nevertheless capable of giving a superior representation at liquid densities when judged by its ability to yield correctly the various thermodynamic functions of state.

Again, systematic improvement of the PY approximation depends upon the inclusion of the class of elementary diagrams, $E(r)$, and the formulation of the correction term $D(r)$. Of course, $D(r)$ is small, and negligible at low densities, and may be shown in fact to be of the order ρ_2 provided the total potential Φ_N can be expressed as a sum of pair potentials(44). However it is apparent that $D(r)$ must eventually be expressed in terms of integrals over triplet distribution functions such as $g_{(3)}$ and $c_{(3)}$. This is a field which is now being studied intensively(45), and whose progress will be critically dependent upon our knowledge of these higher functions. One interesting result which has already emerged(46, 47) is that the leading term in $D(r)$ is *not* of the order ρ^2 but is of the order ρ if there is a triplet potential $\Phi(123)$ acting in the fluid. The leading term in $D(r)$ that arises from $\Phi(123)$ is

$$D_1(12) = \rho \int [1 + f(13)][1 + f(23)] f(123)\, d3, \tag{2.14.18}$$

where

$$f(123) = \exp\left\{-\frac{\Phi(123)}{kT}\right\} - 1. \tag{2.14.19}$$

2.15 Solution of the PY equation for hard spheres

The Percus–Yevick theory has the quite unexpected advantage that the equations may be solved analytically for hard spheres[48]. The equation of state gives no evidence of the hard sphere phase change revealed by machine experiments; but for relative densities beyond about 0.8, $g_{(2)}(r)$ takes on negative values, and so at these high densities the PY solutions can be rejected on physical grounds[49]. This is, however, an impossibly high density since the relative density cannot exceed that of a regular close-packed array of spheres, 0.7405.

Temperley[50] observed that there are other solutions that have additional terms containing functions of the form $\cos(a_i r)$, where the coefficients a_i are reciprocal distances. These oscillatory solutions resemble the form of the distribution function of a solid, but Hutchinson[51] has unfortunately shown that they are physically unacceptable since they imply negative intensity of scattered radiation at certain scattering angles.

We now give a summary of Wertheim's analytic solution in the Percus–Yevick approximation[48].

The pair distribution function is written in the form

$$g_{(2)}(12) = \exp\left(-\frac{\Phi(12)}{kT}\right)\tau(12), \tag{2.15.1}$$

where $\tau(12)$ has the significance

$$\tau(12) = \exp\left(\frac{\Phi(12)-\Psi(12)}{kT}\right) = \exp\left(-\frac{W(12)}{kT}\right), \tag{2.15.2}$$

where $W(12)$ is the potential 'tail' (cf. (2.13.2)) which arises as a modification of the pair potential $\Phi(12)$ by virtue of indirect interactions. The direct correlation in the Percus–Yevick approximation then becomes

$$c_{\mathrm{PY}}(12) = \tau(12)\left\{\exp\left(-\frac{\Phi(12)}{kT}\right)-1\right\} \tag{2.15.3}$$

which, when substituted in the OZ relation yields

$$\begin{aligned} h(12)-c(12) &= \tau(12)-1 \\ &= \rho\int \exp\left(-\frac{\Phi(23)}{kT}\right)\left\{\exp\left(-\frac{\Phi(13)}{kT}\right)-1\right\}\tau(23)\,\tau(13)\,d3 \\ &\quad -\rho\int \tau(13)\left\{\exp\left(-\frac{\Phi(13)}{kT}\right)-1\right\}d3. \end{aligned} \tag{2.15.4}$$

If we now insert the hard sphere potential

$$\Phi(r) = +\infty \quad (r < \sigma),$$
$$\Phi(r) = 0 \quad (r \geqslant \sigma),$$

where σ is the hard sphere diameter, (2.15.4) simplifies to

$$\tau(12) = 1+\rho\int_{13<\sigma}\tau(13)\,d3-\rho\int_{\substack{13<\sigma\\23>\sigma}}\tau(13)\,\tau(23)\,d3. \tag{2.15.5a}$$

Solution of (2.15.5a) for $\tau(12)$ will, from (2.13.3), (2.15.1), yield the direct correlation function and the radial distribution function. If we rewrite (2.15.5a) in vector form explicitly,

$$\tau(r) = 1+\rho\int_{r'<\sigma}\tau(r')\,d\mathbf{r}'-\rho\int_{\substack{r'<\sigma\\|\mathbf{r}-\mathbf{r}'|>\sigma}}\tau(r')\,\tau(|\mathbf{r}-\mathbf{r}'|)\,d\mathbf{r}' \tag{2.15.5b}$$

then we may Laplace transform (2.15.5b) to give

$$p[F(p)+G(p)] = \frac{1}{p}\left[1+\frac{24\eta}{\sigma^3}\int_0^\sigma \tau(t)\,r^2\,dr\right]-12\eta[F(-p)-F(p)]\,G(p), \tag{2.15.6}$$

where

$$\left.\begin{aligned} F(p) &= \frac{1}{\sigma^2}\int_0^\sigma r\tau(r)\,e^{-pr/\sigma}\,dr, \\ G(p) &= \frac{1}{\sigma}\int_\sigma^\infty r\tau(r)\,e^{-pr/\sigma}\,dr, \end{aligned}\right\} \tag{2.15.7}$$

$$\eta = \tfrac{1}{6}\pi\sigma^3\rho. \tag{2.15.8}$$

Study of the virial expansion in r-space[52] and k-space[53] suggests that $c(r)$ for hard spheres is a polynomial in r^3 inside the hard sphere and zero outside. (See (2.13.14):

$$c_{\mathrm{PY}}(r) = g_{(2)}(r)f(r)\exp\left(\frac{\Phi(r)}{kT}\right)$$

and $f(r) = 0$ for $r > \sigma$ in the hard sphere case.) Thus a trial solution

$$\left.\begin{aligned} \tau(r) = -c(r) &= \alpha+\beta\left(\frac{r}{\sigma}\right)+\gamma\left(\frac{r}{\sigma}\right)^2+\delta\left(\frac{r}{\sigma}\right)^3 && r < \sigma, \\ &= 0 && r > \sigma \end{aligned}\right\} \quad (2.15.9)$$

is assumed to hold. Here α, β, γ are functions of the packing density, η. Wertheim[48] shows that the defining relations are in fact

$$\left.\begin{aligned} (1-\eta)^4\alpha &= (1+2\eta)^2, \\ (1-\eta)^4\beta &= -6\eta(1+\eta/2)^2, \\ (1-\eta)^4\delta &= \tfrac{1}{2}\eta(1+2\eta)^2, \end{aligned}\right\} \quad (2.15.10)$$

(γ turns out to be zero, but the other coefficients are finite).

Two analytic equations of state for the PY hard sphere fluid are found, depending upon whether the pressure or compressibility relation is used. The expressions obtained are

$$\left.\begin{aligned} \left(\frac{PV}{NkT}\right)_p &= \frac{1+2\eta+3\eta^2}{(1-\eta)^2} = \frac{1+\eta+\eta^2-3\eta^3}{(1-\eta)^3}, \\ \left(\frac{PV}{NkT}\right)_c &= \frac{1+\eta+\eta^2}{(1-\eta)^3}. \end{aligned}\right\} \quad (2.15.11)$$

The same equations have been obtained for a system of hard spheres by Reiss, Frisch and Lebowitz[54] by a different method. The numerical consequences of (2.15.11) will be discussed in chapter 3; however, some preliminary observations are relevant here. First we notice that the only singularity in the equation of state occurs when

$$\eta = 1,$$

i.e.

$$\pi\sigma^3\rho = 6$$

which, as observed earlier, is a physically impossible density. Next we see that the pressure and compressibility equations are both functionally and numerically different, although both have the same low density limit, and indeed appear to bracket the correct isotherm.

2.16 Functional differentiation

What at first sight seems an entirely different approach to the pair distribution function is that of functional differentiation. The technique seems to have been used in various problems[55], but only relatively recently has it been applied to the liquid state[56, 57]. It is found that the method provides alternative derivations of the BGY, HNC and PY equations, and moreover, yields higher order components to these equations which have been shown to improve the thermodynamic data and virial coefficients obtained from the first-order theory.

The technique has also led to new equations relating the pair potential to the pair distribution.

The singlet distribution of particle 2 in a bulk system of N particles would be given by

$$\rho_{(1)}(2) = \frac{N\int \dots \int w \, \mathrm{d}\mathbf{1} \, \mathrm{d}\mathbf{3} \, \mathrm{d}\mathbf{4} \dots \mathrm{d}\boldsymbol{N}}{\int \dots \int w \, \mathrm{d}\mathbf{1} \, \mathrm{d}\mathbf{2} \dots \mathrm{d}\boldsymbol{N}}, \tag{2.16.1}$$

where $\quad w(1, 2, \dots, N) = \exp[-\rho_N(1, 2, \dots, N)/kT]$

and $\quad \rho_{(1)}(2) = \rho g_{(1)}(2).$

In the absence of any external fields this distribution would, of course, be a constant. However, we may write down the singlet distribution for atom 2 *in the mean field of atom* 1, where the mean field is considered 'external' to the set 2, 3, ..., N particles and arises from the interaction of 1 with its environment of $(N-1)$ particles. Clearly, the singlet distribution of 2 along the radius is a functional of the imposed 'external' potential, and if we denote the mean field of particle 1 as

$$\Phi_1 = \sum_{i=2}^{N} \Phi(1, i), \tag{2.16.2}$$

where we have assumed a pairwise decomposable interaction, then the singlet distribution may be written:

$$\rho_{(1)}(2) = \frac{N\int \dots \int w \exp\left\{-\sum_{i=2}^{N} \Phi(1, i)/kT\right\} \mathrm{d}\mathbf{1} \, \mathrm{d}\mathbf{3} \, \mathrm{d}\mathbf{4} \dots \mathrm{d}\boldsymbol{N}}{\int \dots \int w \exp\left\{-\sum_{i=2}^{N} \Phi(1, i)/kT\right\} \mathrm{d}\mathbf{1} \dots \mathrm{d}\boldsymbol{N}}. \tag{2.16.3}$$

This as it stands brings us no closer to a working relation between the pair potential and $g_{(2)}(r)$. The situation is somewhat similar to that in the derivation of the BGY equation. Here, however, we enquire what the variation in $\rho_{(1)}(2)$ will be as we vary the mean field of 1. This also reminds us of Kirkwood's variation of the coupling parameter ξ.

We now take the functional derivative of (2.16.3) with respect to $\Phi(1,i)$, giving

$$\delta\rho_{(1)}(2) = -\frac{\delta\Phi(12)}{kT}\frac{N\int\ldots\int w\exp\left\{-\sum_{i=2}^{N}\Phi(1,i)/kT\right\}d1\,d3\,d4\ldots dN}{\int\ldots\int w\exp\left\{-\sum_{i=2}^{N}\Phi(1,i)/kT\right\}d1\ldots dN}$$

$$-\frac{N\int\ldots\int w\exp\left\{-\sum_{i=2}^{N}\Phi(1,i)/kT\right\}\sum_{i=3}^{N}\left(\frac{\delta\Phi(1,i)}{kT}\right)d1\,d3\,d4\ldots dN}{\int\ldots\int w\exp\left\{-\sum_{i=2}^{N}\Phi(1,i)/kT\right\}d1\ldots dN}$$

$$+\frac{N\int\ldots\int w\exp\left\{-\sum_{i=2}^{N}\Phi(1,i)/kT\right\}d1\,d3\,d4\ldots dN}{\int\ldots\int w\exp\left\{-\sum_{i=2}^{N}\Phi(1,i)/kT\right\}d1\ldots dN}$$

$$\times\frac{\int\ldots\int w\exp\left\{-\sum_{i=2}^{N}\Phi(1,i)/kT\right\}\sum_{i=2}^{N}\left(\frac{\delta\Phi(1,i)}{kT}\right)d1\ldots dN}{\int\ldots\int w\exp\left\{-\sum_{i=2}^{N}\Phi(1,i)/kT\right\}d1\ldots dN}. \quad (2.16.4)$$

The second term contains $(N-2)$ identical contributions whilst the third contains $(N-1)$. Examination of the integrals and comparison with (2.16.3) enables us to write

$$\delta\rho_{(1)}(2) = -\rho_{(1)}(2)\frac{\delta\Phi(12)}{kT} - \int\rho_{(2)}(23)\frac{\delta\Phi(13)}{kT}d3$$

$$+\int\rho_{(1)}(2)\rho_{(1)}(3)\frac{\delta\Phi(13)}{kT}d3. \quad (2.16.5)$$

This may be written in terms of the three-dimensional Dirac δ-function, since

$$\delta\Phi(12) = \int \delta(\boldsymbol{r}_3-\boldsymbol{r}_2)\,\delta\Phi(13)\,\mathrm{d}3,$$

$$\delta\rho_{(1)}(2) = -\int \rho_{(1)}(2)\,\delta(\boldsymbol{r}_3-\boldsymbol{r}_2)\frac{\delta\Phi(13)}{kT}\,\mathrm{d}3 - \int \rho_{(2)}(23)\frac{\delta\Phi(13)}{kT}\,\mathrm{d}3 + \int \rho_{(1)}(2)\,\rho_{(1)}(3)\frac{\delta\Phi(13)}{kT}\,\mathrm{d}3. \qquad (2.16.6)$$

Thus, the functional derivative of $\rho_{(1)}(2)$ with respect to $\Phi(13)$ is

$$kT\frac{\delta\rho_{(1)}(2)}{\delta\Phi(13)} = -\rho_{(1)}(2)\,\delta(\boldsymbol{r}_3-\boldsymbol{r}_2) - \rho_{(2)}(23) + \rho_{(1)}(2)\,\rho_{(1)}(3). \qquad (2.16.7)$$

Writing $h(23)$ for $\rho_{(2)}(23)/\rho_{(1)}(2)\,\rho_{(1)}(3) - 1$,

$$h(23) = -\frac{kT}{\rho_{(1)}(2)\,\rho_{(1)}(3)}\frac{\delta\rho_{(1)}(2)}{\delta\Phi(13)} - \frac{\delta(\boldsymbol{r}_3-\boldsymbol{r}_2)}{\rho_{(1)}(3)}. \qquad (2.16.8)$$

Substitution of (2.16.8) into the Ornstein–Zernike relation leads at once to the inverse of (2.16.8), namely

$$c(23) = \frac{1}{kT}\frac{\delta\Phi(13)}{\delta\rho_{(1)}(2)} + \frac{\delta(\boldsymbol{r}_2-\boldsymbol{r}_3)}{\rho_{(1)}(3)}. \qquad (2.16.9)$$

Equations (2.16.8) and (2.16.9) were first given by Yvon (1958)[(57)], and are basic to the method. We shall now indicate how the technique of functional differentiation may be used to re-derive the known equations of BGY, PY and HNC.

First let us assume that $\mathscr{F}(\rho_{(1)}(2), \Phi(12))$ is a function of $\eta(3)$. Then we may write in terms of a functional Taylor expansion:

$$\ln g_{(2)}(12) + \frac{\Phi(12)}{kT} = \rho\int [g_{(2)}(13)-1]\frac{\delta\mathscr{F}}{\delta\eta(3)}\,\mathrm{d}3 + \frac{\rho^2}{2!}\iint [g_{(2)}(13)-1] \times [g_{(2)}(14)-1]\frac{\delta^2\mathscr{F}}{\delta\eta(3)\,\delta\eta(4)}\,\mathrm{d}3\,\mathrm{d}4 + \ldots. \qquad (2.16.10)$$

We shall see in a moment that for certain functional forms of $\mathscr{F}(\rho_{(1)}(2), \Phi(12))$ we may regain the HNC, PY and BGY equations depending upon the choice of $\mathscr{F}$, η. At the moment the functional form of $\mathscr{F}$ appears quite arbitrary, and application of the method depends to a large extent upon the skill in choosing a 'successful' functional form: different equations result from different choices of $\mathscr{F}$ and η.

Already (2.16.10) is bearing considerable similarity to the three integro-differential equations thus far derived, (2.5.24, 2.13.16, 2.13.20) especially if the functional Taylor expansion is truncated at the first term, giving a first-order theory. Percus has observed that the choice[58]

$$\mathscr{F}(\rho_{(1)}(2), \Phi(12)) = \ln\left\{\rho_{(1)}(2)\exp\left(\frac{\Phi(12)}{kT}\right)\right\} \qquad (2.16.11)$$

as a functional of $\eta = \rho_{(1)}(3)$ leads directly to the HNC equation. For

$$\begin{aligned}\frac{\delta[\ln\{\rho_{(1)}(2)\exp(\Phi(12)/kT)\}]}{\delta\rho_{(1)}(3)} &= \frac{1}{\rho_{(1)}(2)}\frac{\delta\rho_{(1)}(2)}{\delta\rho_{(1)}(3)} + \frac{1}{kT}\frac{\delta\Phi(12)}{\delta\rho_{(1)}(3)} \\ &= \frac{\delta(\boldsymbol{r}_2-\boldsymbol{r}_3)}{\rho_{(1)}(2)} + c(23) - \frac{\delta(\boldsymbol{r}_2-\boldsymbol{r}_3)}{\rho_{(1)}(2)} \\ &= c(23), \end{aligned} \qquad (2.16.12)$$

where the second line of (2.16.12) comes from the application of the general defining relation (2.16.9). Substituting this expression for the first-order functional derivitive into (2.16.10), and elimination of c from this equation by means of the OZ relation yields the HNC equation (2.13.15).

Similarly, if we chose the form[58]

$$\mathscr{F}(\rho_{(1)}(2), \Phi(12)) = \rho_{(1)}(2)\exp\left(\frac{\Phi(12)}{kT}\right) \qquad (2.16.13)$$

we should obtain the Percus–Yevick equation, whilst if we assume

$$\mathscr{F}(\rho_{(1)}(2), \Phi(12)) = \frac{1}{kT}\rho_{(1)}(2)\,\nabla_1\Phi(12) \qquad (2.16.14)$$

to be a functional of

$$\eta = \ln\left\{\rho_{(1)}(2)\exp\left(\frac{\Phi(12)}{kT}\right)\right\},$$

rather than $\rho_{(1)}(3)$, the BGY equation results. Cole[59] has shown that if we choose (2.16.14) to be a functional of $\rho_{(1)}(3)$ the equation resulting is

$$\nabla_1 g_{(2)}(12) + \frac{1}{kT}\nabla_1 \Phi(12) g_{(2)}(12)$$
$$= -\frac{\rho}{kT}\int [g_{(2)}(23) - 1] g_{(2)}(13) \nabla_1 \Phi(13)\, d3 \quad (2.16.15)$$

which may be obtained directly from the Ornstein–Zernike relation by assuming

$$-kT\frac{\partial c(\boldsymbol{r})}{\partial \boldsymbol{r}} = g_{(2)}(\boldsymbol{r})\frac{\partial \Phi(\boldsymbol{r})}{\partial \boldsymbol{r}}\boldsymbol{r}. \quad (2.16.16)$$

Thus there is a whole range of equations which may be set up whose form depends directly upon the choice of $\mathscr{F}$-function and the function upon which it is functional.

This elegant derivation of the HNC and PY equations seems to place them on a sounder theoretical basis, particularly in the case of the HNC approximation, without requiring extensions of validity from the dilute gas phase. To some extent this is illusory since we know that the theories themselves become less satisfactory as the liquid phase is approached, and moreover, the selection of the functional form of $\mathscr{F}$ appears quite arbitrary, and indeed is more or less the product of inspired guesswork. It would be more satisfactory if some physical basis could be found which would enable us to converge on some form for $\mathscr{F}$, such as minimization of the free energy, for example.

2.17 Second-order theories

The technique of functional differentiation does have the very important advantage that it indicates quite clearly how we might construct a second-order theory of the pair function. Inclusion of the second term in the functional Taylor expansion (2.16.10) gives[60]

$$\ln g_{(2)}(12) + \frac{\Phi(12)}{kT} = \rho\int [g_{(2)}(13) - 1]\frac{\delta\mathscr{F}}{\delta\eta(3)}\, d3$$
$$+\frac{\rho^2}{2!}\iint [g_{(2)}(13) - 1][g_{(2)}(14) - 1]$$
$$\times \frac{\delta^2\mathscr{F}}{\delta\eta(3)\,\delta\eta(4)}\, d3\, d4. \quad (2.17.1)$$

In principle a succession of higher order terms could be included but mathematical difficulties have so far restricted the expansion to the second (quadratic) term. Retaining the same choices (2.16.11) and (2.16.12) for $\mathscr{F}$ and η we proceed to the second-order HNC and PY theories, generally designated HNC2 and PY2. In the case of the second-order HNC theory we have to consider

$$\frac{1}{2!}\iint [g_{(2)}(13)-1][g_{(2)}(14)-1]\frac{\delta^2[\ln\{\rho_{(1)}(2)\exp(\Phi(12)/kT)\}]\,d3\,d4}{\delta\rho_{(1)}(3)\,\delta\rho_{(1)}(4)}. \tag{2.17.2}$$

There is little technical difficulty in evaluating this extra term, and the result is that the OZ equation is retained, but the inclusion of the quadratic term (2.17.2) in the Taylor expansion supplements the HNC1 approximation to give

$$c_{\mathrm{HNC}\,2}(12) = h(12)-\ln g_{(2)}(12)+\frac{\Phi(12)}{kT}+\epsilon(12). \tag{2.17.3}$$

Comparison of this with the exact relation

$$c(12) = h(12)-\ln g_{(2)}(12)+\frac{\Phi(12)}{kT}+E(12) \tag{2.17.4}$$

suggests that some account is being taken of the elementary clusters dropped in the HNC1 approximation. The second-order term may be shown to be

$$\epsilon(12) = \Delta(12)-\tfrac{1}{2}[h(12)-c(12)]^2, \tag{2.17.5}$$

where

$$\Delta(12) = \tfrac{1}{2}\rho^2\iint c(12)c(13)g_{(2)}(34)[g_{(3)}(234)-\ln g_{(2)}(234)$$
$$h(23)+\ln g_{(2)}(23)-g_{(2)}(23)+\ln g_{(2)}(24)]\,d3\,d4. \tag{2.17.6}$$

The most significant thing about these rather cumbersome equations is that they involve the triplet distribution $g_{(3)}(234)$. The theory is no longer self-contained and some closure must be applied. Verlet in his paper[60] suggests a possible first-order theory for $g_{(3)}$ for insertion into (2.17.6). The suggestion is in the spirit of the two-body functional derivative (2.16.12):

$$\ln\left\{\rho_{(2)}(23)\exp\left(\frac{\Phi(12)}{kT}+\frac{\Phi(13)}{kT}\right)\right\}$$

is now taken as a linear functional of $\{\rho_{(1)}(4)\}$. The result is an equation for $g_{(3)}$

$$\ln\left\{\frac{g_{(3)}(123)}{g_{(2)}(23)}\exp\left(\frac{\Phi(12)}{kT}+\frac{\Phi(13)}{kT}\right)\right\} = \rho\int\left\{\frac{g_{(3)}(234)}{g_{(2)}(23)}-1\right\}c(14)\,\mathrm{d}4. \tag{2.17.7}$$

The set of equations, (2.17.7), (2.17.6), (2.17.5), (2.17.3) and the OZ relation then allow $g_{(2)}$ to be determined for a given interparticle potential. This theory, designated HNC2 by Verlet, is cumbersome and difficult to apply. It has, in fact, been applied to dilute systems only, but with some improvement on the HNC1 estimates of the virial coefficients.

Similarly, in the PY case, the corresponding quadratic term in the functional Taylor expansion is

$$\frac{1}{2!}\iint[g_{(2)}(13)-1][g_{(2)}(14)-1]\frac{\delta^2[\rho_{(1)}(2)\exp(\Phi(12)/kT)]\,\mathrm{d}3\,\mathrm{d}4}{\delta\rho_{(1)}(3)\,\delta\rho_{(1)}(4)} \tag{2.17.8}$$

(cf. (2.17.2)). The second-order contribution to the direct correlation function will then be of the form

$$c_{\mathrm{PY2}}(12) = h(12)-\left\{g_{(2)}(12)\exp\left(\frac{\Phi(12)}{kT}\right)-1\right\}+\Delta(12), \tag{2.17.9}$$

where $\Delta(12)$ is defined above (equation (2.17.6)). Clearly some account is being taken of the class of diagrams $D(r)$ (cf. (2.14.6)). The class $D(r)$ includes $E(r)$, and $\Delta(r)$ includes $\epsilon(r)$. In the PY case closure is obtained by taking

$$\left\{\rho_{(2)}(23)\exp\left(\frac{\Phi(12)}{kT}+\frac{\Phi(13)}{kT}\right)\right\}$$

to be a linear functional of $\{\rho_{(1)}(4)\}$. The result for $g_{(3)}$ is the expression

$$g_{(3)}(123)\exp\left(\frac{\Phi(12)}{kT}+\frac{\Phi(13)}{kT}\right) = g_{(23)}(23)+\rho\int\{g_{(3)}(234)-g_{(2)}(23)\}c(14)\,\mathrm{d}4. \tag{2.17.10}$$

The equations (2.17.10), (2.17.9), (2.17.6) and the OZ relation enable $g_{(2)}$ to be determined from a given interparticle potential. This theory is designated PY2 and gives superior results to PY1, although calculations have not yet been extended to liquid densities. In principle HNC3 and PY3 theories could be constructed, but these higher approximations have not been attempted in practice. Note the form of $\Delta(r)$, (2.17.6), under the assumption of pairwise additivity of the interparticle potential. In (2.14.18) we saw that if ***triplet*** potentials are taken into account, $\Delta(r) = \mathcal{O}(\rho)$. This aspect of the problem is being intensively investigated by several workers[45, 46, 47].

We have seen that Verlet's method for passing from the first-order to second-order PY or HNC theories via the functional Taylor expansion involves the triplet distribution $g_{(3)}(123)$. To proceed further some *ad hoc* closure device has to be incorporated, and to avoid this Wertheim[61] has recently derived a second-order PY or HNC equation based on graphical concepts. In particular he has done at triplet correlation level what was done in first-order theories at pair correlation level. Thus, ignoring triplet potentials, instead of (2.14.3) we have

$$\frac{\Phi(123)-\Psi(123)}{kT} = \frac{W(123)}{kT} = C(123)+E(123). \quad (2.17.11)$$

To obtain Wertheim's second-order HNC equation we ignore the elementary triplet class $E(123)$, that is, we replace $W(123)$ in

$$\frac{g_{(3)}(123)}{g_{(2)}(12)g_{(2)}(23)g_{(2)}(31)} = \exp\left(\frac{W(123)}{kT}\right) \quad (2.17.12)$$

by $C(123)$. Equation (2.17.12) may be regarded as a 'second-order superposition approximation'. The corresponding PY expression is obtained by linearizing the exponential:

$$\frac{g_{(3)}(123)}{g_{(2)}(12)g_{(2)}(23)g_{(2)}(31)} = 1+C(123). \quad (2.17.13)$$

Beyond this there are no further approximations, only a topological analysis leading to integral equations.

2.18 Some general comments

It is pertinent to ask what one hopes to achieve from these highly complex integro-differential equations for the pair distribution function. Certainly none other than the BGY, HNC1 and PY1 equations could possibly be employed for the inversion of X-ray data to yield the pair interaction. The second-order theories do lead to appreciable numerical improvement, but this alone cannot justify the manipulation of sets of complex equations such as those developed by Verlet and Wertheim. Nevertheless, these more sophisticated theories can be defended on two grounds. It is important to have numerically reliable theories in the form of analytic prescriptions which have a sound physical basis. Then, perhaps, we might develop some insight into the melting transition: as yet there can be no definite identification of the instability of the first-order solutions with the onset of solidification. It is particularly curious that the most satisfactory theory numerically (PY1) fails to yield a solid–fluid phase transition, whilst the least satisfactory (BGY1) appears to do so. We do not know whether PY2 is capable of describing the phase change since solutions at super-critical densities have only been obtained as yet.

It is in many ways disappointing that none of the theories described in this chapter provide a fully adequate description of a sub-critical dense fluid. However, as we shall see in the next chapter, we are able to provide more than a qualitative description of a dense fluid, but whether future progress in the field will depend upon the inclusion of more terms in the functional Taylor expansion, or what amounts to the same thing, a systematic evaluation of the diagrammatic class $E(r)$, or whether we must re-route our theories to take account of triplet potentials is not yet clear.

2.19 Non-additive effects

There is mounting evidence that the assumption of pairwise additivity of the interparticle potential does not adequately describe the total configurational energy Φ_N. We have instead:

$$\Phi_N = \sum_{i<j}\sum \Phi_{ij} + \sum_{i<j<k}\sum\sum \Phi_{ijk} \tag{2.19.1}$$

and, of course, all equations remain unchanged up to the point where the approximation is introduced. In particular the internal energy becomes

$$E = \tfrac{3}{2}NkT + \tfrac{1}{2}\rho \int \Phi(12)\, g_{(2)}(12)\, d\mathbf{2} + \tfrac{1}{6}\rho \iint \Phi(123)\, g_{(3)}(123)\, d\mathbf{2}\, d\mathbf{3} \quad (2.19.2)$$

and the equation of state contains the three-body virial

$$PV = \rho kT - \tfrac{1}{6}\rho \int \nabla_1 \Phi(12)\, g_{(2)}(12)\, d\mathbf{2} - \tfrac{1}{18}\rho^2 \iint F_{123}\, g_{(3)}(123)\, d\mathbf{2}\, d\mathbf{3}, \quad (2.19.3)$$

where
$$F_{123} = \left(r_{12}\frac{\partial}{\partial r_{12}} + r_{13}\frac{\partial}{\partial r_{13}} + r_{23}\frac{\partial}{\partial r_{23}} \right) \Phi(123). \quad (2.19.4)$$

If the dominant term in $\Phi(123)$ has the simple form of the triple-dipole potential of Axilrod and Teller(62), then (2.19.4) reduces to

$$F_{123} = -9\Phi(123). \quad (2.19.5)$$

We therefore now need to know the triplet distributions $g_{(3)}(123)$ if we are to incorporate the effects of three-body potentials. Direct evidence for triplet potentials comes from the large maximum always found in the third virial coefficient, B_3, at low temperatures. We may suspect therefore that their influence on the equation of state, or the internal structure, of fluids may show up even at those relatively low densities at which the first-order theories HNC1 and PY1 take good account of the pairwise interactions. Any discussion of triplet interactions ought, therefore, to be given against the background of the first-order theories. Rushbrooke and Silbert(47), and Rowlinson(46) have taken the triplet potentials into account, but the theories must be considered incomplete since they include no prescription for $g_{(3)}(123)$.

How may we incorporate the triplet potentials into the PY1 and HNC1 theories? The topological arguments given in §2.14 have been extended by Friedman(63) and Baxter(64) to include new graphs generated by the triplet potential. In the HNC case, for example, all the elementary clusters $E(r)$ are dropped, the first being given by (2.14.7), of order ρ^2. But, as Rushbrooke and Silbert have observed[47] (cf. §2.14), if we include triplet potentials the first elementary cluster is of order ρ, and so its inclusion will presumably represent an improvement on HNC1 even when all subsequent elementary graphs are dropped. The leading triplet term is

$$E(12)^{\text{triplet}} = \rho \int f(123)\,[1+f(13)]\,[1+f(23)]\,d3, \quad (2.19.6)$$

where $$f(123) = \exp\{-\Phi(123)/kT\} - 1. \quad (2.19.7)$$

Rowlinson[46] has determined the triplet modification of the PY1 equation. Here the graphical class dropped is $D(12)$ which includes $E(12)$. Equation (2.14.16) may be taken as a definition of $D(12)$:

$$\begin{aligned} D(12) &= c(12) - h(12) + [y(12) - 1] \\ &= c(12) - y(12)f(12). \end{aligned} \quad (2.19.8)$$

In this case the leading graphical term is still (2.19.6), and it can be shown that the graphs present in y may be constructed from f and $(1+f)\,E^{\text{triplet}}$ bonds.

The inclusion of triplet effects in $c(r)$ in either the HNC1 or PY1 approximations has the effect of replacing the pair potential by an effective pair potential where the two are related by

$$\Phi^*(12) = \Phi(12) - kT\rho E(12)^{\text{triplet}},$$

where $\Phi^*(12)$ is the effective pair potential and $E(12)^{\text{triplet}}$ is the first three-body elementary cluster defined in (2.19.6). Kihara[65] has made a theoretical estimate of the *attractive* contribution to the three-body correction term, but does not estimate the repulsive component. Graben and Present[66] have shown that the attractive component alone could provide corrections of the order of 100 per cent to the third virial coefficient. However, Sherwood and Prausnitz[67] and subsequently Sherwood *et al.*[68] have observed that the repulsive component is of opposite sign and may substantially offset the attractive terms. At present the *net* contribution of the triplet effect is not known either theoretically or experimentally.

Since triplet effects are assumed inoperative at very low densities, we may expect the correction to become more apparent with increasing density. X-ray scattering experiments on liquid argon[69] seem to show that the depth of the pair potential well decreases with increasing density which leads us to assume that $\Phi(123)$ is positive at liquid densities. This concurs with other evidence, but Levesque and Verlet[70] suggest that this experimental technique is not yet sufficiently accurate as to detect triplet effects unambiguously.

2.20 Perturbation theories

The hard sphere model is obviously an idealization, although it has been shown that for purely repulsive soft sphere potentials the quantitative form of $g_{(2)}(r)$ at small r is extremely well described by a hard sphere approximation. This is clearly seen from the k-space computations of Weeks, Chandler and Anderson[80], fig. 2.20.1. The hard sphere approximation constitutes an excellent basis for a perturbation treatment of realistic fluids about the 'ideal' rigid sphere potential. Ashcroft and Lekner[43] have also demonstrated the application of the hard sphere structure factor to the case of liquid metals where the ion–ion interaction is effectively that of a rigid sphere.

Several authors have developed theories to relate hard sphere data to the properties of fluids with other repulsive potentials. Rowlinson[81] considered fluids whose repulsive intermolecular potential varied as the inverse nth power of the intermolecular separation. He expanded the thermodynamic properties in powers of $1/n$, the lowest order result corresponding to $n = \infty$, that is, the hard sphere model. Thus, the Helmholtz free energy may be written as

$$F_n = F^0 + n^{-1}(\partial F/\partial n^{-1})_0 + \mathcal{O}(n^{-2}) \tag{2.20.1}$$

for a repulsive potential varying as the inverse with power of the separation. n is known as the inverse steepness parameter. The value of Rowlinson's expression (2.20.1) lies in the fact that the derivative is an accessible thermodynamic function of the hard sphere fluid:

$$\left(\frac{\partial F}{\partial n^{-1}}\right)_0 = 3(PV - NkT)_0\,[\ln x + F(x)],$$
$$x = \epsilon/kT. \tag{2.20.2}$$

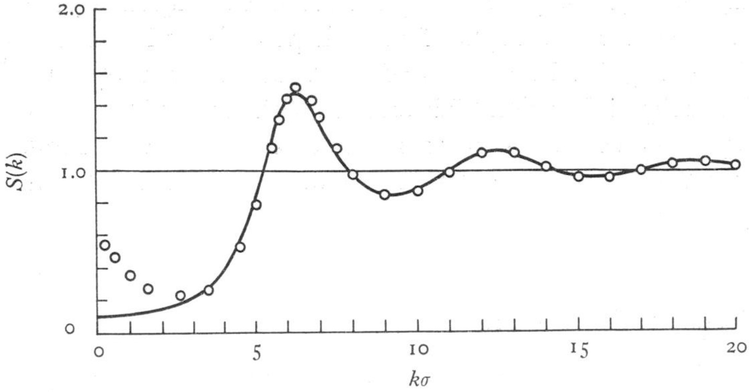

Fig. 2.20.1. The hard sphere structure factor $S(k)$. The circles represent points from a molecular dynamic determination. The full curve is from the perturbation approach of Weeks, Chandler and Anderson. (Redrawn by permission from Weeks *et al.*, *J. Chem. Phys.* **54**, 5237 (1971).)

$F(x)$ is a readily calculable function which depends upon the exact shape of the potential. For the simple case of soft spheres,

$$\Phi^*(x) = \epsilon(r/\sigma)^{-n},$$

$F(x)$ may be shown to reduce to Euler's constant $\gamma = 0.57722$. If we now introduce a reduced length s, which is a function of temperature defined as

$$s = x^{1/n}[1+n^{-1}F(x)], \tag{2.20.3}$$

then we have $$F_n(V,T) = F^0(V/s^3, T) - 3NkT\ln s. \tag{2.20.4}$$

Thus, the fluid with a soft potential ($n \neq 0$) behaves as if it were one of hard spheres whose diameters are, by (2.20.3), explicit functions of temperature.

Barker and Henderson[82] have generalized Rowlinson's method to more realistic potentials. First, however, we consider their perturbation treatment of the square-well fluid. They expand the free energy of the system as a Taylor series about the hard sphere value F^0:

$$\frac{F}{kT} = \frac{F^0}{kT} - \frac{\epsilon\langle n\rangle}{kT} - \frac{1}{2}\left(\frac{\epsilon}{kT}\right)^2 \langle n^2\rangle + \ldots, \tag{2.20.5}$$

where ϵ is the (attractive) perturbation and $\langle n\rangle$ is the mean number of molecules in the range σ to 1.5σ *in the unperturbed (hard sphere) distribution*, $g^0_{(2)}(r)$. Equation (2.20.5) may be written

$$\frac{F-F^0}{kT} = -\frac{\epsilon\langle n\rangle}{kT} - \frac{1}{2}\left(\frac{\epsilon}{kT}\right)^2 \{\langle n^2\rangle - \langle n\rangle^2\} + \dots \quad (2.20.6)$$

It is clear that this expression will converge best at high densities (since then the fluctuation about the mean $\{\langle n^2\rangle - \langle n\rangle^2\}$ is at its minimum), and high temperatures. Equation (2.20.6) is known as the high temperature approximation (HTA). Now,

$$\langle n^2\rangle - \langle n\rangle^2 = NkT(\partial\rho/\partial P)_T \quad (2.20.7)$$

where $(\partial\rho/\partial P)_T = \chi_T$ the macroscopic isothermal compressibility. An even better approximation would be to choose the *local* compressibility, i.e. $(\partial/\partial P)[\rho g^0_{(2)}(r)]$ instead of $(\partial\rho/\partial P)_0 g^0_{(2)}(r)$. We may obtain $(\partial\rho/\partial P)_0$ from the analytic PY hard sphere isotherm, (2.15.11), whereupon

$$\frac{F-F^0}{NkT} = -2\pi\rho\left(\frac{\epsilon}{kT}\right)\int_0^{1.5\sigma} g^0_{(2)}(r)\,r^2\,\mathrm{d}r$$
$$-\pi\rho\left(\frac{\epsilon}{kT}\right)^2 \frac{(1-\eta)^4}{1+4\eta+4\eta^2}\frac{\partial}{\partial\rho}\left\{\rho\int_0^{1.5\sigma} g_{(2)}(r)\,r^2\,\mathrm{d}r\right\}. \quad (2.20.8)$$

This procedure has been generalized by Barker and Henderson to the case of realistic potentials with soft cores. For an arbitrary potential $\Phi(r)$ they define a modified potential by the equations

$$\begin{aligned} v(\alpha,\gamma,d,\sigma;r) &= \Phi\left(d+\frac{r-d}{\alpha}\right) && \text{for} \quad d+\frac{r-d}{\alpha} < \sigma \\ &= 0 && \text{for} \quad \sigma < d+\frac{r-d}{\alpha} < d+\frac{\sigma-d}{\alpha} \\ &= \gamma\Phi(r) && \text{for} \quad r > \sigma. \end{aligned}$$

When $\alpha = \gamma = 0$, v becomes the hard sphere potential with diameter d, while for $\alpha = \gamma = 1$ the original potential $\Phi(r)$ is recovered. α is an inverse steepness parameter for the repulsive region and a depth parameter for the attractive region. Barker and Henderson find[82]

$$\frac{F}{NkT} = \frac{F^0}{NkT} - \frac{2\pi N\alpha d^2}{V} g^0_{(2)}\left(\frac{Nd^3}{V}, 1\right)\left\{\int_0^{\sigma}\exp\left(-\frac{\Phi(r)}{kT}\right)\mathrm{d}r - (\sigma-d)\right\}$$
$$+\frac{2\pi N\gamma}{V}\int_{\sigma}^{\infty} g^0_{(2)}\left(\frac{Nd^3}{V}, \frac{r}{d}\right)\frac{\Phi(r)}{kT} r^2\,\mathrm{d}r + \dots, \quad (2.20.9)$$

where F^0 refers to the hard sphere function. For $\alpha = \gamma = 1$ (2.20.9) gives the free energy corresponding to the potential $\Phi(r)$. In this generalized Rowlinson method the perturbation in the repulsive term is about a hard sphere *of the appropriate diameter*, and d is therefore temperature dependent. If we plot the difference between $\exp[-\Phi_r(r)/kT]$ for the realistic repulsive potential $\Phi_r(r)$, and the corresponding function for the hard sphere potential we obtain the 'blip' function shown in fig. 2.20.2 (*b*). d is chosen so as to make the two areas equal, giving

$$d = \int_0^\infty \{1 - \exp[\Phi_r(r)/kT]\}\, dr. \tag{2.20.10}$$

It is seen from (2.20.9) that Barker and Henderson split the potential up into two regions about $r = \sigma$: that is, into the repulsive and attractive components. However, they have more recently[83] chosen to split the integration into two regions $r < \mu$, $r > \mu$ where $\mu < \sigma$. This was found to be necessary since the deviation from the hard sphere interaction in the region $\mu < r < \sigma$ (see fig. 2.20.3 (*a*)) was spoiling the convergence. The region $r < \mu$ is dealt with by the generalized Rowlinson inverse steepness method, whilst the contribution to the thermodynamic properties for $r > \mu$ is given by the first term of the high temperature approximation (2.20.6). μ is determined variationally so as to minimize the excess free energy per particle as determined by this combination of methods.

Anderson, Weeks and Chandler[84] obtain a functional Taylor expansion in terms of $\Delta\phi$ about a hard sphere reference system for the free energy of a system of soft spheres:

$$F_s = F^0 + \int \frac{\partial F^0}{\partial \phi(\mathbf{r})} \Delta\phi(\mathbf{r})\, d\mathbf{r} + \frac{1}{2} \iint \frac{\partial^2 F^0}{\partial \phi(\mathbf{r})\, \partial \phi(\mathbf{r}')} \Delta\phi(\mathbf{r})\, \Delta\phi(\mathbf{r}')\, d\mathbf{r}\, d\mathbf{r}' + \ldots, \tag{2.20.11}$$

where $\quad \phi(r) = \exp[-\beta\Phi(r)], \quad \Delta\phi = \phi_s - \phi^0.$

From F we can, of course, calculate the pressure, entropy, and other thermodynamic functions by straightforward differentiations with

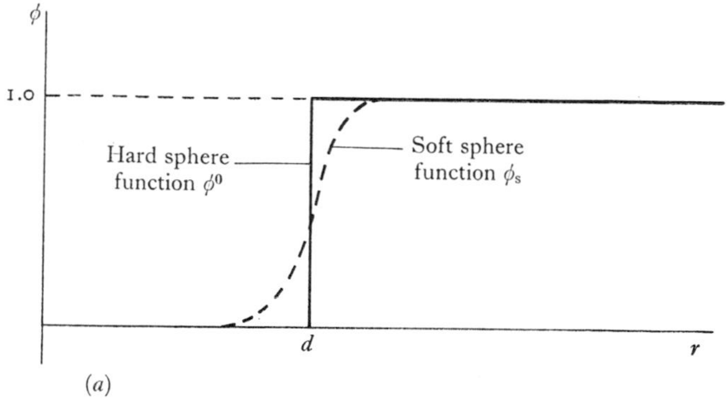

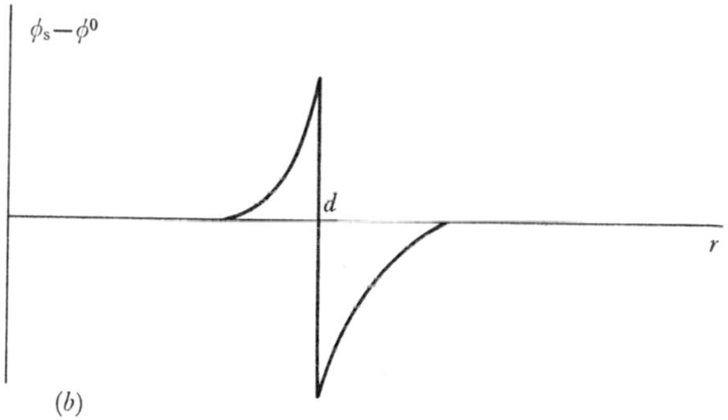

Fig. 2.20.2. (*a*) The function $\phi = \exp[-\Phi(r)/kT]$ for a hard sphere and soft sphere interaction, shown by the full and broken lines, respectively. (*b*) The 'blip'-function, $\phi_s - \phi^0$, representing the difference between the soft and hard sphere functions shown in (*a*).

respect to ρ and β. Also, the radial distribution function can be determined by functional differentiation with respect to $\phi(r)$:

$$\tfrac{1}{2}\rho^2 g_{(2)}(\boldsymbol{r}) = \phi(\boldsymbol{r})\frac{\partial F}{\partial \phi(\boldsymbol{r})}. \qquad (2.20.12)$$

Anderson *et al.* deal in terms of a function $y(r)$ rather than $g(r)$, defined as

$$y(\boldsymbol{r}) = e^{\beta\Phi(r)} g_{(2)}(r) \qquad (2.20.13)$$

giving, from (2.20.12)

$$y(\boldsymbol{r}) = \frac{2}{\rho^2}\frac{\partial F}{\partial \phi}. \qquad (2.20.14)$$

Further, defining a 'blip' function (fig. 2.20.3(*b*))

$$B(r) = y^0(r)\,[\phi_s(r) - \phi^0(r)] = y^0(r)\,\Delta\phi \qquad (2.20.15)$$

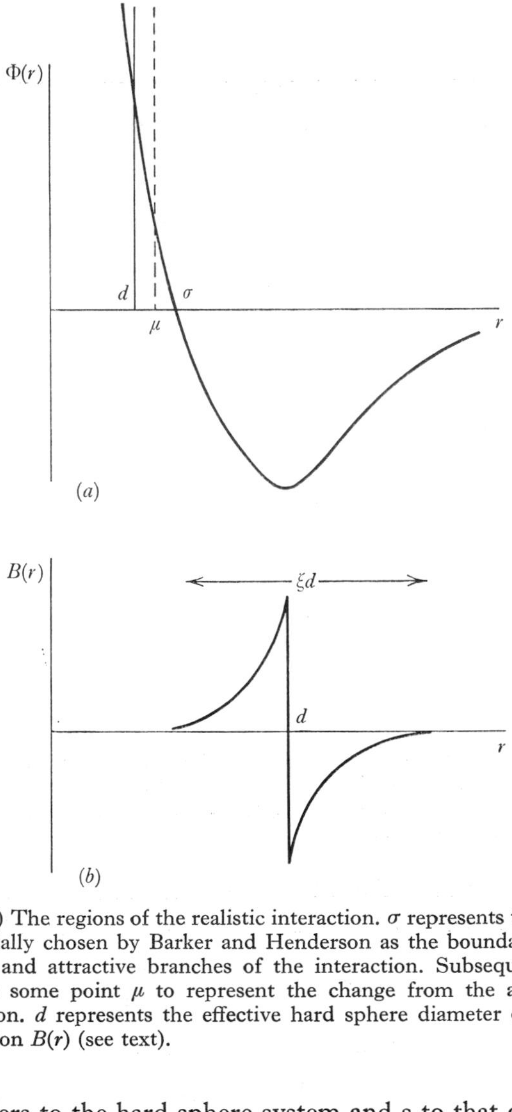

Fig. 2.20.3. (*a*) The regions of the realistic interaction. σ represents the collision diameter, initially chosen by Barker and Henderson as the boundary between the repulsive and attractive branches of the interaction. Subsequently these authors chose some point μ to represent the change from the attractive to repulsive region. d represents the effective hard sphere diameter of the core. (*b*) The function $B(r)$ (see text).

where 0 refers to the hard sphere system and s to that of the soft sphere, (2.20.11) can immediately be written in terms of $B(r)$:

$$F_s = F^0 + \tfrac{1}{2}\rho^2 \int B(r)\, \mathrm{d}\boldsymbol{r} + \dots. \tag{2.20.16}$$

The range of $B(r)$ is very small and depends, of course, upon the softness of the sphere. It is reasonable to chose d such that

$$\int B(r)\,\mathrm{d}\boldsymbol{r} = 0 \tag{2.20.17}$$

and is therefore a decreasing function of both temperature and density. This choice causes the first functional derivative (and potentially the largest term) to vanish identically. If the range of the function B is ξd, then it may be shown that the second derivative term appears to order ξ^4 for this choice of d, and, indeed, all higher terms are at least of order ξ^4:

$$F_s = F^0[1+\mathcal{O}(\xi^4)]. \tag{2.20.18}$$

Since ξ is very small it is clear that the series (2.20.16) converges rapidly, and provides a direct connection between the thermodynamic properties of soft and hard spheres. Similarly, it may be shown that

$$g_s(r) = \phi_s(r)\,y^0(r)\,[1+\mathcal{O}(\xi^2)]. \tag{2.20.19}$$

This equation gives a relationship between the structure of a soft and a hard sphere fluid. All that is now required for the application of these formulae are analytic expressions for F^0 and $y^0(r)$ for a hard sphere fluid. Verlet and Weis[85] have presented analytic expressions for these functions which accurately summarize the results of computer calculations on hard sphere systems. It can be seen that the thermodynamic relationship (2.20.18) is inherently more accurate than the structural one, (2.20.19). One might expect the former to be accurate even for quite soft potentials since the first correction is of fourth order in the softness. Andersen *et al.* restrict their attention to intermolecular potentials which are positive and repulsive. In real fluids, of course, the intermolecular forces are repulsive for some ranges of intermolecular separations. Nevertheless, at high densities (and at high temperatures for all densities) the structure is dominated by the repulsive forces, and so an accurate theory of repulsive forces can provide a foundation for an equilibrium theory of liquids.

The results of the Andersen, Weeks and Chandler perturbation treatment are a substantial improvement on the generalized

Rowlinson model which breaks down badly at high density. This is to be expected from a theory which attempts to describe soft spheres with a density-*independent* effective hard sphere diameter. The Barker–Henderson variational method allows the hard sphere diameter to be density dependent and hence gives consistently better results over a range of density (except for low densities where apparently the variational method fails to give the correct second virial coefficient). The results of the blip-function method are better still, being within 1 per cent of the Monte Carlo results[86]. Henderson[88] observes that the differences between the Anderson–Weeks–Chandler model and the Monte Carlo results are due to the use of the PY theory to calculate $y^0(r)$, rather than to any fundamental problems in the theory. The main reason why their perturbation theory is sensitive to errors in the PY theory is that values of $y^0(r)$ for r near the hard sphere diameter are required, where the PY theory is appreciably in error. More recently Andersen *et al.*[84] have used the Taylor series

$$\ln y^0(r) = \ln y^0(d) + \left(\frac{\partial \ln y^0(r)}{\partial r}\right)_{r=d} (r-d)$$

for $r < d$ together with the hard sphere Monte Carlo results of Barker and Henderson[90] for $r > d$ to calculate $y(r)$. The agreement is then excellent.

2.21 Quantum liquids

Thus far discussion has been centred on essentally classical liquids – liquids at relatively high temperatures and high atomic mass. In such a system the wavelength of the particle is small in comparison to the interatomic spacing and it is meaningful to speak of an individual particle. For quantum liquids, however, we can no longer treat the atoms as individual particles, but must instead speak of the atomic density. Moreover, quantum liquids belong to a different statistical system and the conditions of exchange symmetry of the wave function under interchange of particle coordinates establishes whether the system obeys Bose–Einstein or Fermi–Dirac statistics.

The only quantum fluid of practical importance is liquid helium

^{3}He and its isotope ^{4}He. These fluids constitute a *fermion* and *boson* system respectively. We shall assume that the fluid is in its ground state and has a uniform density $\rho = N/V$. Furthermore, the Hamiltonian of the system is now written[71]

$$\mathscr{H} = -\frac{\hbar^2}{2m}\sum_{i=1}^{N}\nabla_i^2 + \sum_{i<j=1}^{N}\Phi(r_{ij}), \tag{2.21.1}$$

where a pairwise additive interaction potential has been assumed. Expression (2.21.1) represents the quantum mechanical counterpart of the classical Hamiltonian. The eigenfunctions of (2.21.1) are usually approximated by (symmetrized or anti-symmetrized) products of one-particle functions

$$\psi \sim \prod_{i=1}^{N} \phi_i(r_i). \tag{2.21.2}$$

However, such an expression does not represent the correlation of the particles amongst themselves, i.e. does not depend explicitly on the interparticle separations r_{ij}. Of such functions, the simplest is the product over all $N(N-1)/2$ pairs

$$\Psi(1 \dots N) = \mathscr{S}(\mathbf{1} \dots \mathbf{N}) \prod_{i<j=1}^{N} f(r_{ij}) \tag{2.21.3}$$

with $f(r_{ij})$ defined to vanish for $r < r_0$ and to approach unity for $r \gg r_0^3$. This form was suggested by Mott for the hard sphere Bose gas. $\mathscr{S}(\mathbf{r}_1 \dots \mathbf{r}_N)$ is the Slater determinant of one particle wave functions, and is unity for a Bose system.

The two-particle distribution function is given by

$$\rho^2 g_{(2)}(12) = \frac{N(N-1)\int \dots \int \Psi(1 \dots N)\Psi^*(1 \dots N)\,\mathrm{d}3 \dots \mathrm{d}N}{\int \dots \int \Psi(1 \dots N)\Psi^*(1 \dots N)\,\mathrm{d}1 \dots \mathrm{d}N}. \tag{2.21.4}$$

We may proceed further by assuming a wave function of the Bijl–Dingle–Jastrow[72] type for the ground state of the system:

$$\Psi(1 \dots N) = \prod_{i<j} \exp\left[\tfrac{1}{2}u(r_{ij})\right] = \prod_{i<j} (f_{ij}), \tag{2.21.5}$$

where

$$u(r) = -\infty \quad \text{at} \quad r = 0, \qquad g_{(2)}(r) = 0 \quad \text{at} \quad r = 0,$$
$$u(r) = 0 \quad \text{at} \quad r = \infty, \qquad g_{(2)}(r) = 1 \quad \text{at} \quad r = \infty.$$

The BGY equation in the form[73]

$$\nabla_1 g_{(2)}(12) = g_{(2)}(12)\nabla_1 u(12) + \rho \int [g_{(3)}(123) - g_{(2)}(12) g_{(2)}(13)] \cos(12, 13)\, d3, \tag{2.21.6}$$

can be used to derive a $u(r)$ from a given $g_{(2)}(r)$. For $g_{(3)}(123)$ we may insert the Kirkwood superposition approximation. Equation (2.21.6) is a linear inhomogeneous integral equation for $\nabla u(r)$. Abe[74] obtains a formal solution in the form of an infinite series (closely related to (2.11.5)) by a process of iteration. Each term in the series can be represented by a cluster diagram of a simple closed loop and a complete set of cross links radiating from particle 1. Wu and Feenberg[73] have computed essentially the exact solution by setting up the iteration procedure on a high speed computer. Abe establishes an approximate closed relation between the correlation function $u(r)$ and $g_{(2)}(r)$. A recent improvement on this relation has been given by Davison and Lee[87]. The Hamiltonian of the ground state can be expressed as

$$\langle \mathscr{H} \rangle = \langle \mathscr{T} \rangle + \langle \mathscr{V} \rangle, \tag{2.21.7}$$

where $\qquad \mathscr{T} = -\dfrac{\hbar^2}{2m}\sum_i \nabla_i^2, \quad \mathscr{V} = \sum_{i<j} \Phi(r_{ij}),$

are the kinetic energy and potential energy operators, respectively. It follows that for a Bose system

$$\langle \mathscr{T} \rangle = \frac{N\rho\hbar^2}{8m} \int \nabla_1 g_{(2)}(12)\nabla_1 u(12)\, d\mathbf{2}, \tag{2.21.8}$$

$$\langle \mathscr{V} \rangle = \tfrac{1}{2} N\rho \int g_{(2)}(12)\Phi(12)\, d\mathbf{2}. \tag{2.21.9}$$

The pressure of the system is now given by

$$P = \frac{2}{3}\frac{\rho\hbar^2}{8m} \int \nabla_1 g_{(2)}(12)\nabla_1 u(12)\, d\mathbf{2} - \frac{\rho^2}{6} \int g_{(2)}(12)\nabla_1 \Phi(12)\, r_{12}\, d\mathbf{2}. \tag{2.21.10}$$

Clearly, the 'correlation' function $u(12)$ must be determined. Equation (2.21.10) may be written in classical form:

$$PV = Nk\tau - \frac{N\rho}{6}\int g_{(12)}(12)\,\nabla_1\Phi(12)\,r_{12}\,d\mathbf{2}, \qquad (2.21.11)$$

where $$\tau = \frac{2}{3}\frac{\hbar^2}{8km}\int \nabla_1 g_{(2)}(12)\,\nabla_1 u(12)\,d\mathbf{2}.$$

For classical fluids τ is equal to the thermodynamic temperature T, but for quantum fluids it may be considerably greater than T, due to a high zero-point kinetic energy. In a similar manner, the molal internal energy E, of the fluid may be expressed as

$$E = \tfrac{3}{2}Nk\tau + N\rho\int g_{(2)}(12)\,\Phi(12)\,d\mathbf{2}. \qquad (2.21.12)$$

The quantities $\rho_{(2)}kT_i$ represent the components of the average momentum current density tensor in the configuration space of subsets of atomic pairs. For classical systems in thermodynamic equilibrium all τ_i are equal to $\mathbf{1}_i T$ where T is the thermodynamic temperature and $\mathbf{1}_i$ is the unit dyad in the configuration space of molecule i. For quantum mechanical fluids the τ_i may depart widely from the classical value and will in general be functions of position. This raises interesting problems in regions of inhomogeneity, for example, at the liquid surface (§4.10).

Kirkwood and Mazo[75] observe that in the classical superposition approximation a quantum fluid may be described by the RDF of a classical fluid at a temperature τ. All quantum effects, both dynamical and statistical, are transferred to the determination of τ. Kirkwood and Mazo estimate τ (at $T = 0$) to be 12.6 °K, and this compares well with the semi-empirical results of Henshaw and Hurst[76].

The boson formulation given above must be modified for antisymmetrical systems by evaluating to Slater determinant explicitly[71]. Lado[91] has recently shown how symmetry effects may be incorporated into an effective potential for low density systems – thereafter the problem may be treated classically. We may nevertheless choose plane waves for the one-particle wave functions appearing in $\mathscr{S}(\mathbf{1}, \ldots, \boldsymbol{N})$. The potential energy again takes the form:

$$\langle \mathscr{V} \rangle = \tfrac{1}{2} N\rho \int g_{(2)}(12)\, \Phi(12)\, \mathrm{d}\mathbf{2}, \tag{2.21.13}$$

where now we must make a density expansion for the pair distribution

$$g_{(2)}(12) = \exp\,[u(12)]\,[g^{(2)}(12) + \rho g^{(3)}(12) + \ldots]. \tag{2.21.14}$$

The coefficients $g^{(2)}, g^{(3)}, \ldots$ are simple functions of the Slater 2-body, 3-body, ... determinants. The kinetic energy operator produces three terms

$$\langle \mathscr{T} \rangle = -\frac{N\hbar^2}{2m} \langle \mathscr{S} \nabla^2 (\Pi f) \rangle + \langle (\Pi f)\, \nabla^2 \mathscr{S} \rangle + \langle 2 \nabla \mathscr{S} . \nabla (\Pi f) \rangle. \tag{2.21.15}$$

These terms, in order of appearance, are identified as the correlational kinetic energy (which is identical to the boson case), the Fermi energy (arising from the antisymmetric condition on the wave functions) and a cross term.

CHAPTER 3

NUMERICAL SOLUTION OF THE INTEGRAL EQUATIONS

3.1 Introduction

Numerical evaluation and comparison of the various integro-differential equations developed in chapter 2 may be conveniently divided into three sections. First the low density solutions will be discussed. In this case comparison is not made in terms of the form of the pair distribution function but rather by evaluating the equation of state, based on the pressure and compressibility relations, and comparing the virial coefficients so determined with the *exact* results of chapter 1. Of course, if the theory were self-consistent the equation of state and virial coefficients would be independent of the method of approach – the pressure and compressibility equations would yield the same results. Some effort has gone into *forcing* self-consistency by choosing a form for $c(r)$ which ensures identical results regardless of the approach (the self-consistent approximation: SCA). This device of enforcing thermodynamic consistency cannot be regarded as an advance of physical understanding; nevertheless, the results are excellent at least to the sixth virial coefficient for hard spheres.

At liquid densities direct comparison of the radial distribution function may be made. It will be seen that all the theories discussed in the previous chapter agree in the qualitative form of the pair distribution, but the quantitative discrepancy is large. Again, appeal to the equation of state is made. The extreme sensitivity of the pressure equation to the precise form of $g_{(2)}(r)$ (and, indeed, to the assumed form of the pair potential) provides a severe test of the theory. It is computationally convenient to work in terms of an idealized potential such as the hard sphere or square-well interactions – these models have the important advantage that the resulting equation of state may be compared directly with machine simulations. In fact the Monte Carlo and molecular dynamic equations of state may be taken to represent the exact isotherm.

There is some restriction placed on the form of the model potential in that it is required to be physically representative of a real system: it is generally acknowledged that it is the repulsive component of the interaction which is responsible for the essentially geometric problem of atomic packing whilst the attractive component governs the detailed form of the distribution. There are forms of the model potential, the Gaussian interaction for example, which have little bearing upon physical reality, but which do have the advantage of mathematical expediency. We shall not be concerned with these here.

For direct comparison with real fluids, computations are made on the basis of real interactions. Numerical solution of the integral equations is naturally much more difficult. One of the great problems is knowing to what extent the Lennard-Jones pair interaction gives an adequate representation of the effective pair potential. Also to what extent can we depend upon the experimental scattering experiments for our knowledge of the 'true' radial distribution function? As often as possible direct comparison between the theoretical prediction in k-space and the scattered distribution is made, thereby eliminating the Fourier inversion of the scattering data. But to what extent any discrepancy may be attributed to the assumed form of the pair potential, the integral equation or the experimental determination remains an open question.

Finally an important role of the first-order theories is their inversion to yield the effective pair potential. As we saw in §2.19, the inverted result is likely to incorporate triplet effects which cannot as yet be eliminated. We can therefore expect the pair potential of the inert gases, for example, to be density dependent. We know that the liquid metal pair potentials are density dependent for other reasons. However, this route to the pair interaction is an important one, and raises questions not only on the precision of the inversion, but also on the quality of the scattering data from which $g_{(2)}(r)$ is initially determined. Not surprisingly this aspect of the subject is somewhat controversial in that there is considerable difficulty in ascribing the final form of the pair potential to the form of the equation, the data, the inversion process or any shortcoming in any or all of these. This situation is exacerbated by the strikingly different forms of the pair potential in the liquid metal and inert gas cases.

3.2 Low density solutions

The exact results for the first few virial coefficients have been discussed in chapter 1. Here we consider the estimates of the virial coefficients on the basis of the various approximate integral equations. There are two distinct routes to these coefficients, via the pressure and compressibility relations, designated (p) and (c) respectively:

$$(p)\quad \frac{PV}{NkT} = 1 - \frac{2\pi\rho}{3kT}\int_0^\infty g_{(2)}(r)\frac{\mathrm{d}\Phi(r)}{\mathrm{d}r}r^3\,\mathrm{d}r, \tag{3.2.1}$$

$$(c)\quad \frac{1}{kT}\left(\frac{\partial P}{\partial \rho}\right)_T = 1 - 4\pi\int_0^\infty c(r)\,r^2\,\mathrm{d}r. \tag{3.2.2}$$

Ideally these two equations should, of course, yield identical virial expansions, but, by the nature of the approximations made for $g_{(2)}(r)$ and $c(r)$ in the various integral equations the pressure and the compressibility estimates remain inconsistent. The self-consistent approximation (SCA) of Hurst(14) and of Rowlinson(12) does ensure consistency, but this is gained at the expense of some accuracy.

Now, if we take the PY approximation we have

$$c(r)_{\mathrm{PY}} = f(r)\,y(r), \tag{3.2.3}$$

where

$$y(r) = \exp[\beta\Phi(r)]\,g_{(2)}(r). \tag{3.2.4}$$

$c(r)_{\mathrm{PY}}$ may be expressed diagrammatically as in (2.14.5) and inserted in the compressibility relation (3.2.2). This will yield a diagrammatic expansion in density whose coefficients may be immediately identified with the virial coefficients. Similarly, inverting (3.2.4) to read

$$g_{(2)}(r) = y(r)\exp[-\beta\Phi(r)]$$

the integrand of the pressure integral (3.2.1) may be formed as

$$g_{(2)}(r)\,r\frac{\mathrm{d}\Phi(r)}{\mathrm{d}r} = y(r)\,r\frac{\mathrm{d}f}{\mathrm{d}r}. \tag{3.2.5}$$

Again, in the PY approximation, we may easily form the diagrammatic expansion of $y(r)$ from (3.2.3), and having inserted the series in the pressure equation, identify the virial coefficients. Since both the PY and HNC expansions for $c(r)$ are identical up to but not including the four-body cluster integrals, it follows that the coefficients $B_3(\text{PY})_p$, $B_3(\text{HNC})_p$ will be identical, as will $B_3(\text{PY})_c$, $B_3(\text{HNC})_c$, although thereafter they will be internally inconsistent, i.e.

$$B_4(\text{PY})_p \neq B_4(\text{PY})_c.$$

Proceeding in essentially this manner, Kim, Henderson and Oden[1] obtain the following diagrammatic expansions for the fifth virial coefficient:

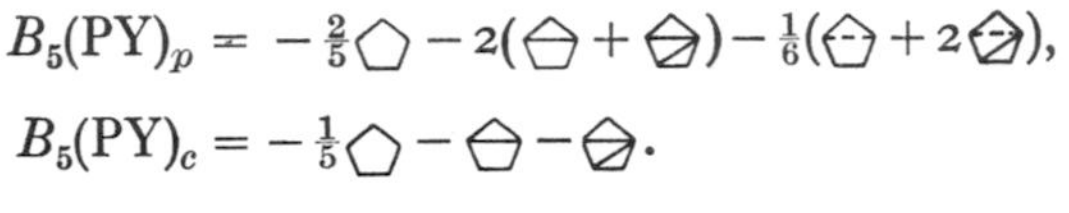

$$B_5(\text{PY})_p = -\tfrac{2}{5}\bigcirc - 2(\ominus + \ominus) - \tfrac{1}{6}(\ominus + 2\ominus), \quad (3.2.6)$$

$$B_5(\text{PY})_c = -\tfrac{1}{5}\bigcirc - \ominus - \ominus. \quad (3.2.7)$$

The presence of a solid line linking particles i and j denotes the f_{ij}-bond, and the broken line represents the factor rf_{ij}. It is clear that the coefficients (3.2.6) and (3.2.7) are inconsistent. This may be directly associated to the approximate nature of (3.2.3).

Similarly, for the HNC approximation

$$c(r)_{\text{HNC}} = f(r)y(r) + y(r) - 1 - \ln y(r). \quad (3.2.8)$$

The resulting fifth virial coefficients are

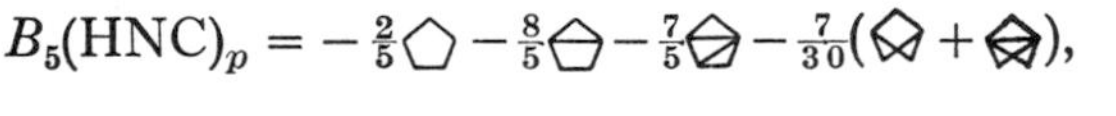

$$B_5(\text{HNC})_p = -\tfrac{2}{5}\bigcirc - \tfrac{8}{5}\ominus - \tfrac{7}{5}\ominus - \tfrac{7}{30}(\ominus + \ominus), \quad (3.2.9)$$

and $$B_5(\text{HNC})_c = -\tfrac{2}{5}\bigcirc - 2(\ominus + \ominus) - \tfrac{1}{3}(\ominus + \ominus). \quad (3.2.10)$$

Again internal consistency is lost because of the approximations which have been introduced.

Rowlinson[12] has proposed the self-consistent approximation;

$$c(r)_{\text{RSCA}} = f(r)y(r) + \alpha(\rho, T)[y(r) - 1 - \ln y(r)]. \quad (3.2.11)$$

Following Rowlinson we write α in the form

$$\alpha(\rho, T) = \sum_{n=0}^{\infty} \rho^n \alpha_{n+4}(T). \quad (3.2.12)$$

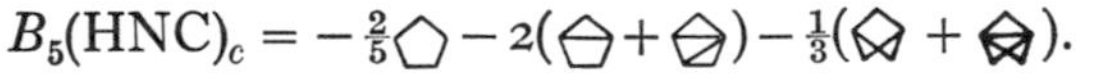

The resulting fifth virial coefficients are

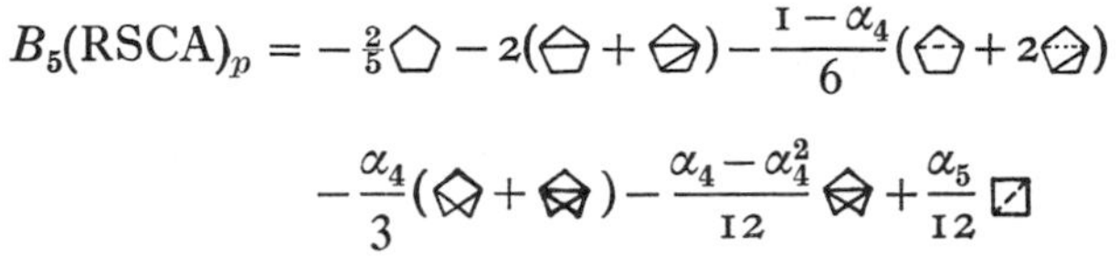

$$B_5(\text{RSCA})_p = -\tfrac{2}{5}\bigcirc - 2(\ominus + \ominus) - \frac{1-\alpha_4}{6}(\ominus + 2\ominus)$$

$$-\frac{\alpha_4}{3}(\diamondsuit + \diamondsuit) - \frac{\alpha_4 - \alpha_4^2}{12}\diamondsuit + \frac{\alpha_5}{12}\boxempty \qquad (3.2.13)$$

and

$$B_5(\text{RSCA})_c = -\frac{1+\alpha_4}{5}\bigcirc + \frac{5+3\alpha_4}{5}\ominus - \frac{5-2\alpha_4}{5}\ominus$$

$$-\frac{4\alpha_4 + 3\alpha_4^2}{30}(\diamondsuit + \diamondsuit) - \frac{\alpha_5}{10}(\square + \boxslash). \qquad (3.2.14)$$

The coefficient α_l is chosen so that $B_l(\text{RSCA})_p$ and $B_l(\text{RSCA})_c$ are consistent.

Hurst[14] has proposed a very similar relation:

$$c(r)_{\text{HSCA}} = f(r)y(r) + y(r) - 1 - y^m(r)\ln y(r)$$

$$= c(r)_{\text{HNC}} + [1 - y^m(r)]\ln y(r). \qquad (3.2.15)$$

Following Henderson[71], we expand m in the series

$$m(\rho, T) = \sum_{n=0}^{\infty} \rho^n m_{n+4}(T). \qquad (3.2.16)$$

The resulting expressions for the fifth virial coefficient has been given by Henderson. The coefficients m_l are chosen so that $B_l(\text{HSCA})_p$ and $B_l(\text{HSCA})_c$ are self-consistent.

Verlet[20] has extended the PY and HNC theories by means of a functional Taylor expansion, discussed in §2.17. The PY2 theory then becomes based on the relation (2.17.9)

$$c(12)_{\text{PY2}} = f(12)y(12) + \Lambda(12).$$

The resulting expressions for the fifth virial coefficient are

$$B_5(\text{PY2})_p = -\tfrac{2}{5}\pentagon - 2(\ominus + \ominus) + \tfrac{1}{3}(\diamondsuit + \diamondsuit) - (\diamondsuit + \diamondsuit)$$

$$-\tfrac{1}{6}(3\diamondsuit + \diamondsuit) - \tfrac{1}{36}(\diamondsuit + 3\diamondsuit) - \tfrac{1}{24}(2\diamondsuit + \diamondsuit) \qquad (3.2.17)$$

and

$$B_5(\text{PY2})_c = -\tfrac{2}{5}\bigcirc - 2(\ominus + \ominus) \quad \tfrac{3}{10}(\diamondsuit + \diamondsuit) - \tfrac{4}{5}(\diamondsuit + \diamondsuit)$$

$$-\tfrac{1}{10}(3\diamondsuit + \diamondsuit). \qquad (3.2.18)$$

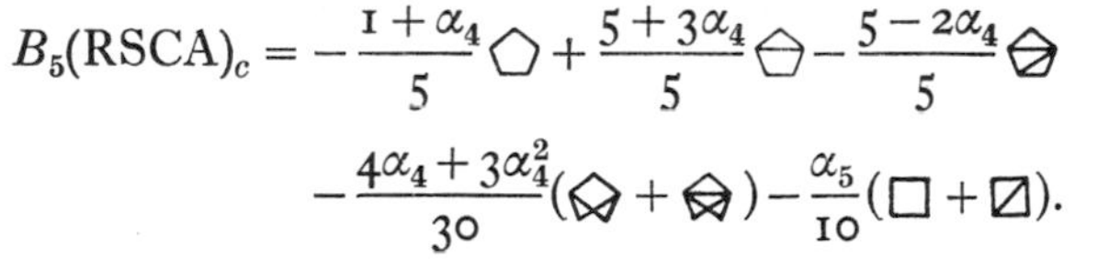

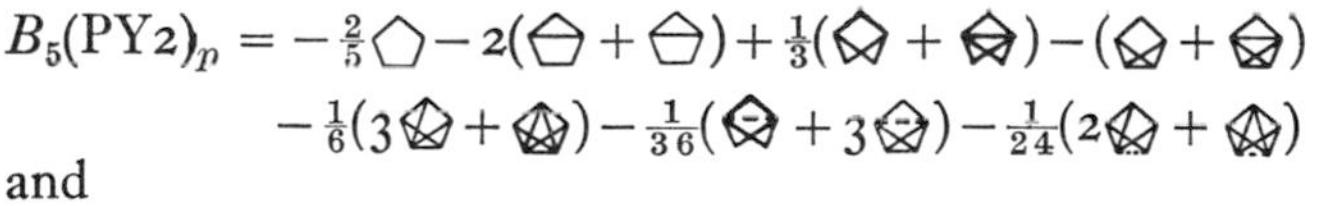

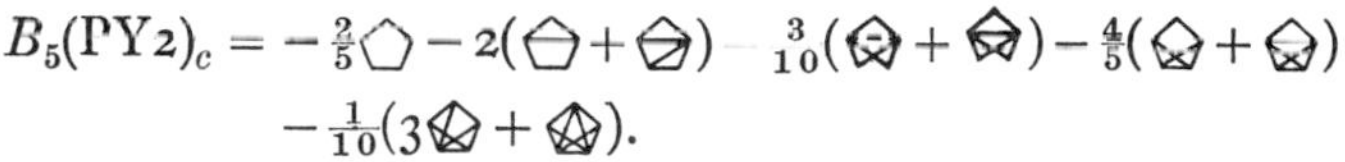

Again the consistency of the pressure and compressibility equations is lost because of the approximations made.

The HNC2 theory is based on the relation (2.17.3)

$$c(12)_{\mathrm{HNC2}} = f(12)y(12) + y(12) - 1 - \ln y(12) + \epsilon(12)$$

and the resulting expansions for the fifth virial coefficient are

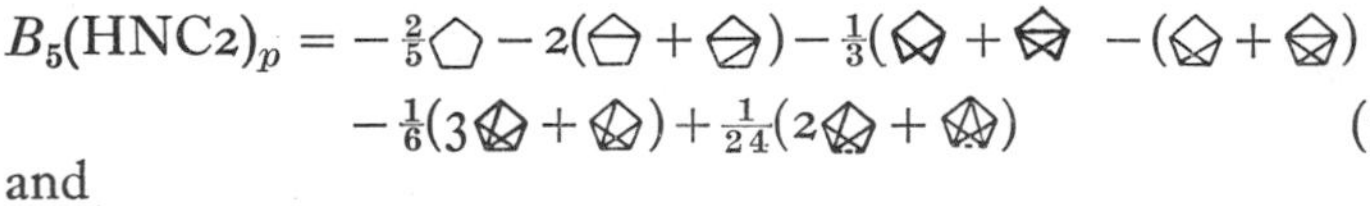

$$B_5(\mathrm{HNC2})_p = -\tfrac{2}{5}\pentagon - 2(\diagram + \diagram) - \tfrac{1}{3}(\diagram + \diagram) - (\diagram + \diagram)$$
$$- \tfrac{1}{6}(3\diagram + \diagram) + \tfrac{1}{24}(2\diagram + \diagram) \qquad (3.2.19)$$

and

$$B_5(\mathrm{HNC2})_c = -\tfrac{2}{5}\pentagon - 2(\diagram + \diagram) - \tfrac{1}{3}(\diagram + \diagram) - \tfrac{9}{10}(\diagram + \diagram)$$
$$- \tfrac{1}{10}(3\diagram + \diagram). \qquad (3.2.20)$$

Consistency is again lost.

We consider the virial coefficients B_l in the density expansion of the equation of state

$$\frac{PV}{NkT} = 1 + B_2\rho + B_3\rho^2 + B_4\rho^3 + \dots. \qquad (3.2.21)$$

The majority of work has been done for the hard sphere potential, and this we now consider in one, two and three dimensions.

Hard spheres

One dimension. The equation of state of an array of rods constrained to move on a line is

$$\frac{PV}{NkT} = (1 - \alpha)^{-1}, \quad \alpha = \frac{\rho b}{N}, \qquad (3.2.22)$$

where b/N is the length of one rod[2]. The exact results for the virial coefficients are, from (3.2.22),

$$B_2/b = B_3/b^2 = B_4/b^3, \text{ etc.} = 1. \qquad (3.2.23)$$

The results for one-dimensional spheres are compared in table 3.2.1, where p and c refer to the pressure and compressibility solutions respectively.

Notice that all one-dimensional virial coefficients in the PY approximation are exact since there can be no indirect correlation and $c(r) = f(r)$ exactly.

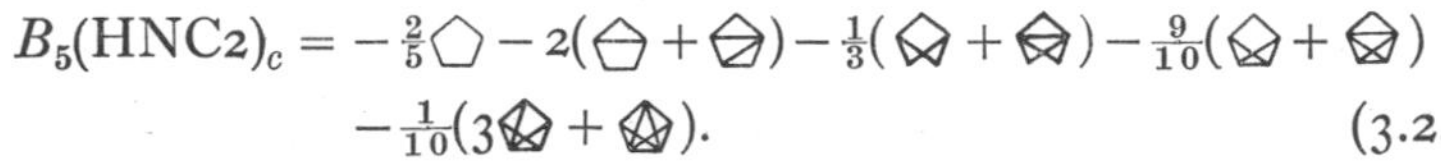

TABLE 3.2.1. *Virial coefficients for hard spheres in one dimension*

	B_4/b^3	B_5/b^4	Reference
HNC_p	3/2	11/6	(3)
HNC_c	11/12	4/5	(3)
PY_p	1	1	
PY_c	1	1	

TABLE 3.2.2. *Virial coefficients for hard spheres in two dimensions*

	B_2/b	B_3/b^2	B_4/b^3	B_5/b^4	B_6/b^5
Exact	1	0.7820	0.5322	0.3338	0.1992
HNC_p			0.8066		
HNC_c			0.4423		
PY_p			0.5008		
PY_c			0.5377		

Two dimensions. The equation of state is not known. The third virial coefficient[(4)] and the fourth[(5)] are known exactly (cf. § 1.4) and the fourth, fifth and sixth have been obtained by Monte Carlo calculations[(6)]. The results are shown in table 3.2.2.

Three dimensions. Not surprisingly most effort has been concentrated on the three dimensional hard sphere system. The results are displayed in table 3.2.3, and we see immediately that none of the approximate theories is consistent beyond the third virial coefficient. It is possible to develop procedures which ensure self-consistency (self-consistent approximation: SCA) for both the pressure and the compressibility estimates up to and including the fourth coefficient, but the fifth remains inconsistent and incorrect. Thus, by extending the PY and HNC equations to include extra terms neglected in the initial approximation B_4 becomes exact, but

$$(B_5/b^4)_p\,(\text{PY}) = 0.493 \quad \text{and} \quad (B_5/b^4)_p\,(\text{HNC}) = 0.398:$$

these results are even poorer than the first-order PY and HNC theories.

TABLE 3.2.3. *Virial coefficients for hard spheres in three dimensions*

	B_2/b	B_3/b^2	B_4/b^3	B_5/b^4	B_6/b^5
Exact[6]	1	0.625	0.2869	0.1103	0.0386
BGY_p	1	0.625	0.2252	0.0475[9]	
BGY_c	1	0.625	0.3424	0.1335[9]	
K_p	1	0.625	0.1400		
K_c	1	0.625	0.4418		
$HNC1_p$[7]	1	0.625	0.4453	0.1447	0.0382
$HNC2_p$	1	0.625		0.066[11]	
$HNC1_c$[7]	1	0.625	0.2092	0.0493	0.0281
$HNC2_c$	1	0.625		0.123[10]	
$PY1_p$[8]	1	0.625	0.2500	0.0859	0.0273
$PY2_p$	1	0.625		0.124[11]	
$PY1_c$[8]	1	0.625	0.2969	0.1211	0.0449
$PY2_c$	1	0.625		0.107[10]	

We recall that the analytic equation of state in the PY approximation is given by (2.15.11):

$$\left(\frac{PV}{NkT}\right)_p = \frac{1+2\eta+3\eta^2}{(1-\eta)^2},$$

$$\left(\frac{PV}{NkT}\right)_c = \frac{1+\eta+\eta^2}{(1-\eta)^3},$$

where the reduced density $\eta = \frac{1}{6}\pi\sigma^3\rho$. The virial coefficients of these equations may be shown directly to be

$$(B_l)_p = 2(3l-4)(b/4)^{l-1},$$

$$(B_l)_c = (1+\tfrac{3}{2}l(l-1))(b/4)^{l-1},$$

$$l \geqslant 2. \tag{3.2.24}$$

$(B_l)_p$ and $(B_l)_c$ are seen to be identical for $l = 2, 3$ and to differ slightly thereafter, fig. (3.2.1).

Whilst the inequality has not been proved, it is apparent from table 3.2.3 that the pressure and compressibility isotherms tend to bracket the exact curve. This is seen to be true for all the theories considered in table 3.2.3. It is clear throughout that the PY theory in the first and second order retains its superiority. We have dis-

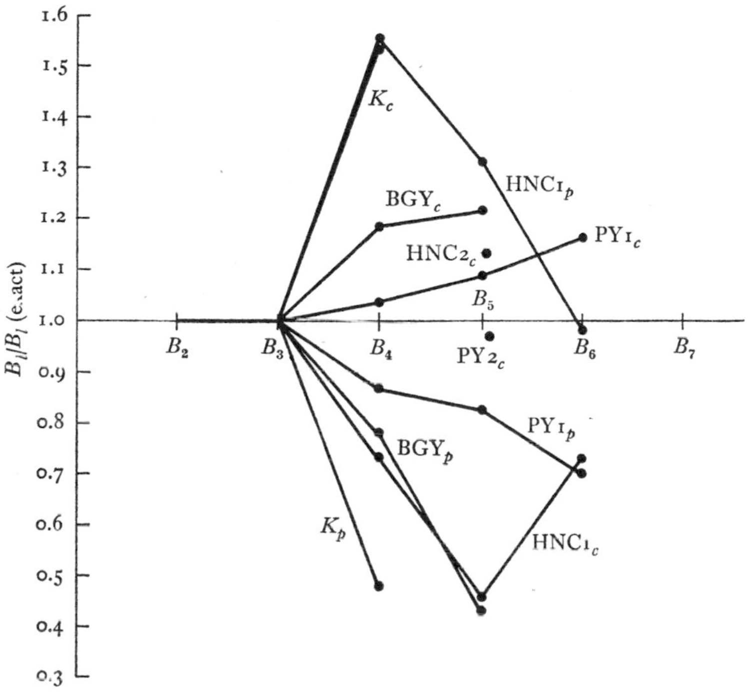

Fig. 3.2.1. The ratio of the hard sphere virial coefficients B_l/B_l (exact) on the basis of various theories.

cussed the reasons for its superior representation for hard spheres in terms of the cluster expansion (§ 2.14). There is nevertheless some motivation to press for consistency between the two equations of state, even at the expense of a slight loss of accuracy.

Rowlinson[12] considers the direct correlation function defined by

$$c(r) = f(r)y(r) + \alpha(\rho, T)[y(r) - 1 - \ln y(r)], \qquad (3.2.25)$$

where α is a pure number, that is, a function of density and temperature. It will depend upon the ratio Φ/kT, but will be independent of r. Equation (3.2.25) is seen to reduce to the PY approximation for $\alpha = 0$ and to the HNC approximation for $\alpha = 1$. At low densities an optimum value of α may be determined by iteration which yields a self-consistent equation of state. Rowlinson's results for hard spheres are shown in table 3.2.4. An alternative SCA, due to Hutchinson and Rushbrooke[7] is consistent up to and including B_4.

TABLE 3.2.4. *SCA hard sphere virial coefficients*

	B_4/b^3	B_5/b^4	B_6/b^5
Exact	0.2869	0.1103	0.0386
Rowlinson	0.2824	0.1041	0.0341

Rodriguez[13] has tried modifying the Kirkwood superposition so as to obtain consistency for B_4. The result was

$$B_4/b^3 = 0.2773.$$

The method is inconsistent beyond this coefficient.

Hurst[14] has suggested a new form of direct correlation function for substitution in the Ornstein–Zernike equation. The new approximation may be related to the HNC expression as follows (cf. 3.2.15)

$$c_{\mathrm{HSCA}}(r) = c_{\mathrm{HNC}}(r) + [1 - y^m(r)] \ln y(r). \qquad (3.2.26)$$

Hurst[14] suggests the name 'generalized HNC' (GHNC) for this approximation. m is a simple numeric. It can be shown that for an optimum value $m = 0.4372$, the coefficients B_2, B_3 are exact and consistent, $B_4/b^3 = 0.2824$ and is consistent, whilst

$$\left.\begin{aligned} (B_5/b^4)_c &= 0.1102 \quad \text{Exact,} \\ (B_5/b^4)_p &= 0.0915 \quad 0.1103, \\ (B_6/b^5)_c &= 0.0386 \quad \text{Exact,} \\ (B_6/b^5)_p &= 0.0353 \quad 0.0386. \end{aligned}\right\} \qquad (3.2.27)$$

These values are seen to be virtually exact. The direct correlation function $c_{\mathrm{GHNC}}(r)$ apparently yields the first six virial coefficients which are both consistent and exact. No application of Hurst's direct correlation function to liquid systems have as yet been made. Henderson[15] has shown that a minor modification in the value of m in (3.2.26) ensures consistency in B_4, B_5 and B_6 simultaneously.

We see from fig. 3.2.1 that the ratio B_l/B_l(exact) on the basis of the various theories diverges for $l > 3$, although the pressure and compressibility estimates *bracket* the correct isotherm. The curious behaviour of the ratio for $(\mathrm{HNC})_p$ and $(\mathrm{HNC})_c$ is presumably to be attributed to a particularly felicitous cancellation

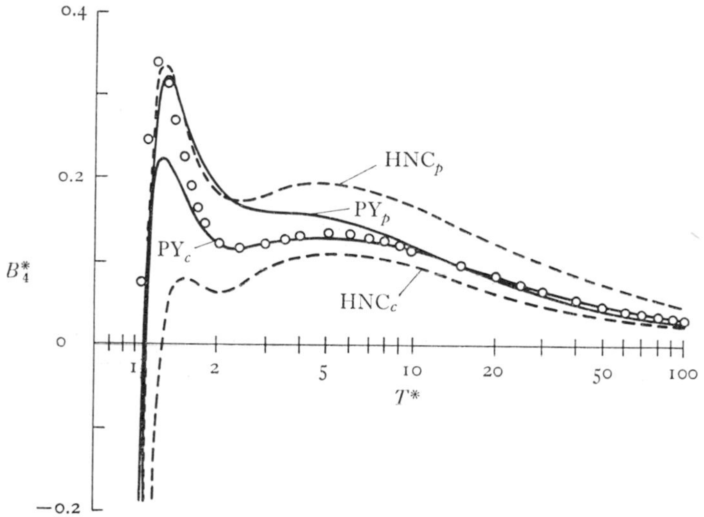

Fig. 3.2.2. The reduced fourth virial coefficient at high temperatures. The points give the exact values and the curves give the results of the PY and HNC theories. (Redrawn by permission from Henderson and Oden, *Mol. Phys.* **10**, 405 (1966).)

amongst the higher order diagrams. In fig. 3.4.1 we show the equation of state of a rigid sphere fluid as a function of density for the various theories.

The square-well potential

Cole[16] summarizes the results for the square-well potential. Unlike the hard sphere potential, the square-well interaction shows a Boyle temperature and to this extent is more realistic. Cole[16] and McQuarrie[17] have given an extensive discussion of this interaction at low densities. For all the equations there is consistency between the second and third virial coefficients B_2 and B_3, and they are, of course, temperature dependent. Thereafter there is the usual discrepancy amongst the various coefficients. Kihara[18] has generalized the square-well potential so that the attractive range is characterized by the parameter $g\sigma$. Cole and McQuarrie use $g = 2$, whereas a value of 1.8 is, in fact, a more reasonable estimate of the attractive range for many simple molecules. Kihara's analysis leads to a generalized expression for the Boyle temperature

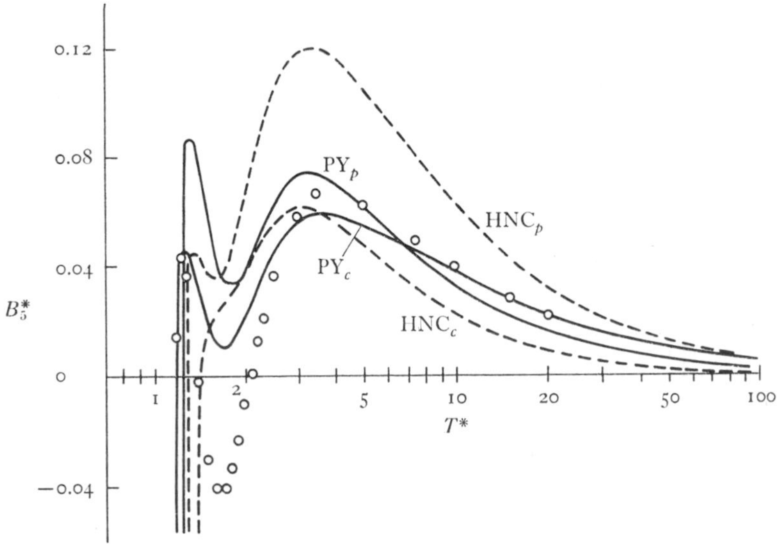

Fig. 3.2.3. Fifth virial coefficient at high temperatures. The points give the exact values of Barker, and the curves give the results of the PY and HNC theories. (Redrawn by permission from Henderson and Oden, *Mol. Phys.* **10**, 405 (1966).)

which reduces to that of Cole and McQuarrie when $g=2$. There is the usual discrepancy amongst the higher virial coefficients. Katsura[19] has given $B_4(T)/b^3$ ($g = 2$) derived on the basis of cluster integrals evaluated by a combination of analytic and numerical techniques.

The Lennard-Jones potential

There are relatively few results for the Lennard-Jones potential. However, the PY approximation retains its superiority at the fourth virial coefficient[20] (fig. 3.2.2). Of course, any theory which is satisfactory for hard spheres is likely to be satisfactory for a Lennard-Jones fluid at high temperatures.

The fifth virial coefficient determined on the basis of the PY, PY2, HNC, HNC2 and SCA approximations are shown in figs. 3.2.3, 3.2.4 and 3.2.5. These have been determined by Kim, Henderson and Oden[1] on the basis of the cluster expansions given at the beginning of this section. In fig. 3.2.3 we have compared the

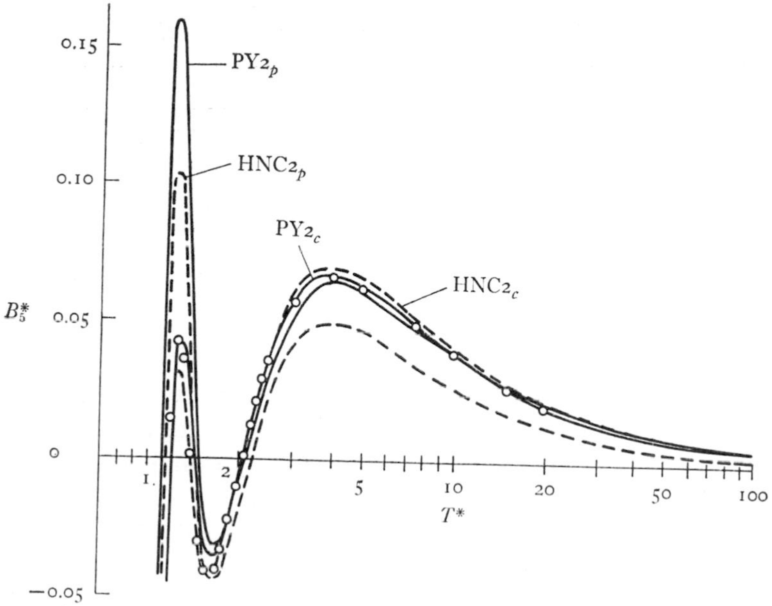

Fig. 3.2.4. Fifth virial coefficient at high temperatures. The points give the exact values of Barker, and the curves give the results of the PY2 and HNC2 theories. (Redrawn by permission from Henderson and Oden, *Mol. Phys.* **10**, 405 (1966).)

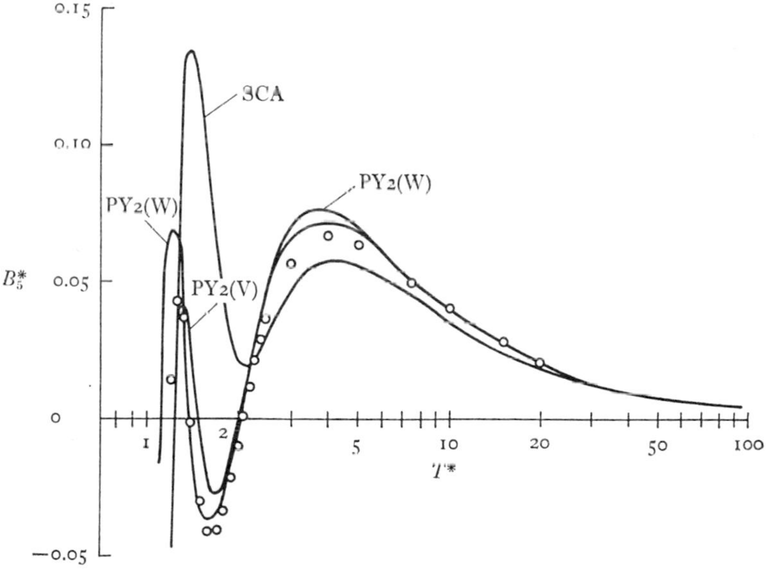

Fig. 3.2.5. Fifth virial coefficient for a Lennard-Jones potential (○) and the PY2 (Verlet), PY2 (Wertheim) approximations to it. Also shown is the self-consistent approximation (SCA). (Redrawn by permission from Henderson and Oden, *Mol. Phys.* **10**, 405 (1966).)

TABLE 3.2.5. *Critical ratios for a Lennard-Jones fluid*

	kT_c/ϵ	b/V_c	$(PV/kT)_c$
$(HNC)_p$	1.25	0.98	0.35
$(HNC)_c$	1.39	0.91	0.38
$(PY)_p$	1.25	0.88	0.30
$(PY)_c$	1.32	0.91	0.36
$(PY2(V))_p$	1.36	0.73	0.31
$(PY2(V))_c$	1.33	0.77	0.34

PY and HNC values for B_5/b^4 with Barker's exact values[70]. The PY values are quite satisfactory at high temperatures but are unsatisfactory at low temperatures. The HNC values are not good at any temperature, and the HNC compressibility equation is not even qualitatively correct since it only has one maximum. In fig. 3.2.4 we show the results of Kim *et al.* for the PY2 and HNC2 B_5: the agreement is seen to be very good, all predicting negative values of B_5 in the region $1.4 > T^* > 2$. The two compressibility curves PY2(V) and PY2(W) are estimates on the basis of the Verlet[20] and Wertheim[72] second-order PY theories. Both are good, but only PY2(V) accurately reproduces the first peak, although both follow the exact curve at low temperatures. The discrepancy may be ascribed to the different 'closure' expressions for $g_{(3)}(123)$ adopted by Verlet and Wertheim. Finally, the SCA estimate is shown in fig. 3.2.5, and the result is seen to be disappointing: the Hurst and Rowlinson curves are virtually indistinguishable.

Only the more accurate PY2 compressibility results are shown for B_5/b^4[21] (fig. 3.2.3). The two approximate curves PY2(V) and PY2(W) are estimates on the basis of the Verlet and Wertheim second-order Percus–Yevick theories. Again, the high temperature limit is good, and both reproduce the maximum at about 5, but only PY2(V) accurately produces the peak at ~ 1.6. Both follow the exact curve at low temperatures.

The critical constants for a Lennard-Jones fluid have been estimated from the equation of state in the HNC, PY and PY2(V) approximations by Levesque[22]. His results are summarized in table 3.2.5. These values may be compared with Verlet and

Levesque's[23] molecular dynamic result of $kT_c/\epsilon = 1.35 \pm 0.01$ for a Lennard-Jones potential. Wood finds a similar critical ratio to that of Verlet, but believes that T_c cannot be determined precisely for such a fluid. The observed critical ratio is found to be 1.26: Verlet and Levesque interpret the discrepancy as evidence of a slight inadequacy of the Lennard-Jones potential, but observe that it becomes *larger* when presumably better interatomic potentials are used in the PY2 calculation.

3.3 Liquid-density solutions

It is important to emphasize that failure of the various theories to predict accurate and self-consistent values for the virial coefficients does not necessarily imply that the approximate theories will be inapplicable at liquid densities. The approximations made for $g_3(123)$ or $c(12)$ contain contributions to all orders of density as we have seen from the classes of diagrams retained in the PY and HNC approximations. The equation of state at liquid densities estimated on the basis of the approximate theories may well prove to be superior to a five-term virial representation.

The inadequacy of using the comparison of the approximate and exact virial coefficients as a criterion of excellence for a theory of liquids is shown clearly in the work of Hoover and Poirier[24]. An estimate of the hard sphere virial coefficients is made on the basis of a relationship between the potential of mean force and the excess chemical potential. It is found that only the Kirkwood equation leads to the correct third virial coefficient, and again, this theory yields the best value of the fourth virial coefficient being in error by about 30 per cent. The HNC and PY estimates are in error by a factor of three or four. The various estimates of the third coefficient are given in table 3.3.1.

McQuarrie[25] has computed the virial coefficients of a square-well fluid from a comparison of the virial equation of state and a density expansion of the internal energy of the fluid. The coefficients are then compared with the known third and fourth virial coefficients, and it appears that the HNC and BGY equations give the least inconsistent results over most of the temperature range.

TABLE 3.3.1

	Exact	K	BGY	HNC	PY
B_3/b^2	5/8	5/8	5/6	5/4	−1/12

We should therefore exercise some caution in evaluating the performance of the integral equations at liquid densities on the basis of the low-density solutions.

A few attempts have been made to effect an analytic solution of the integral equations for the pair distribution function. We have discussed Wertheim's analytic solution[26] of the PY equation for the case of hard spheres in §2.15. Thiele has also obtained an analytic solution of the hard sphere PY equation. He observed that the equation for $[r(\Phi(r)/kT)\ln g_{(2)}(r)]$ required discontinuities in the derivatives of this function at $r = \sigma, 2\sigma, 3\sigma$, etc., and surmised that the function was analytic over each interval. This is the case, and he was able to obtain a solution for $[(\Phi(r)/kT)\ln g_{(2)}(r)]$ in the form of a cubic in r over the range $0 < r < \sigma$. Reiss, Frisch and Lebowitz[28] have also obtained the equation of state of hard spheres though not on the basis of the solution of any integral equation discussed here, and we pursue this aspect no further.

Green[29] has obtained an analytic solution of the BGY equation in the Kirkwood superposition approximation. Green's procedure was to linearize the non-linear BGY integral equation on the assumption that

$$g_{(2)}(r) = \exp\left(-\frac{\Phi(r)}{kT}\right)\exp[\alpha(r)]$$

may be written as $$\exp\left[-\frac{\Phi(r)}{kT}\right][1+\alpha(r)],$$

where $\alpha(r)$ is some unknown function of the separation distance. This, and one or two other approximations are introduced which are difficult to evaluate physically. The final explicit expression for the pair distribution appears to agree reasonably well with experiment[29], and it seems that no essential features are lost in the course of the mathematical simplification. However, we cannot place too much confidence in such a procedure: the non-linearity of the initial integral equation may have important physical consequences which are not necessarily apparent in the form of $g_{(2)}(r)$.

Kirkwood and Boggs[30] have obtained an analytic solution of the Kirkwood equation in the superposition approximation. Again various mathematical approximations are made which are physically obscure, but nevertheless a pair distribution in qualitative agreement with observation is obtained. With the advent of large-scale computers and numerical methods the approximate analytic techniques can no longer be justified. The power of numerical, if not mathematical, techniques has been vastly improved and we are now in a position to make a critical evaluation of the various theories.

3.4 The hard sphere fluid

We first consider the various integral equations for a system of hard spheres. The hard sphere potential is of particular importance since it has been shown to be the effective short range repulsive interaction which governs the geometric problem of packing at high densities. For a pair potential whose repulsive component is adequately described by the hard sphere interaction, such as the liquid metals[31], the conclusions should be particularly important. Furthermore we have the 'experimental' data of machine calculations with which to compare the various pair distributions and the equations of state: in this case there is no problem arising from uncertainty in the form of the pair potential.

The equation of state of the rigid sphere fluid is particularly simple:

$$\frac{PV}{NkT} = 1 + \frac{2\pi\sigma^3}{3}\rho g_{(2)}(\sigma), \tag{3.4.1}$$

where $g_{(2)}(\sigma)$ is the pair distribution at $r = \sigma$. Equation (3.4.1) arises from the insertion of the hard sphere potential in

$$\frac{PV}{NkT} = 1 - \frac{2\pi N}{3VkT}\int_0^\infty \nabla_1\Phi(r)\, g_{(2)}(r)\, r^3\, dr. \tag{3.4.2}$$

Notice in (3.4.1) that $g_{(2)}(\sigma)$ is itself a function of density.

The mathematics by which the BGY and K equations are reduced to a form suitable for numerical solution has been treated in detail by Kirkwood, Maun and Alder[32], and reviewed by Hill[33].

Unfortunately the density parameter λ is defined differently in the BGY and K theories and are not immediately comparable. However for a ratio of volume to close-packed volume of spheres = 1.24 on the Kirkwood theory, and 1.48 on the Born–Green–Yvon theory, the integral equations cease to be integrable. Kirkwood *et al.*[32] surmise that this represents the boundary of stability of the fluid phase, and that the stable phase at higher densities is crystalline[34]. We shall see that this aspect of the K and BGY equations is unique amongst the integral equations developed so far. The BGY and K equations are effectively subject to the 'boundary condition' which may be taken as a criterion for the existence of a fluid phase:

$$r[g_{(2)}(r)-1] \to 0 \quad \text{when} \quad r \to \infty. \tag{3.4.3}$$

Clearly a long range oscillatory solution for $g_{(2)}(r)$ does violate the boundary condition, and although the BGY and K integrals still yield a solution, it must be rejected as not being of the fluid phase.

Whether this singularity can be identified as a solid–fluid phase transition, or whether in fact purely hard spheres can have such a transition, has been debated ever since with no rigorous answer, in spite of the almost quantitative agreement with the molecular dynamic prediction of a phase change[35].

To the extent that the statistical geometry of the packing of hard spheres is characteristic of a real liquid[31], we should expect the excess entropy of the hard sphere fluid to approximate that of a real liquid. The excess entropy is defined as

$$S_{\text{excess}} = S_{\text{fluid}} - S_{\text{ideal gas}} \tag{3.4.4}$$

and is clearly always negative. In table 3.4.1 the excess entropy of a hard sphere fluid in the K and BGY approximations is compared as a function of density. In fig. 3.4.1 we reproduce the equation of state of a hard sphere fluid as a function of density in the various approximations[36]. We see that the BGY and K equations are by far the worst numerically, but have the qualitative advantage of

TABLE 3.4.1. *The excess entropy of a hard sphere fluid*

$\rho\sigma^3$	(S_{ex}/Nk) K	(S_{ex}/Nk) BGY
0.169	−0.03	−0.03
0.299	−0.12	−0.11
0.407	−0.24	−0.23
0.500	−0.39	−0.37
0.585	−0.55	−0.56
0.658	−0.73	−0.76
0.729	−0.92	−1.00
0.790	−1.14	−1.23
0.862	−1.37	−1.49
0.924	−1.60	−1.77
0.982	−1.80	
1.032	−2.07	
1.089	−2.32	
1.141	−2.60	

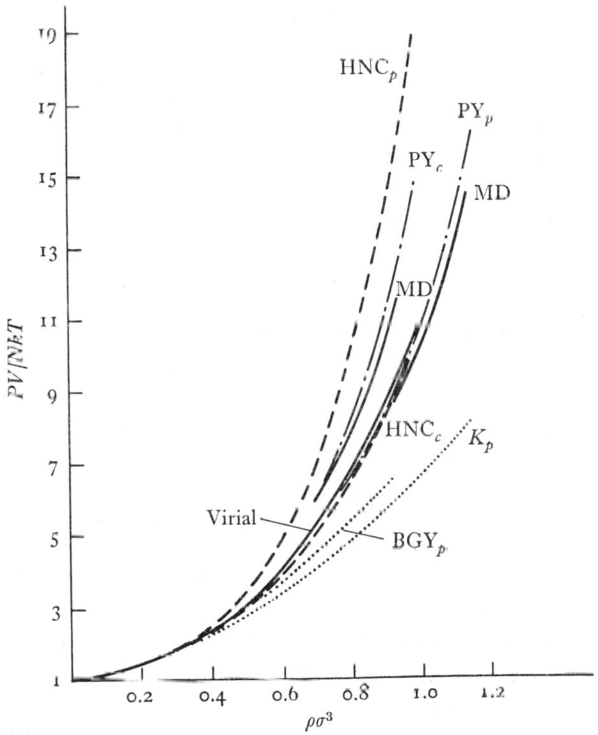

Fig. 3.4.1. The equation of state of the hard sphere fluid in the various approximations. (Redrawn by permission from Klein, *J. Chem. Phys.* **39**, 1388 (1963).)

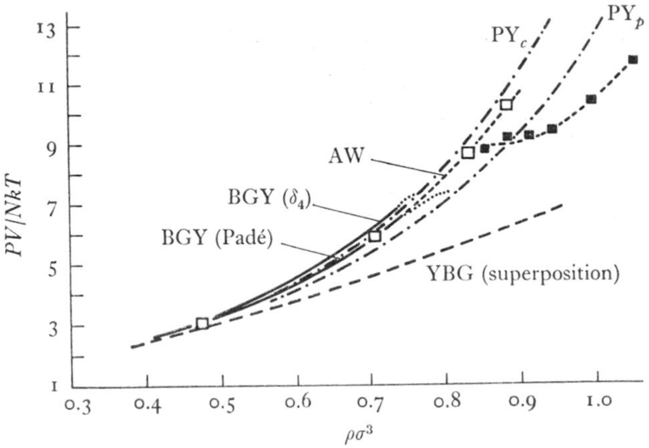

Fig. 3.4.2. Equation of state for a system of hard spheres, showing the BGY (Padé) isotherm to be in essential agreement with the Alder and Wainwright data (fluid □ 108 particles (smoothed); solid ■ 108 particles). (Redrawn by permission from Rice and Lekner, *J. Chem. Phys.* **43**, 3559 (1965).)

ceasing to be integrable at or about the molecular dynamic phase change density.

The Rice–Lekner[51] modification of the BGY equation in terms of a Padé approximant to the triplet distribution retains the singularity but gives better agreement with the molecular dynamics results than any other of the first-order theories (fig. 3.4.2).

In table 3.4.2 we compare PV/NkT estimated from a five-term hard sphere virial Z_5, a six-term hard sphere virial Z_6, and the BGY–Padé approximant with the molecular dynamics calculation (fluid branch).

Since each of the approximations gives a different representation of the fourth and fifth virial coefficients, one expects them to separate as the density increases. The pressure and compressibility estimates tend to bracket the 'experimental' isotherm, and Klein[36] observes that the HNC deviations are quite large. Thus at $\rho\sigma^3 = 0.9$, the compressibility curve differs by 40 per cent, and the pressure curve by 20 per cent from the machine calculations. However, the arithmetic mean of the two curves in the HNC case gives a good representation of the reference isotherm.

Klein[36] fails to find any evidence of a fluid–solid transition of

TABLE 3.4.2

ρ_0/ρ	Z_5	Z_6	BGY-Padé	Mol. dyn.
1.6	8.11	8.95	10.11	10.17*
1.7	7.17	7.79	8.55	8.59
2.00	5.31	5.59	5.83	5.89
3.00	2.98	3.01	3.03	3.05
10.00	1.36	1.36	1.36	1.36

* Density transition occurs at $\rho_0/\rho = 1.63$.

the BGY–K type, although the radial distribution function begins to take on a strongly oscillatory long range form at high densities. The general conclusion seems to be that the PY equation is better than the HNC equation for repulsive potentials, in particular for solid spheres. However, as we shall see in the next section, the HNC equation appears to give better results than the PY equation at low temperature and moderate density[40] for more realistic pair potentials having an attractive component. This may be understood as follows. Consider first the form of the Mayer f-function for a *realistic* potential as a function of temperature. We see that as the temperature decreases the initially hard sphere-like f-function develops a positive component as the attractive forces play an increasing role. Thus, in performing integrations over the field points linked by f-bonds it is clear that, unlike the hard sphere f-function, there can be extensive cancellation between the positive and negative region of the bond. And so it can be that the net contribution of the diagram  dropped in the HNC approximation would have been virtually zero for a realistic potential. Of course, this would not be the case at high temperatures when the positive region of the f-function is effectively washed out, nor can this approximation be satisfactory at very high densities when more than two field points have to be taken into account. The HNC approximation quite clearly will be inadequate for hard-sphere interactions, or those effectively so, and so we might anticipate that the theory would find better application in the case of Lennard-Jones and Guggenheim–McGlashan potentials. In fig. 3.4.3 we show the

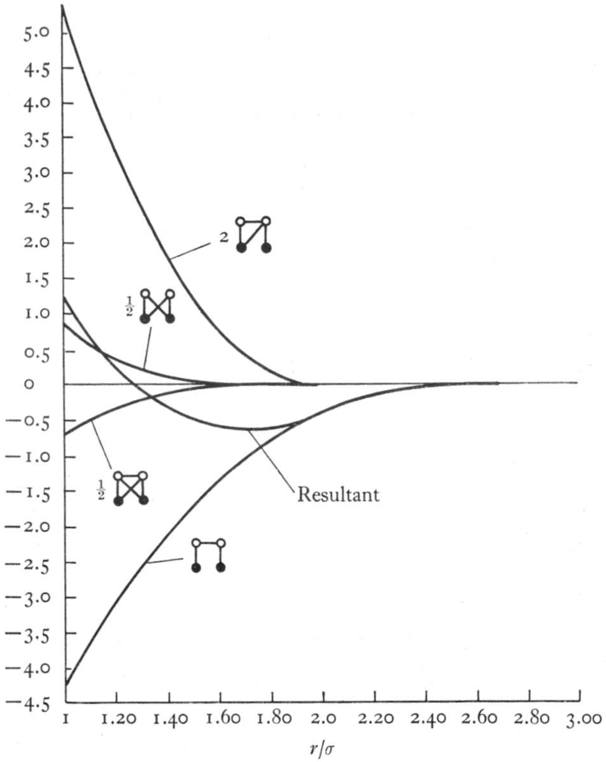

Fig. 3.4.3. Hard sphere cluster integrals involved in the first-order HNC and PY approximations. (Redrawn by permission from Klein, *J. Chem. Phys.* **39**, 1388 (1963).)

contributions of the various cluster diagrams entering with order ρ^2 for a hard sphere interaction. It is clear that  is far from zero in the vicinity of the atomic diameter.

In the PY approximation, however, the diagrams  and  are set to be equal and opposite. In other words PY equate

$$-\tfrac{1}{2}\rho^2\iint f_{13}f_{23}f_{14}f_{24}\,\mathrm{d}\mathbf{3}\,\mathrm{d}\mathbf{4} \quad \left(\tfrac{1}{2}\;\cdot\right) \qquad (3.4.5)$$

with

$$\tfrac{1}{2}\rho^2\iint f_{13}f_{23}f_{14}f_{24}f_{34}\,\mathrm{d}\mathbf{3}\,\mathrm{d}\mathbf{4} \quad \left(\tfrac{1}{2}\;\cdot\right). \qquad (3.4.6)$$

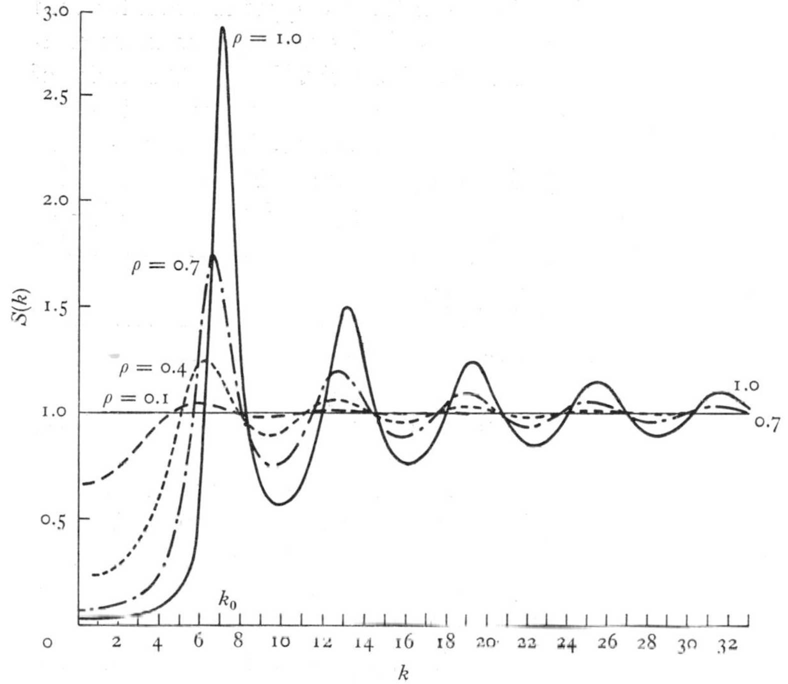

Fig. 3.4.4. Hard sphere structure factor as a function of density in the HNC approximation, as calculated by Klein[36]. Note the development of the first peak at k_0 with increasing density. (Redrawn by permission from Klein, *J. Chem. Phys.* **39**, 1388 (1963).)

This, of course, effectively sets

$$f_{34} = \exp\left[-\frac{\Phi(r_{34})}{kT}\right] - 1 = -1 \qquad (3.4.7)$$

which means that $\Phi(r_{34})$ has been equated to infinity for all values of r_{34}. Now in the case of hard spheres this is a good approximation because when field points 3 and 4 move far apart, one or more of the other bonds f_{13}, f_{14}, f_{23}, f_{24} become zero, and thus the error which would have been introduced in assuming f_{34} equal to -1 for all values of r_{34} is small. As is shown in fig. 3.4.3 this is very good in the case of hard spheres. As the density is increased, however, the approximation becomes inadequate for the same reason as for HNC: we cannot neglect the contribution due to diagrams with more than two field points, and the PY approximation for hard spheres will start to fail. This conclusion is confirmed in the computations of Broyles, Chung and Sahlin[37]. The HNC equation of state is nevertheless an improvement over those of the five-term virial series and the BGY and K equations. Klein calculates the scattering function $S(k)$ for hard spheres in the HNC approximation and observes the development of the curve with density at k_0 – the wave number corresponding to the smallest interparticle spacing for close-packed spheres (fig. 3.4.4).

We look for singularities in k-space since from the Fourier transformation of the Ornstein–Zernike equation we have

$$S(k) = 1 + \frac{c(k)}{1 - \rho c(k)}. \tag{3.4.8}$$

Singularities will occur when the denominator is zero. For hard spheres the singularity is purely density dependent, whilst for more realistic potentials there will be singularities for a given combination of density and temperature values, presumably defining the limit of stability of the fluid in the ρ–T plane. $c(k)$ is a decreasing function of k for small k so that no poles in (3.4.8) appear as long as $\rho c(k) < 1$. Comparison of (3.4.8) with the compressibility relation shows that the latter condition is equivalent to

$$\left(\frac{\partial P}{\partial \rho}\right)_T > 0. \tag{3.4.9}$$

Klein[36] concludes that there is a tendency for a singularity to develop. He bases his conclusion on the tendency toward singularity of $S(k_0)$, and the absence of marked broadening of successive peaks in the RDF such as are found in liquids. Kugler[69] gives some interesting inequalities for $c(k)$ and $S(k)$ in classical fluids.

A more recent application of the generalized HNC theory of Hurst[14] and of Rowlinson[12] has been made for hard spheres by Lado[67] at liquid densities. Lado writes the direct correlation as

$$c(r) = g_{(2)}(r) - g_{(2)}(r)\, e^{\beta\Phi(r)} + \alpha\{g_{(2)}(r)\, e^{\beta\Phi(r)} - 1 - \ln[g_{(2)}(r)\, e^{\beta\Phi(r)}]\} \tag{3.4.10}$$

which is inserted into the Ornstein–Zernike equation. With $\alpha = 0$

and $\alpha = 1$ we recover the PY and HNC approximations respectively. This expression is identical to that of Rowlinson[12], (3.2.11). α is again chosen to yield consistency between the pressure and compressibility isotherms. It is found that this approach leads to an improvement over the PY and HNC approximations for the hard sphere thermodynamic and distribution functions. Limited comparisons with Verlet's PY2 equation indicate comparable accuracy with that approximation, which is computationally more difficult. Lado has also reported the results of this approach for hard discs[68].

An intermediate situation which contains many of the features of a realistic potential is the square-well fluid. Barker and Henderson[56] calculate the equation of state by treating the attactive well as a perturbation on the hard sphere potential. The agreement is good even at the lowest temperature $T^* = 0.5$, which is far below the critical temperature. These authors generalize their perturbation approach to the 6–12 potential, and compare their results at $T^* = 2.74$ to the Monte Carlo equation of state of Wood and Parker[38]. The results appear to be quite satisfactory.

3.5 Realistic fluids

There exists a considerable body of published data on the measurements of the equations of state and distribution functions of gases and liquids[39]. However, comparison of theory with experiment is, in the case of real systems, necessarily a comparison of the theory and the assumed form of potential with experiment. This method does not qualify as a complete test of either theory or potential since a deficiency in one can often be compensated for by a change in the other. Comparison is therefore obscured by our inadequate knowledge of the pair potential and the development of triplet contributions to the interaction at liquid densities. However, in as far as the Lennard-Jones and Guggenheim–McGlashan potentials are analytically convenient and nevertheless representative of the class of simple insulating fluids, we shall attempt a comparison and find that agreement is quite good.

Direct comparison can legitimately be made with the various machine calculations which have been made for a Lennard-Jones

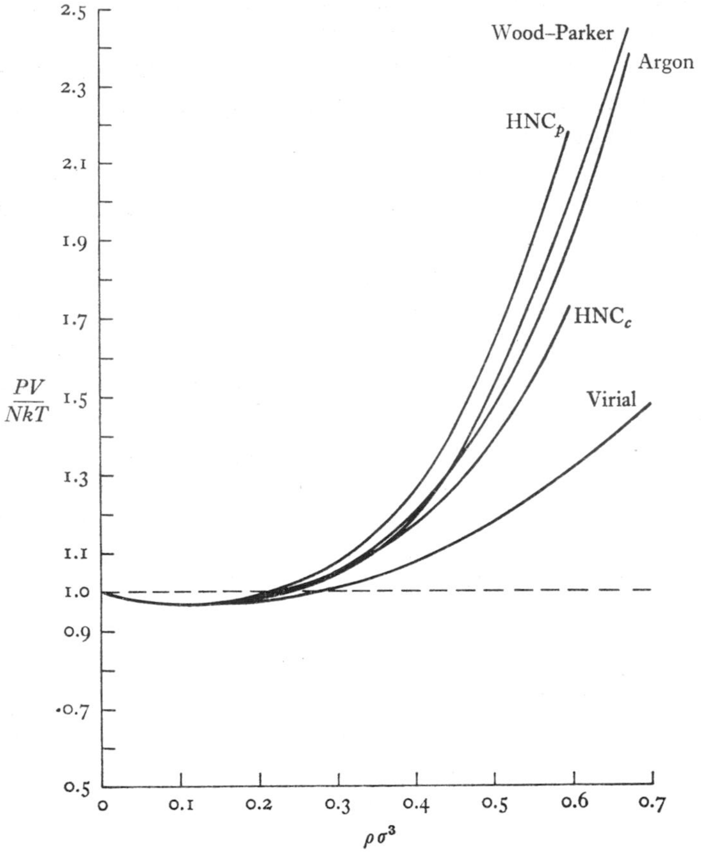

Fig. 3.5.1. Comparison of the experimental and theoretical equations of state for liquid argon (ϵ/k = 119.8 °K, σ = 3.405 Å, $T^* = kT/\epsilon = 2.74$). (Redrawn by permission from Klein and Green, *J. Chem. Phys.* **39**, 1367 (1963).)

fluid(38), and these we shall compare first. In fig. 3.5.1 we show the comparison of the experimental(39) equation of state for argon (ϵ/k = 119.8 °K, σ = 3.405 Å) with the Monte Carlo calculations of Wood and Parker(38) at a reduced temperature $T^* = kT/\epsilon = 2.74$. It is seen that there is striking agreement between the two sets of data, and either can be taken as the reference isotherm. Broyles, Chung and Sahlin(37) have solved the PY, HNC and BGY equations at four particle densities on the Monte Carlo $T^* = 2.74$ isotherm. They compute the thermodynamic quantities PV/NkT and

TABLE 3.5.1. *Thermodynamic functions for argon*

$\rho\sigma^3 =$	0.40	0.83	1.00	1.11
$(PV/NkT)_{MC}$	1.2–1.5	4.01	7.0	7.8
$(PV/NkT)_{PY}(p)$	1.24	4.01	6.8	9.2
$(PV/NkT)_{HNC}(p)$	1.28	5.11	9.1	13.2
$(PV/NkT)_{BGY}(p)$	1.26	2.3	3.1	3.8
$(E/NkT)_{MC}$	−0.86	−1.58	−1.60	−1.90
$(E/NkT)_{PY}$	−0.865	−1.61	−1.67	−1.59
$(E/NkT)_{HNC}$	−0.859	−1.40	−1.19	−0.78
$(E/NkT)_{BGY}$	−0.85	−1.8	−2.2	−2.6

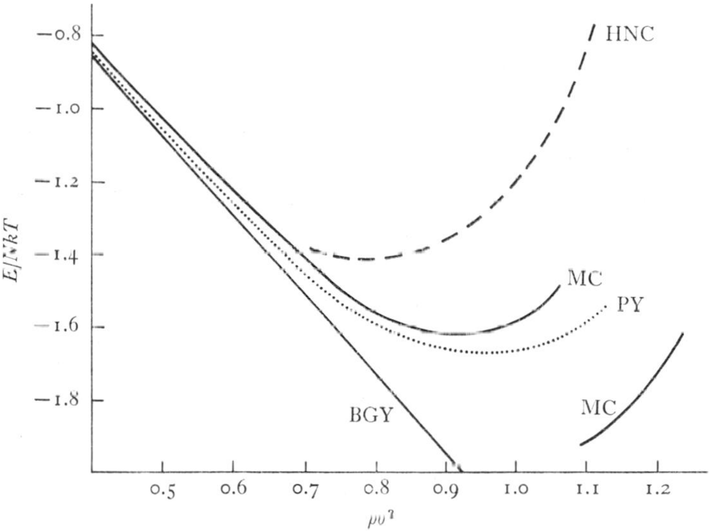

Fig. 3.5.2. Internal energy of a Lennard-Jones fluid as a function of density as computed from the several theories discussed. (Redrawn by permission from Klein and Green, *J. Chem. Phys.* **39**, 1367 (1963).)

E/NkT from the various radial distribution functions. They conclude that the PY quantities agree best with the Monte Carlo results, HNC next best, and BGY most poorly, particularly at high densities. Certainly for the case of highest density, $\rho = 1.111$, PY yields the best pair distribution function. Bearing in mind the great sensitivity of the thermodynamic functions to the detailed form of the first peak of $g_{(2)}(r)$ it is clear that the HNC and BGY equations are

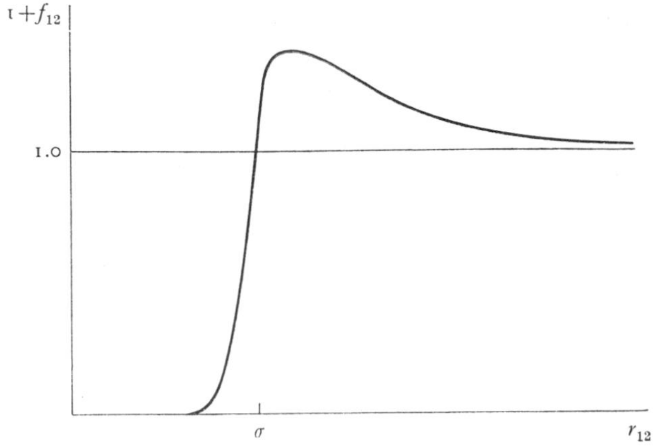

Fig. 3.5.3. The 'exclusion' factor $(1+f_{12})$ for a realistic pair interaction.

inadequate at these densities. In table 3.5.1[37] the thermodynamic quantities are compared, and the order of superiority of the theories is preserved. The results are presented graphically in figs. 3.5.1 and 3.5.2. It is interesting to see that the PY curves for PV/NkT and E/NkT fit the Monte Carlo points quite well, except that the theory fails to yield the transition discontinuity. The PY equation continues to give an analytic solution right through to physically impossible densities. We may understand this by identifying the classes of diagrams dropped in the PY approximation. All elementary diagrams (E) and bundles without the f_{12}-bond (B') are dropped, plus those diagrams formed by inserting the f_{12}-bond into these sets. This may be shown from (2.14.6), (2.14.16). Explicitly, the diagrams omitted are

$$E+B'+f_{12}E+f_{12}B' = (1+f_{12})(E+B'). \qquad (3.5.1)$$

For realistic potentials at low temperatures the function $1+f_{12}$ has the form shown in fig. 3.5.3. Quite clearly the factor $(1+f_{12})$ eliminates contributions to the integrand arising from overlap configurations, since $(1+f_{12}) \to 0$ as $r_{12} \to 0$. It is precisely this contribution which prevents molecules from interpenetrating (which the PY approximation allows) and which leads to a phase transition which the PY equation does not predict. The same reasoning applies to the hard sphere fluid, of course.

Extensive comparison[36, 40, 41] of the HNC results with the experimental data of Michels *et al.*[39] again shows the superiority of the PY approximation, although the HNC results seem to be at least equally good at low temperatures and moderate densities. Khan[40] reviews the HNC situation extensively for the Lennard-Jones and Guggenheim–McGlashan potentials. His results show a disconcerting sensitivity of the pair distribution and thermodynamic results upon the precise form of the pair potential.

Klein and Green[36] find that in the HNC integral equation singularities develop at certain temperature–density points. From (3.4.8) singularities will occur when $1-\rho c(k) = 0$. At such singularities $S(k)$ is infinite. Since the isothermal compressibility χ_T is given by

$$\chi_T = \frac{1}{\rho}\left(\frac{\partial \rho}{\partial P}\right) = \frac{S(0)}{\rho k T} \tag{3.5.2}$$

it follows that, when they occur at $k = 0$, such singularities predict infinite compressibility. All singularities found by Klein and Green were for $k = 0$, which implies $(\partial P/\partial \rho)_T = 0$. With the single exception of the critical point $(\partial P/\partial \rho)_T \neq 0$ at all points on the experimental coexistence curve. Unfortunately for this reason the theoretical locus of singularities cannot be identified with the experimental coexistence curve (except at the critical point).

The PY2 and HNC2 theories remain untested.

The Padé extension of the BGY approximation has been applied to a Lennard-Jones system at $T^* = 2.74$[50]. Again, the BGY equation is retained since it is believed to contain the qualitative features necessary to describe the limits of stability of the fluid phase.

The same Padé approximant to the Salpeter expression of the triplet distribution evaluated by Rice and Lekner[51] for hard spheres is now determined for a Lennard-Jones interaction. Rice and Young[50] calculate the equation of state of the fluid in the superposition approximation, for the one term and the two term Salpeter series, and for the Padé approximant to the entire Salpeter series. The BGY–Padé approximation is in excellent agreement with the fluid branch of the Monte Carlo calculations[35]. The internal energy data for the Monte Carlo 6–12 fluid are compared with the BGY–Padé results. Again, the agreement is seen to be very good.

Kirkwood[44] has investigated the stability of solutions of the one-dimensional BGY equation with respect to small perturbations in the singlet density. We may write

$$-\beta^{-1}\nabla_1 \ln \rho_{(1)}(\boldsymbol{x}_1) = \int \frac{\nabla_1 \Phi(x_{12})\,\rho_{(2)}(\boldsymbol{x}_1\boldsymbol{x}_2)}{\rho_{(1)}(\boldsymbol{x}_1)}\,\mathrm{d}\boldsymbol{x}_2 + \nabla_1 \Phi_0(\boldsymbol{x}_1), \quad (3.5.3)$$

where Φ_0 is an external potential. The following perturbations to the singlet and pair density distributions are now made, and Kirkwood incorrectly assumes here that these may be formed independently;

$$\left.\begin{aligned} \rho_{(1)}(\boldsymbol{x}_1) &= \rho_l[1+\phi(\boldsymbol{x}_1)], \\ \rho_{(2)}(\boldsymbol{x}_1\boldsymbol{x}_2) &= \rho_{(1)}(\boldsymbol{x}_1)\,\rho_{(1)}(\boldsymbol{x}_2)\,g_{(2)}(\boldsymbol{x}_1\boldsymbol{x}_2), \\ g_{(2)}(\boldsymbol{x}_1\boldsymbol{x}_2) &= g_l(x_{12}) + \chi(\boldsymbol{x}_1\boldsymbol{x}_2), \end{aligned}\right\} \quad (3.5.4)$$

where ϕ and χ are small perturbations to the (uniform) liquid distributions ρ_l and g_l. Linearizing (3.5.3) with respect to ϕ and χ we obtain

$$\begin{aligned} -\beta^{-1}\nabla_1 \phi(\boldsymbol{x}_1) \\ = \nabla_1 \Phi_0(\boldsymbol{x}_1) + \rho_l \int \nabla_1 \Phi(x_{12})\,[g_l(x_{12})\,\phi(\boldsymbol{x}_2) + \chi(\boldsymbol{x}_1\boldsymbol{x}_2)]\,\mathrm{d}\boldsymbol{x}_2. \end{aligned} \quad (3.5.5)$$

It is then easy to solve for ϕ by Fourier transformation

$$\tilde{\phi}(k) = M(k)/[1-G(k)], \quad (3.5.6)$$

$$\left.\begin{aligned} \text{where} \quad M(k) &= \frac{\mathrm{i}}{k^2}\int k\cdot\Delta(\boldsymbol{x}_1)\exp(-\mathrm{i}\boldsymbol{k}\cdot\boldsymbol{x}_1)\,\mathrm{d}\boldsymbol{x}_1, \\ \Delta(\boldsymbol{x}_1) &= \beta\nabla_1\Phi(\boldsymbol{x}_1) + \beta\rho_l\int \nabla_1\Phi(x_{12})\,\chi(\boldsymbol{x}_1\boldsymbol{x}_2)\,\mathrm{d}\boldsymbol{x}_2 \\ \text{and} \quad G(k) &= \frac{\mathrm{i}\beta\rho_l}{k^2}\int \frac{\boldsymbol{k}\cdot\boldsymbol{x}}{x}\Phi'(x)\,g_l(x)\exp(-\mathrm{i}\boldsymbol{k}\cdot\boldsymbol{x})\,\mathrm{d}\boldsymbol{x}. \end{aligned}\right\} \quad (3.5.7)$$

Thus, an instability occurs when $G(k)-1$ vanishes, and this is determined solely by unperturbed quantities. At a zero of $G(k)-1$, k_c, the equation $G(k_c)-1=0$ describes a curve in the ρ–β plane.

Now, specializing to hard spheres of diameter a, we have in one-dimension:

$$1+\lambda(\sin z/z) = 0, \quad \lambda = 2\rho_l a g(a), \quad (z = ka)$$

and the first real zero occurs at

$$z = 4.49, \quad \lambda = 4.60.$$

In two-dimensions:

$$1+(\lambda/z)J_1(z) = 0 \quad (J_1 = \text{first-order Bessel function})$$

$$\lambda = 2\pi\rho_1 a^2 g(a) \quad (z = ka)$$

and the first real zero occurs at

$$z = 5.135, \quad \lambda = 15.12.$$

In three-dimensions

$$1+(\lambda/z^3)(\sin z - z\cos z) = 0, \quad \lambda = 4\pi a^3\rho_1 g(a) \quad (z = ka)$$

and the first real zero occurs at

$$z = 5.76, \quad \lambda = 34.8.$$

In each case there exists a critical λ above which the system cannot be stable with respect to small perturbations in the singlet distribution. Kirkwood was able to show that the limit of stability of solutions to the BGY equation was given by the same condition as found by Kirkwood and Boggs(30). Rice and Lekner(51) observe that this treatment of the problem requires no assumption about the triplet distribution function: it was at one time thought that Kirkwood's original stability criterion was dependent upon the superposition approximation and the linearization of the equations.

Kirkwood's assumption that perturbations in $\rho_{(1)}$ and $g_{(2)}$ may be formed independently is even less justified in two dimensions, since more recent work with hard discs shows a phase transition far above the two-dimensional Kirkwood critical pressure, namely, $(P/\rho kT)_K = 4.78$(63). Finally, we know a solid–fluid phase transition cannot occur(65) in a one-dimensional liquid contrary to the above prediction. Kunkin and Frisch(64) have attempted to modify the Kirkwood stability criterion in the light of its several shortcomings, but without success.

A recent analytic approach to the theory of phase transitions has been given by Weeks, Rice and Kozak(42). A criterion for the uniqueness of the solution to the singlet Kirkwood integral equation is given, and on this basis the region of ρ–T space is found over which the one-dimensional Kirkwood solution is unique. Multiple solutions of the non-linear equation are associated with instability of the single phase and thus signal a phase transition. A *bifurcation equation* which can be related to Kirkwood's instability condition describes the initiation of multiple solutions: the periodic singlet density falls naturally out of the theory. No phase transition is found for a system of hard spheres which concurs with Meeron's[43] recent suggestion that systems having purely repulsive forces have no phase transitions, contrary to Kirkwood's prediction[44].

Kerber[45] approaches the melting problem by adopting the PY equation for the description of the liquid phase and an anharmonic model for the solid. The melting points are then computed in a formal way by locating the intersection of the chemical potential curves for the solid and liquid states.

Stell has shown that the direct correlation function may be written as a linear combination of the bundle (B) and elementary (E) diagrams, thus

$$\left.\begin{aligned} c &= B+E && (a),\\ &= f(1+C+B'+E)+B'+E && (b).\end{aligned}\right\} \tag{3.5.8}$$

This expression may be written as a combination of the long range and short range components:

$$c = f(1+C)+(1+f)(B'+E). \tag{3.5.9}$$

C represents the class of chain diagrams and B' the bundle diagrams without the f_{12}-bond. On the basis of the definition of class B, the second expression (3.5.8*b*) follows directly from (3.5.8*a*) by the addition of the f_{12}-bond. The HNC approximation consists in dropping the elementary diagrams from (3.5.8*b*) resulting in the prescription

$$c_{\mathrm{HNC}} = f(1+C+E)+(1+f)B' \tag{3.5.10}$$

written in long and short range form. We note that in the HNC approximation the short range form differs from the exact form below $r_{12} = 1.0$ (cf. (3.5.9)) by the inclusion of the diagrams fE. Moreover, the HNC form for the direct correlation exhibits a long range tail beyond $r_{12} = 1.0$.

The PY approximation, however, consists in dropping the long range component $(1+f_{12})(B'+E)$, but preserving the *exact* short range form:

$$c_{\mathrm{PY}} = f(1+C). \tag{3.5.11}$$

Thus, provided the long range tail $(1+f_{12})(B'+E)$, generally designated $D(r)$, is negligible with respect to the short range component, the PY approximation is superior.

The PY theory however, predicts no phase change even beyond physically impossibly high densities: there is no geometrical exclusion within the volume of integration of the field points in the cluster integrals retained in the PY approximation. If we examine the tail, $D(r)$, in a little more detail we find that the first two members of the set $B'+E$ dropped in the PY approximation ( + ) may be written $(1+f_{34})$ , so that to a first approximation the tail may be written $(1+f_{12})(1+f_{34})$ . Furthermore, for hard spheres  is everywhere positive. We may represent this graphically as

 (3.5.12)

where the wiggly bonds represent the factor $(1+f)$ and effect an exclusion on the pair interactions 12 and 34. Thus the first evidence of geometric packing effects becomes evident in the long range component of the direct correlation function.

Croxton[73] has shown that many of the elementary class of diagrams E may be constructed from the class B' by the addition of l Mayer f-bonds ($l = 1 \ldots \hat{l}$) between the n unbonded field points. Then the class of diagrams dropped in the PY approximation becomes

$$(1+f_{12})(B'+E) \simeq (1+f_{12}) \sum_{n=2}^{\infty} \left[1 + \sum_{l=1}^{\hat{l}} \frac{(-1)^l \hat{l}!}{l!\,(\hat{l}-l)!} \right] B'_n, \quad (3.5.13)$$

where
$$\hat{l} = \sum_{j=0}^{n-1} [(n-j)-1],$$
$$B'_n = n\text{th order } B' \text{ diagram},$$

provided the range of all such l bonds < 1. $\hat{l}$ represents the maximum number of l bonds which may be applied to an nth order diagram. For one or all l bonds > 1

$$(1+f_{12})(B'+E) \simeq (1+f_{12}) \sum_{n=2}^{\infty} B'_n, \quad (3.5.14)$$

that is, the elementary diagrams make no contribution when the range of any l bond > 1. Furthermore, it may be very easily demonstrated that

$$\sum_{l=1}^{\hat{l}} \frac{(-1)^l \hat{l}!}{l!(\hat{l}-l)!} = -1 \quad (l \text{ bonds} < 1), \quad (3.5.15)$$

where $\hat{l}$ is given above. Thus, for l bonds < 1 the B' and E diagrams cancel, and we may write

$$(1+f_{12})(B'+E) \simeq (1+f_{12})\hat{B}', \tag{3.5.16}$$

where $\hat{B}'$ represents the evaluation of the class B' *subject to the condition that all field points in that class are separated by unity or more.* This is, of course, a general statement of the geometrical exclusion property of the cluster integrals dropped in the PY approximation. Furthermore, the class $\hat{B}'$ is very small indeed. The stringent requirement that all field points should be separated by an atomic diameter or more very rapidly diminishes the contribution of the higher-order diagrams, and indeed, a fundamental geometrical limit is placed on the class when twelve particles are in simultaneous interaction: then the exclusion condition has diminished the contribution of the integral to zero.

The non-nodal subclass of rooted B' clusters retained in this approximation are termed *small watermelon diagrams*, and are characterized by having only one field point per chain and, of course, no direct f_{12}-bond. The remaining diagrams in the watermelon class, consisting of parallel chains of field points together with their associated elementary derivatives, are not considered in this approximation: these diagrams, moreover, do not satisfy the exclusion conditions (3.5.13), (3.5.16). Thus, from (3.5.16), only those *physically accessible* configurations of field points contribute to the long range form of the direct correlation function, i.e.

$(1+f_{12})(B'+E) \sim (1+f_{12})(\rho^2$  $+\rho^3$  $+\ldots)$, (3.5.17)

where the wiggly bond represents the *exclusion* $(1+f) = \tilde{f}$ which prevents the field points approaching closer than an atomic diameter. In as far as the diagram  completes the diagrammatic contribution to four-particle configurations, we may anticipate that (3.5.17) ensures that the fourth virial coefficient is given exactly.

We need to sum (3.5.17), but if we consider only the exclusion components of the series, we have the series of *infinite range* diagrams

ρ^2  $+\rho^3$ 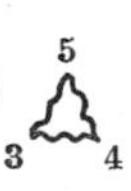 $+\rho^4$ 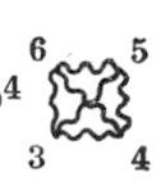 $+\ldots$ (3.5.18) which may be represented approximately by

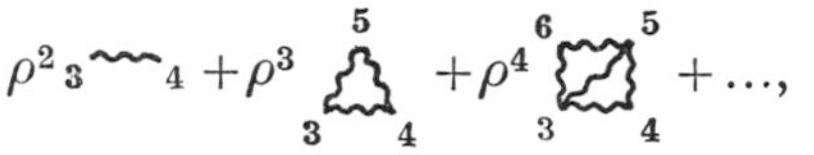

$$\rho^2 \, _3\!\!\sim\!\!\sim\!\!\sim_4 + \rho^3 \, \underset{3 \quad 4}{\overset{5}{\triangle}} + \rho^4 \, \underset{3 \quad 4}{\overset{6 \quad 5}{\boxtimes}} + \ldots, \qquad (3.5.19)$$

i.e. as a netted ring of exclusion bonds. Whilst this series does not *completely* preserve the exclusion characteristics, the field point accessibility remains much the same, and more particularly satisfies a simple convolution relation whose Fourier transform is

$$\sim \frac{\rho^2 \tilde{f}(k)}{1 - \rho \tilde{f}(k)} \qquad (3.5.20)$$

where $\tilde{f}(k)$ is the transform of $\tilde{f}(r)$. If now we artificially *truncate* the range of the $\tilde{f}$-bond at some reduced radius R we partially recover the finite-ranged diagrams (3.5.17) whilst preserving their exclusion characteristics. Designating this truncated function $\tilde{\phi}(r)$, (3.5.20) becomes

$$\sim \frac{\rho^2 \tilde{\phi}(k)}{1 - \rho \tilde{\phi}(k)} \qquad (3.5.21)$$

$\tilde{\phi}(k)$ has a maximum at $k = 0$ when $\phi(0) = 4\pi(R^3 - 1)/3$. ($R = 1$ corresponds to the PY approximation.) We see that (3.5.21) shows singular behaviour at $k = 0$ at a density which is sensitively dependent upon R.

Now,
$$S(0) = \rho k_{\mathrm{B}} T \chi_T = k_{\mathrm{B}} T (\partial \rho / \partial P)_T = \frac{1}{1 - \rho c(0)} \qquad (3.5.22)$$

and since $c(0)_{\mathrm{PY}}$ becomes progressively more negative with increasing ρ, it follows that PY inevitably yields a monotonic equation of state with no evidence of a solid-fluid phase transition. In the present approximation however, (3.5.22) exhibits singular behaviour as $\rho c(0) \to 1$, the isothermal compressibility χ_T becoming infinite at this point, yielding the isotherm shown in figure 3.5.4. The real space form of the direct correlation function in this case shows a long range monotonically decreasing positive tail for $r > 1.0$ and a PY core for $r < 1.0$. From the compressibility relation

$$\frac{1}{k_{\mathrm{B}} T} \left(\frac{\partial P}{\partial \rho} \right)_T = 1 - 4\pi\rho \left\{ \int_0^{1.0} c(r) r^2 \, dr + \int_{1.0}^{\infty} c(r) r^2 \, dr \right\} \qquad (3.5.23)$$

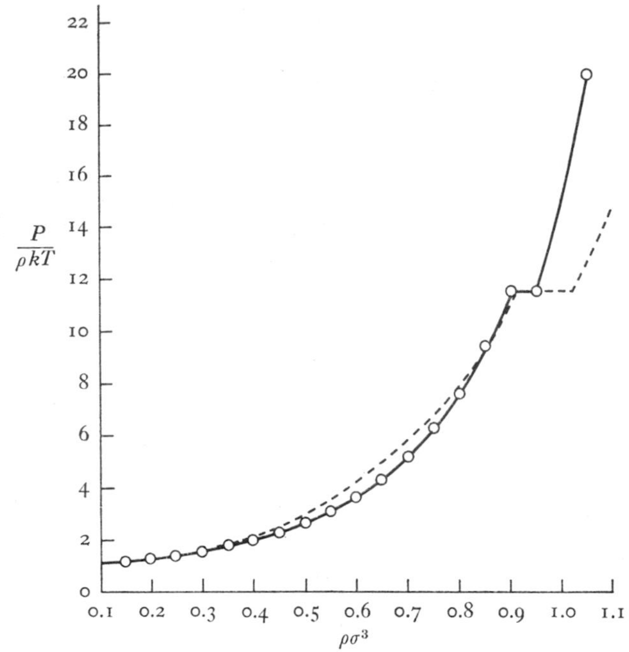

Fig. 3.5.4. The compressibility equation of state for hard spheres as determined by Croxton (solid curve). The broken curve shows the molecular dynamics results of Alder.

we see that the positive tail corrects the PY isotherm onto the molecular dynamics curve (fig. 3.4.1). In the vicinity of the transition, however, the singular long wavelength behaviour of $c(k)$ results in a dramatic development in amplitude of the long range component of $c(r)$, and from (3.5.23), accounts for the zero slope of the compressibility isotherm. We reiterate that the diagrammatic contributions to the long range component arise only from geometrically accessible configurations in the cluster integrals, i.e. those satisfying the exclusion condition.

The structure of the system is given by the dominant poles in $c(k)$ and $h(k)$, and above the transition density the RDF develops 'kinks' which are generally taken to be indicative of a solid phase (cf. (5.7.3)). It may be shown in the present approximation that

$$\frac{\Psi}{kT} = \frac{\Psi_{\mathrm{PY}}}{kT} - \hat{B}',$$

where $\hat{B}'$ is the known long range positive tail of $c(r)$. The diagrams $\hat{B}'$ evidently enhance the oscillations in the RDF, which is precisely what is required to correct the pressure curve on to the reference isotherm (fig. 3.4.1).

It must be emphasized that it is the *cancellation* amongst the watermelon diagrams and their elementary derivatives that is responsible for the features reported here. The HNC approximation, whilst incorporating the entire watermelon class, may be shown not only to have an incorrect short and long range form (3.5.10), but in dropping the entire E class, forfeits the exclusion characteristics which seem to be associated with geometric packing.

3.6 Inversion of integral equations

It is quite evident that given the experimental data for $g_{(2)}(r)$, the various first-order integro-differential equations developed in this book may be inverted to yield the effective pair potential, $\Phi(r)$. Of course, this involves first of all the Fourier inversion of the experimental structure factor $S(k)$ which is in itself a hazardous business. Rather more direct information on the pair potential may be obtained from the Fourier transform of the Ornstein–Zernike equation, for then we have

$$c(k) = \frac{S(k)-1}{S(k)}. \qquad (3.6.1)$$

We have already seen that there is an intimate relationship between the long range form of the direct correlation function and the pair potential (equation (2.12.10)):

$$c(r) \simeq -\frac{\Phi(r)}{kT} \quad \text{for} \quad c(r) \ll 1, \qquad (3.6.2)$$

i.e. the long range form of the pair potential is particularly sensitive to the small k region of $c(k)$[46]. In fig. 3.6.1 we show $c(k)/c(0)$ for lead (600 °K) and argon (84 °K), and conclude from the localized form that the potential has a long range form[47], and indeed the real space transform suggests a long range oscillatory potential.

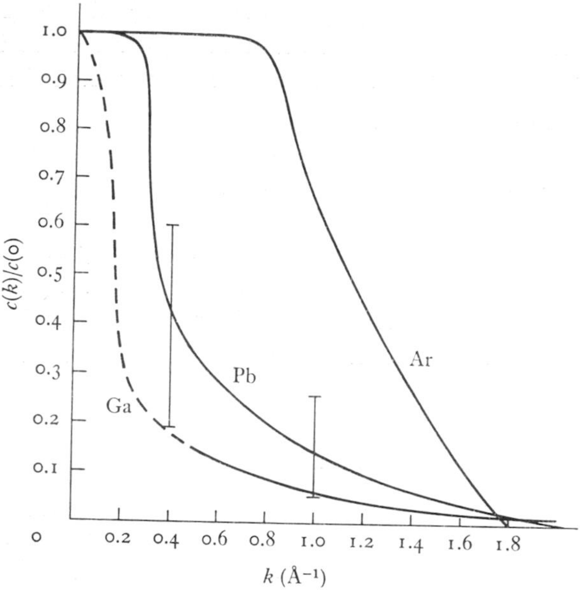

Fig. 3.6.1. Measured values of $c(k)/c(0)$ at small scattering angles: argon at 84 °K; lead at 600 °K; gallium at 320 °K, using results of Ascarelli. (Redrawn by permission from March, *Liquid Metals* (Pergamon Press).)

If we rearrange the HNC and PY approximations (2.14.17, 2.14.15) we find immediately

$$\Phi_{PY}(r) = kT \ln\left\{1 - \frac{c(r)}{g_{(2)}(r)}\right\}, \tag{3.6.3}$$

$$\Phi_{HNC}(r) = kT\{h(r) - c(r) - \ln[1 + h(r)]\}. \tag{3.6.4}$$

Gaskell[(49)] has pointed out that starting from the same experimental data for $c(r)$ and $g_{(2)}(r)$, then it follows that $\Phi_{HNC}(r) \geqslant \Phi_{PY}(r)$. Further, at the nodes of the total correlation function

$$h(r), c(r) = -\Phi_{HNC}/kT.$$

This latter relation holds asymptotically (i.e. $r \to \infty$) for both approximations (cf. (3.6.2)).

Ascarelli[(48)] has obtained the effective pair potential for gallium by determining the PY inversion. The resulting curves show a pronounced repulsive maximum beyond the first minimum, but there appears to be no further oscillatory long range structure. March[52] ascribes this to an inadequacy of the PY theory used by Ascarelli, applied to liquid metals, and suggests the calculations should be repeated using the BGY inversion. Enderby and March[53] have, in fact, shown that the PY inversion applied to bismuth and thallium yields results similar to those of Ascarelli.

Gehlen and Enderby[62] have investigated the *k*-space form of the BGY equation numerically, and find it to be very sensitive to the low angle experimental $S(k)$ for lead. They compare the resulting Φ_{BGY} with Φ_{PY} determined from the same scattering data, and both are quite dissimilar to the pseudopotential determination.

Hutchinson[54] reports the inversion of Rahman's molecular dynamics radial distribution function for argon at 85.5 °K, i.e. just above the triple point. This investigation is important in that the precise form of the pair potential is known. Rahman[55] used a 6-exp Buckingham potential truncated at 7.65 Å. First the exact potential is compared with the 'experimentally' determined $c(r)$ (fig. 3.6.2(*a*)). The most notable feature of $c(r)$ is that it has the same general shape as $\Phi(r)$, but dies away rather more rapidly. This is contrary to the HNC and PY expectations which suggest the asymptotic relationship $c(r) \simeq -\Phi(r)/kT$ for r large. In fig. 3.6.2(*b*) we compare the HNC and PY inversions with the exact potential, and it is seen that the HNC potential is much the better of the two, and to a certain degree concurs with the findings of Khan[40] for low temperature systems with a relatively soft repulsive core. It is, however, contrary to the conclusions of Johnson *et al.*[46] who consider that for the short range forces in argon the PY method was probably more appropriate. Returning to Hutchinson's findings, he observes that the slight displacement of the well depth in Φ_{HNC} to the right accounts for the pressures several orders of magnitude too large estimated by Gaskell[49] for the HNC virial equation of state. The comparison of well depths gives

$$\Phi_{exact}(3.8)/kT = -1.40, \quad \Phi_{HNC}(4.0)/kT = -1.28,$$
$$\Phi_{PY}(4.6)/kT = -4.01.$$

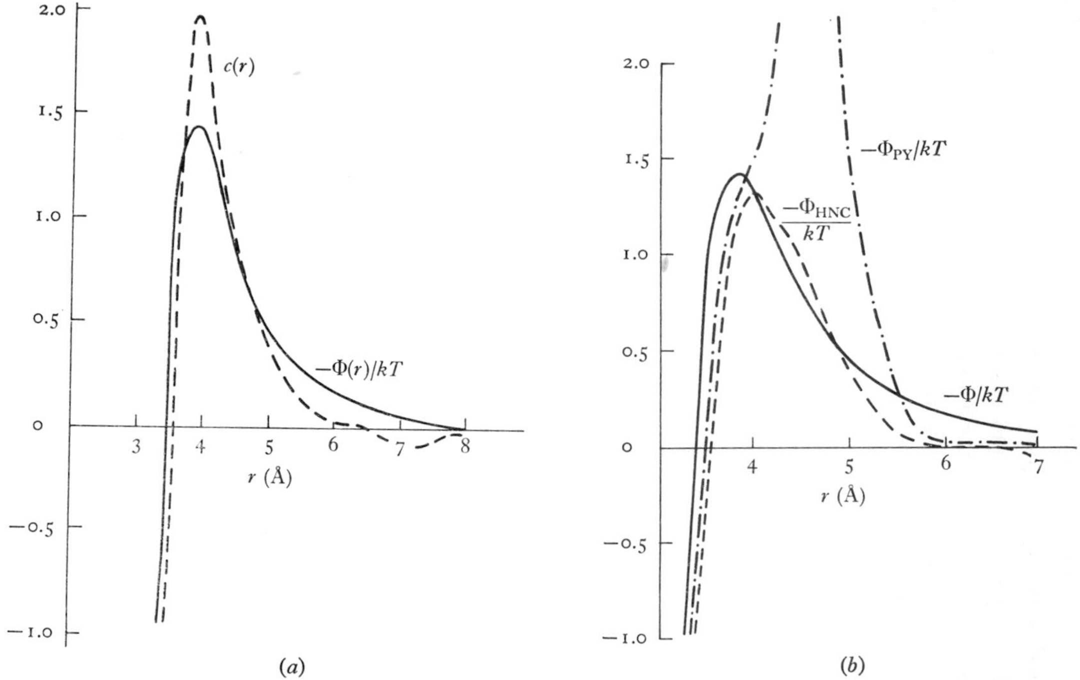

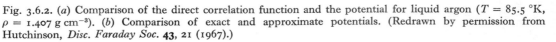

Fig. 3.6.2. (*a*) Comparison of the direct correlation function and the potential for liquid argon ($T = 85.5$ °K, $\rho = 1.407$ g cm^{-3}). (*b*) Comparison of exact and approximate potentials. (Redrawn by permission from Hutchinson, *Disc. Faraday Soc.* **43**, 21 (1967).)

Inversion of the PY2 equation is not possible as such, but using this approximation at low densities and a combination of Monte Carlo and molecular dynamics results at high densities, the equation of state from any two-body potential can be determined to within 2 per cent. Levesque and Vieillard–Baron[66] attempt to find that pair potential most in accord with experiment. They conclude that the Lennard-Jones 6–12 potential gives the best results in the non-critical region whilst a 6–9 interaction appears to give better agreement in the vicinity of the critical point.

In a very recent piece of work Ballentine and Jones[74], using the PY and HNC theories, investigate the sensitivity of the pair potential $\Phi(r)$ to errors in the measured structure factor $S(k)$. They study the linear relation between small changes in $S(k)$ and $\Phi(r)$ by means of a generalized eigenvector analysis. Only those combinations of errors in $S(k)$ which correspond to large eigenvalues will lead to serious uncertainties in $\Phi(r)$. The method is illustrated for sodium, and some qualitative conclusions are: the small k-region of $S(k)$ is by far the most sensitive feature of the structure factor with regard to the determination of pair potentials; the radius of the repulsive core is not sensitive to errors; the depth of the attractive well is the parameter most sensitive to errors in $S(k)$; and the existence of a second repulsive region beyond the first minimum is confirmed.

3.7 Quantum fluids

Wu and Feenberg[57] evaluated the Bijl–Dingle–Jastrow (BDJ)[58] wave function $\Psi = \Pi \exp\{\frac{1}{2}u(r_{ij})\}$ for the ground state of a boson system. Taking an experimentally determined radial distribution function they numerically evaluate the homogeneous non-linear integral equation relating $g_{(2)}(r)$ and $u(r)$ by Abe's[59] iteration scheme. The BDJ wave function is essentially a product of single-particle wave functions which incorporates correlation. Thus $u(r)$ satisfies the boundary conditions (2.21.5). Wu and Feenberg's effectively exact solution is shown in fig. 3.7.1 for ^{4}He at 2.06 °K.

Massey[60] approaches the problem slightly differently. He uses

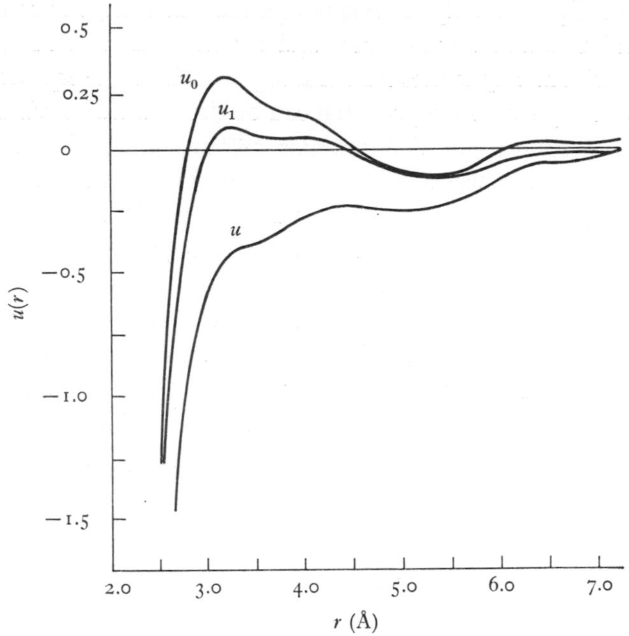

Fig. 3.7.1. The 'correlation' function $u(r)$ for the ground state of liquid ^{4}He, showing the zeroth, first-order and essentially exact functions.

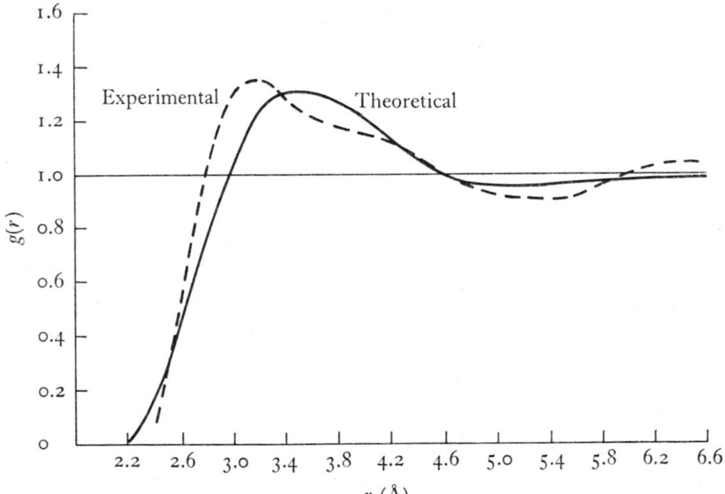

Fig. 3.7.2. Comparison of the experimental and theoretical radial distribution functions of liquid ^{4}He.

a set of trial functions $\{g_{(2)}(r)\}$ and determines $u(r)$. The expectation value of the Hamiltonian may then be calculated (cf. (2.21.7)) and he selects that radial distribution function which minimizes $\langle\mathscr{H}\rangle$. Comparison of Massey's result to the experimental form of Goldstein and Reekie(61) is shown in fig. 3.7.2.

Provided the core of the potential is sufficiently repulsive, it follows that short-range exchange effects in the form of Pauli repulsion act as only a small correction beyond the atomic diameter, and to this extent we may describe the ground state of an assembly of fermions as a cluster expansion of the exchange effects around the boson ground-state properties. The fermions are, of course, assumed to be of the same mass and interacting through the same potential as the bosons. We may therefore use a ground-state wave function Ψ_F of BDJ form (2.21.3, 2.21.5), whereupon the expectation energy may be expressed as a cluster expansion

$$E_F = \frac{\langle\Psi_F|\mathscr{H}|\Psi_F\rangle}{\langle\Psi_F|\Psi_F\rangle} = E_B + \tfrac{3}{5}\left(\frac{\hbar^2 k_F^2}{2m}\right) + \epsilon(12) + \epsilon(123), \quad (3.7.1)$$

where E_F and E_B are the Fermi and Bose ground-state energies, $k_F = (3\pi^2\rho)^{\frac{1}{3}}$ and $\epsilon(12)$ and $\epsilon(123)$ arise in the antisymmetrization of clusters of two and three particles respectively. Wu and Feenberg(75) have proposed such a 'pseudopotential' approach to properties of ground-state fermion assemblies, and Schiff(76), in calculating the crystallization density of cold, dense neutron matter in cores of neutron stars observes that $\epsilon(12)$ is considerably less, and $\epsilon(123)$ very much less, than E_B thereby justifying *a posteriori* the Wu–Feenberg perturbation scheme. Of course, the ground-state boson properties are a prerequisite, and Kalos, Levesque and Verlet(77) have succeeded in solving numerically the Schrödinger equation for a system of 256 hard sphere bosons in its ground state. This is essentially a perturbation method in which the attractive forces are regarded as the perturbation about a hard sphere reference system.

CHAPTER 4

THE LIQUID SURFACE

4.1 Introduction

A statistical mechanical theory of the liquid surface presents special problems in that the distribution functions and the integral relations describing them are no longer scalar functions of the interatomic separation, but instead depend upon the vector field $\boldsymbol{r}$ and the distance of the origin molecule from the surface $\boldsymbol{z}$. We might anticipate that far from the liquid–vapour interface the general vector expressions reduce to their scalar counterparts. Alternatively we may say that the atomic inhomogeneity is restricted to the region of the transition zone between liquid and vapour.

There is another serious difficulty. The apparatus of the correlation function and its associated theoretical basis developed in the earlier chapters contains no means of establishing the structure or location of the surface. However, the method of correlation functions does contain within itself the liquid–vapour phase transition in as far as there exist two different values of the density of the system, that of the gas and that of the liquid which in the virial equation of state will yield identical pressures along a sub-critical isotherm:

$$P = \rho kT - \frac{2\pi\rho^2}{3}\int_0^\infty \nabla_1 \Phi(r)\, g_{(2)}(r)\, r^3\, \mathrm{d}r. \qquad (4.1.1)$$

Clearly this is quadratic in ρ, ignoring for the moment the density dependence of $g_{(2)}(r)$. Similarly the bulk liquid and bulk vapour densities yield identical chemical potentials for an equilibrium system, and indeed, the continuity of these two quantities across the liquid–vapour transition zone indicates the stability of the two-phase coexistence. Nevertheless, not until an external field is applied will the system resolve itself into homogeneous liquid and vapour phases. It seems an auxilliary condition is required to establish the free surface: then we may begin to enquire what the correlation functions can tell us about the thermodynamic functions of the liquid surface.

4.2 A formal theory of liquid surface

The first statistical mechanical analysis of the liquid surface in terms of the intermolecular forces acting at the interface between two fluid phases was due to Fowler[1]. However, almost at the outset, Fowler introduces the approximation of a mathematical density discontinuity to describe the liquid–vapour density transition. Quite clearly this cannot be a satisfactory model of the liquid surface, especially as the critical point is approached. Kirkwood and Buff (KB)[2] develop a formal statistical mechanical theory of the liquid surface free from this mathematical expedient, although for the purposes of calculation these authors are forced to shrink the transition zone back to one of density discontinuity.

Following Kirkwood and Buff we consider a heterogeneous system consisting of two fluid phases, and define a rectangular coordinate system, the Gibbs dividing surface defining the xy-plane. The z-axis is normal to the liquid surface and directed from the liquid into the vapour phase (see fig. 4.2.1). We consider a mechanical definition of the surface tension as the stress transmitted across a strip of unit width. If we take the stress normal to the unit strip at depth z to be $P(z)$, then the surface tension, γ, is defined as

$$\gamma = \int_{-\infty}^{\infty} \{P - P(z)\}\,\mathrm{d}z, \tag{4.2.1}$$

where P is the uniform system pressure, (4.1.1). Of course, far from the surface $P(z) \to P$ and clearly it is only the region of interphasal inhomogeneity which contributes to the surface tension. We now attempt to determine $P(z)$.

The average force exerted by molecules in a volume element $\mathrm{d}\mathbf{2}$ on molecules in a volume element $\mathrm{d}\mathbf{1}$ is

$$\rho_{(2)}(z_1, r)\,\nabla\Phi(r)\,\mathrm{d}\mathbf{1}\,\mathrm{d}\mathbf{2} \tag{4.2.2}$$

with a component in the x-direction:

$$\boldsymbol{e}_x \cdot \nabla\Phi(r)\,\rho_{(2)}(z_1, \boldsymbol{r})\,\mathrm{d}\mathbf{1}\,\mathrm{d}\mathbf{2}, \tag{4.2.3}$$

where $\boldsymbol{e}_x$ is the unit vector in the x-direction. Integration over all

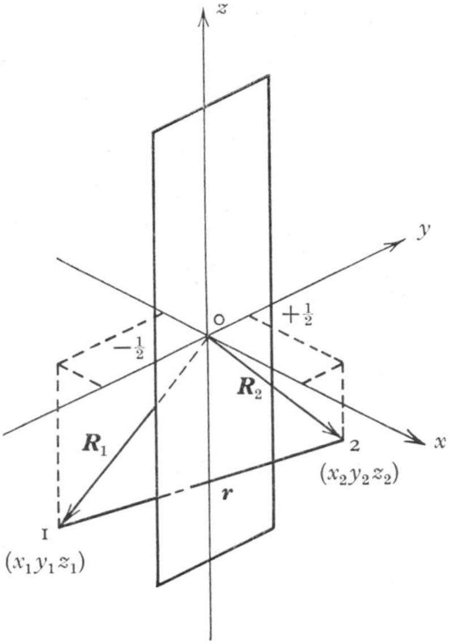

Fig. 4.2.1. Geometry for the determination of the stress transmitted across a plane normal to the liquid surface. The origin of coordinates is located on the Gibbs dividing surface, this defining the xy-plane.

positions of **1** and **2** subject to the transmission of stress across the strip yields

$$\int_{-\frac{1}{2}l}^{+\frac{1}{2}l}\int_{-\frac{1}{2}}^{+\frac{1}{2}}\int_{-\infty}^{0}\int_{-x_1}^{\infty}\int_{-\infty}^{\infty}\int_{-\infty}^{\infty}\frac{x_2}{r}\nabla\Phi(r)\,\rho_{(2)}(z_1,\boldsymbol{r})\,\mathrm{d}y_2\,\mathrm{d}z_2\,\mathrm{d}x_2\,\mathrm{d}x_1\,\mathrm{d}y_1\,\mathrm{d}z_1 \tag{4.2.4}$$

which, after partial integration with respect to x_1 and extension of the residual integration over x_2 to the interval $-\infty \leqslant x_2 \leqslant +\infty$ by virtue of the symmetry requirement that the integrand be an even function of x_2, becomes

$$\frac{1}{2}\int_{-\frac{1}{2}l}^{+\frac{1}{2}l}\int\frac{x_2^2}{r}\nabla\Phi(r)\,\rho_{(2)}(z_1,\boldsymbol{r})\,\mathrm{d}\mathbf{2}\,\mathrm{d}z_1. \tag{4.2.5}$$

l is chosen to be much greater than the transition region. Expression (4.2.5) represents the molecular force transmitted across the strip.

To this we must add the momentum transport across the strip – the kinetic contribution to the stress. This is

$$kT\int_{-\frac{1}{2}l}^{+\frac{1}{2}l} \rho_{(1)}(z_1)\,\mathrm{d}z_1, \tag{4.2.6}$$

where $\rho_{(1)}(z_1)$ is the density profile across the transition zone. Thus, finally

$$P(z) = kT\rho_{(1)}(z_1) - \frac{1}{2}\int \frac{x_2^2}{r}\nabla\Phi(r)\,\rho_{(2)}(z, \boldsymbol{r})\,\mathrm{d}\mathbf{z}. \tag{4.2.7}$$

The above equation for $P(z)$ reduces to the bulk value P when $\rho_{(1)}$ becomes independent of z and equal to the number density of the liquid (vapour), and when the pair density $\rho_{(2)}$ also becomes independent of z, and a function of the scalar separation r appropriate to the liquid (vapour) phase. The surface tension then follows from (4.2.1).

It is quite evident that for the evaluation of the surface tension integral (4.2.1) we must determine the singlet density profile $\rho_{(1)}(z)$ and the anisotropic pair distribution function $\rho_{(2)}(z, \boldsymbol{r})$. Furthermore, it is clear that these functions are intimately related thus:

$$\left.\begin{aligned} \rho_{(2)}(z_1, \boldsymbol{r}) &= \rho_{(1)}(z_1)\,\rho_{(1)}(z_2)\,g_{(2)}(z_1, \boldsymbol{r}), \\ \rho_{(1)}(z) &= \rho g_{(1)}(z) \end{aligned}\right\} \tag{4.2.8}$$

for a plane interfacial zone.

As yet there has been no need to specify precisely the location of the origin of coordinates: (4.2.7) is independent of the position of the origin. However, for the determination of the surface energy of the liquid we shall need to locate our coordinates on the *Gibbs surface*. First let us suppose the origin of coordinates is located on some arbitrary plane within the liquid surface. Clearly we may choose an origin of coordinates such that

$$\int_{-\infty}^{0} [\rho_{\mathrm{L}} - \rho_{(1)}(z)]\,\mathrm{d}z = \int_{0}^{\infty} [\rho_{(1)}(z) - \rho]\,\mathrm{d}z, \tag{4.2.9}$$

and in fact this is the location of the Gibbs discontinuity. The Gibbs surface is often defined as that surface at which the superficial excess density vanishes. In other words, it is a hypothetical density discontinuity chosen such that the liquid and vapour densities may be assumed constant up to the discontinuity, subject to the conservation of matter.

The surface *excess* energy, u_S may be written as the sum of the liquid and vapour components:

$$u_S = \frac{\rho_L^2}{2}\int_{-\infty}^{0}\int_{-\infty}^{0}\{g_{(1)}(z_1)g_{(1)}(z_2)-1\}\int_0^\infty g_{(2)}^L(z_1,\boldsymbol{r})\,\Phi(r)\,\mathrm{d}\boldsymbol{r}\,\mathrm{d}z_1\,\mathrm{d}z_2$$

$$+\frac{\rho_V^2}{2}\int_0^\infty\int_0^\infty\{g_{(1)}(z_1)g_{(1)}(z_2)-1\}\int_0^\infty g_{(2)}^V(z_1,\boldsymbol{r})\,\Phi(r)\,\mathrm{d}\boldsymbol{r}\,\mathrm{d}z_1\,\mathrm{d}z_2, \tag{4.2.10}$$

where $g_{(2)}^L$ and $g_{(2)}^V$ represent the liquid and vapour pair distribution functions respectively. It is quite apparent from (4.2.10) that the surface excess energy is highly sensitive to the location of the origin of coordinates.

In the absence of an accurate and explicit theory of the functions $\rho_{(1)}(z)$ and $\rho_{(2)}(z_1, \boldsymbol{r})$, Fowler, as an initial approximation, and Kirkwood and Buff, for the purposes of numerical evaluation, resort to the expedient of shrinking the transition zone to a mathematical surface of density discontinuity coincident with the Gibbs dividing surface. In this case the F and KB formulations become identical, and both assume that the liquid remains homogeneous right up to the surface of separation. In this approximation KB obtain

$$\gamma = \frac{\pi\rho_L^2}{8}\int_0^\infty \nabla_1\Phi(r)\,g_{(2)}^L(r)\,\mathrm{d}r \tag{4.2.11}$$

for the surface tension at an interface between a liquid phase and vapour phase of negligible density. For the surface excess energy they obtain

$$u_S = -\frac{\pi\rho_L^2}{2}\int_0^\infty g_{(2)}^L(r)\,\Phi(r)\,r^3\,\mathrm{d}r. \tag{4.2.12}$$

These expressions allow relatively simple calculation of the thermodynamic observables γ and u_S from the potential of intermolecular force of the liquid phase. However, Kirkwood and Buff found that (4.2.11) could not be immediately satisfied with the LJ potential and the experimental RDF – in fact the wrong sign was obtained for the surface tension. KB made an analytical approximation to the experimental $g_{(2)}(r)$ to secure the results given in table 4.5.1.

It is found that in the KB formulation, agreement between theoretical and experimental values for the surface tension deteriorates with increasing temperature. This is undoubtedly due to inadequate account being taken of the delocalization of the liquid–vapour interface as $T \to T_c$. The KBF theory retains the density discontinuity throughout: whilst this might be a reasonable approximation at the triple point, it is clearly untenable as the temperature rises. In all cases the inequalities $\gamma_{(\text{step})} > \gamma_{(\text{expt})}$ for the surface tension, and $u_{S(\text{step})} < u_{S(\text{expt})}$ for the surface energy on the KB model. Kirkwood and Buff's original calculation[2] for liquid argon at 90 °K using the $g_{(2)}(r)$ data of Eisenstein and Gingrich was about 25 per cent above the experimental value. There have been attempts to compute the effect of an exponential type of transition zone using the KB analysis, but the discrepancy is increased rather than decreased. Berry *et al.*[77] obtain an expression for γ in the KB formalism with an exponential transition in which the single adjustable parameter, the relaxation length, has the significance of the thickness of the transition zone. These authors conclude that the transition is complete within two or three atomic diameters for a number of systems (N_2, O_2, CH_4, Ne, Ar) at the triple point. Linear[66, 89] and cubic [89] density gradients have also been tried with some moderate success, but these cannot, of course, afford much physical insight into the problem. It would probably be true to say that the theoretical emphasis has shifted from the numerical estimate of the thermodynamic parameters of the surface more toward the determination of the transition profile, to which the thermodynamic parameters are somewhat insensitive.

Shoemaker *et al.*[3] (SPC) have utilized recent X-ray scattering determinations of the RDF for the evaluation of the KBF surface tension and surface energy relations (table 4.5.1), whilst Freeman and McDonald[88] have evaluated these expressions by a Monte Carlo technique. These results may be taken as the best possible estimate of γ and u_S on the basis of the KBF model.

It is clear that an equilibrium spatial delocalization of the liquid–vapour interface develops subject to the thermodynamic constraints of constancy of the chemical potential and normal pressure of the transition zone. Not any density profile will satisfy this restriction,

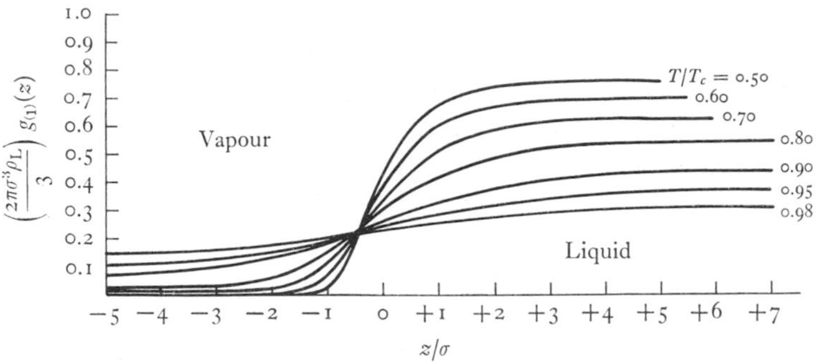

Fig. 4.2.2. Variation of the density transition profile with temperature according to Hill[4] on the basis of the continuity of chemical potential across the interface. Any model of the liquid surface must be able to account for the relaxation of the density transition as the critical point is approached. (Redrawn by permission from Hill, *J. Chem. Phys.* **20**, 141 (1952).)

but equally, neither is the singlet distribution uniquely defined in satisfying the condition on the chemical potential alone. However, several attempts have been made to solve the non-linear integral equation expressing the constancy of the chemical potential across the transition zone so as to determine the equilibrium density profile $\rho_{(1)}(z)$. In these quasi-thermodynamic developments the Kirkwood–Buff stress-tensor definition of the surface tension is retained, where $P(z)$ is now identified as the macroscopic pressure of the fluid appropriate to the local density $\rho_{(1)}(z)$ in an elemental stratum parallel to the planar interface. No account is taken of the dependence of $P(z)$ on $\rho_{(2)}(z, \boldsymbol{r})$ in these treatments and, of course, an equation of state is required to link $\rho_{(1)}(z)$ to $P(z)$. This inevitably restricts the analysis to the simplest pair interactions (hard sphere, square-well, etc.) with suitable account taken of the attractive branch of the interaction.

In the vicinity of the critical point the transition zone is gradual and extended, and under these circumstances certain thermodynamic functions of the interphasal region may be calculated[9]. The implicit assumption is made that the density does not change significantly over a correlation length within the liquid. However, near the triple point the quasi-thermodynamic approach must be abandoned and a more rigorous statistical mechanical formulation adopted. This is currently a point of contention, but the dangers in applying macroscopic thermodynamics in regions where molecular inhomogeneity varies by a factor $\sim 10^3$ over a few molecular diameters enforces an approach at the microscopic rather than macroscopic level. Unfortunately the two approaches yield quite distinct results, particularly as the triple point is approached. To generalize, the quasi-thermodynamic analyses tend to yield monotonic density transition profiles, whilst the statistical mechanical approach usually results in an oscillatory transition. As we might anticipate, the quasi-thermodynamic approach cannot hope to reproduce structural features appropriate to the molecular structure of the liquid, just as thermodynamics can provide no structural information in the bulk fluid and the quasi-thermodynamic theories generally result in a smoother density profile. This is not to say, of course, that these theories cannot yield accurate values of the thermodynamic parameters of the liquid surface γ and u_S. Indeed, these quantities are obtained as integrations over density distributions, and to this extent are relatively insensitive to the detailed features of the density transition, etc. It is precisely because of this that the emphasis has shifted to the determination of $g_{(1)}(z)$.

4.3 The distribution function $g_{(1)}(z)$

The density profile $\rho_{(1)}(z)$ must be determined in both the statistical mechanical and quasi-thermodynamic approaches. In the former approach we then go on to determine the anisotropic pair distribution $\rho_{(2)}(z, \mathbf{r})$ – and as we shall see, there are a number of approximate means of doing this (§4.4). Then we have all the components of the exact Kirkwood–Buff expression for $P(z)$ given in (4.2.7). In the quasi-thermodynamic development we first determine the transition profile $\rho_{(1)}(z)$ which satisfies the condition on the chemical potential, and subsequently relate $P(z)$ to $\rho_{(1)}(z)$ through an idealized equation of state. Thus, Hill[4] and Plesner and Platz[5] consider a system of hard spheres of diameter σ interacting with an attractive potential

$$\left.\begin{aligned}\Phi(r) &= -\epsilon(\sigma/r)^6, & g_{(2)}(r) &= 1.00 & (r \geqslant \sigma),\\ &= +\infty, & &= 0 & (r < \sigma),\end{aligned}\right\} \tag{4.3.1}$$

whereupon the bulk equation of state is given as

$$P = P_{\mathrm{hs}}(\rho) - \frac{24\epsilon\eta^2}{kT}, \tag{4.3.2}$$

where $\eta = \pi\sigma^3\rho/6$, $\rho = N/V$. Whilst the first term in (4.3.2) represents a purely density-dependent hard sphere equation of state the second term represents the van der Waals correction arising from the attractive branch of the potential, assuming the pair distribution (4.3.1). Of course, an immediate improvement could be obtained in (4.3.2) in setting $g_{(2)}(r) = \exp(-\Phi(r)/kT)$. For $P_{\mathrm{hs}}(\rho)$ Hill utilizes the Tonks[6] equation of state, whilst Plesner and Platz use the Reiss–Frisch–Lebowitz (RFL)[7] equation of state for a hard sphere fluid. Now, for either bulk phase it is straightforward to write down the appropriate expression for the chemical potential: in the case of the RFL expression of Plesner and Platz we have

$$\ln\frac{\eta}{1-\eta} + \frac{7\eta}{1-\eta} + \frac{15\eta^2}{2(1-\eta)^2} + \frac{3\eta^3}{(1-\eta)^3} - 8\left(\frac{\epsilon}{kT}\right)\eta = \text{constant}, \tag{4.3.3}$$

where the last term is the van der Waals term representing the potential energy of interaction of one molecule with the rest of the fluid. We actually need the expression analogous to (4.3.3), but applicable to the transition region. The interaction energy (in units of kT) in the anisotropic region then becomes:

$$\Psi(z) = -\frac{3\epsilon}{kT}\left\{\int_{-\infty}^{-1} t^{-4}\eta(z+t)\,\mathrm{d}t + \int_{-1}^{1} \eta(z+t)\,\mathrm{d}t + \int_{1}^{\infty} t^{-4}\eta(z+t)\,\mathrm{d}t\right\} \tag{4.3.4}$$

so that the non-linear integral equation

$$\ln\frac{\eta(z)}{1-\eta(z)} + \frac{7\eta(z)}{1-\eta(z)} + \frac{15\eta^2(z)}{2(1-\eta(z))^2} + \frac{3\eta^3(z)}{(1-\eta(z))^3} + \Psi(z) = \text{constant} \tag{4.3.5}$$

may be solved numerically for the transition profile

$$\rho_{(1)}(z) = 6\eta(z)/\pi\sigma^3.$$

The constant is determined by the bulk equilibrium between gas and liquid at the temperature of interest

$$(P_{\mathrm{gas}} = P_{\mathrm{liquid}};\ \mu_{\mathrm{gas}} = \mu_{\mathrm{liquid}}).$$

It now only remains to relate $\rho_{(1)}(z)$ to $P(z)$ via the assumed equation of state and we have estimates of the surface tension. Similarly, we may determine the surface energy.

The equilibrium thermodynamic properties of simple fluids can be obtained by a perturbation technique, the Zwanzig expansion[78], where the potential function is split up into a short range repulsive part $\Phi^0(r)$ and a weaker long range part $\Phi^1(r)$. This approach has been extensively developed by Barker and Henderson for a uniform fluid (§ 2.20). Toxvaerd has extended the Barker–Henderson perturbation scheme to include non-uniform fluids. In particular, perturbation expansions are obtained for the chemical potential and for the free energy. The constancy of the chemical potential across the transition zone enables us to write:

$$\text{constant} = \ln \gamma^0(z) + 2\pi \int_{-\infty}^{\infty} \rho(z+z')\, dz' \int_{|z'|}^{\infty} \beta\Phi^1(r) \times \{g^0_{(2)}[r, \rho(z)] + \tfrac{1}{2}\beta\Phi^1(r)\, g^1_{(2)}[r, \rho(z)] + \dots\}\, r\, dr, \quad (4.3.6)$$

where γ^0 is the fugacity in a uniform fluid of particles which interact through the pair potential $\Phi^0(r)$. The superscript 0 refers to the uniform reference fluid at $\rho = \rho(z)$. This expression is interesting in that it explicitly involves the local structure through the correlations $g^0_{(2)}$, $g^1_{(2)}$. Toxvaerd[81] iteratively solves (4.3.6) for a square-well fluid for which the hard sphere system represents the reference fluid. Verlet and Weiss[79] have given a parametrized expression for $g^0_{(2)}(r, \rho)$ obtained from computer simulations, whilst $g^1(r, \rho)$ is tabulated by Smith, Henderson and Barker[80]: these local (isotropic) functions inserted in (4.3.6) finally yield a monotonic density transition. Toxvaerd[82] also obtains a perturbation expression of the per-particle Helmholtz free energy in the vicinity of the density transition:

$$a(z) = a^0(z) + \frac{1}{2}\int_V \beta\Phi^1(r')\rho(z+z')\, g^0_{(2)}[r', \rho(z)]\, d\mathbf{r}' - \frac{1}{4}\int_V \beta^2(\Phi^1(r'))^2 \rho(z+z')\, g^0_{(2)}[r', \rho(z)]\, \frac{\partial \rho(z+z')}{\partial \beta p_0}\, d\mathbf{r}' \quad (4.3.7)$$

in which case the equilibrium density $\rho(z)$ for the interface zone can be determined as the function which minimizes the interfacial free energy, subject to the boundary conditions $\rho(+\infty) = \rho_{\text{gas}}$, $\rho(-\infty) = \rho_{\text{liquid}}$. In this formulation the excess free energy per unit area is identical to the surface tension, provided the origin of co-ordinates is located on the Gibbs surface, and so we have

$$\gamma = \lim_{h\to\infty}\left\{\int_{-h}^{+h} \rho(z)\, a(z)\, dz - (\rho_l a_l + \rho_g a_g)\, h\right\}.$$

Toxvaerd takes the PY hard sphere system as the reference fluid in evaluating (4.3.7) which inevitably leads to a systematic error, particularly as the triple point is approached when the hard sphere representation of $\Phi^0(r)$ is least appropriate. A two-parameter trial function is used to describe $\rho(z)$:

$$\rho(z) = \rho_l\{\exp\,[a(b-z)] + \rho_g\}/\{\exp\,[a(b-z)] + 1\},$$

where a and b are varied to minimize (4.3.7). The usual structureless transition profile is obtained, although Toxvaerd[82] reports a $\gamma(T)$ characteristic in excellent agreement with experiment.

In an essentially statistical mechanical treatment, the first stage of the development is the determination of the density transition profile $g_{(1)}(z)$ across the liquid vapour interface. This is approached in terms of the exact BGY single-particle integro-differential equation which, generalized to the inhomogeneity of the liquid surface, now becomes

$$kT\nabla_1 \rho_{(1)}(\boldsymbol{r}_1) + \int_2 \nabla_1 \Phi(|\boldsymbol{r}_1 - \boldsymbol{r}_2|)\, \rho_{(2)}(\boldsymbol{r}_1, \boldsymbol{r}_2, |\boldsymbol{r}_1 - \boldsymbol{r}_2|)\, d\boldsymbol{2} = 0. \quad (4.3.8)$$

This as it stands is exact. However, in the absence of a knowledge of the anisotropic pair distribution at the liquid surface it is necessary to resort either to the expedient of assuming that $\rho_{(2)}$ retains its bulk form right up to a discontinuity of density, or to form some interpolated[66] function of the two limiting bulk values. Toxvaerd, for example, solves (4.3.8) iteratively having set

$$g_{(2)}(\boldsymbol{r}_1, \boldsymbol{r}_2) = \alpha g_{(2)}(r_{12}, \rho(\boldsymbol{r}_1)) + (1-\alpha)\, g_{(2)}(r_{12}, \rho(\boldsymbol{r}_2)), \quad (4.3.9)$$

where $g_{(2)}(r_{ij}, \rho(\boldsymbol{r}_i))$ is the RDF in an *unstable fluid* at a uniform density equal to the local density $\rho(z_i)$ at $\boldsymbol{r}_i$. $g_{(2)}(r_{ij}, \rho(\boldsymbol{r}_i))$ is itself

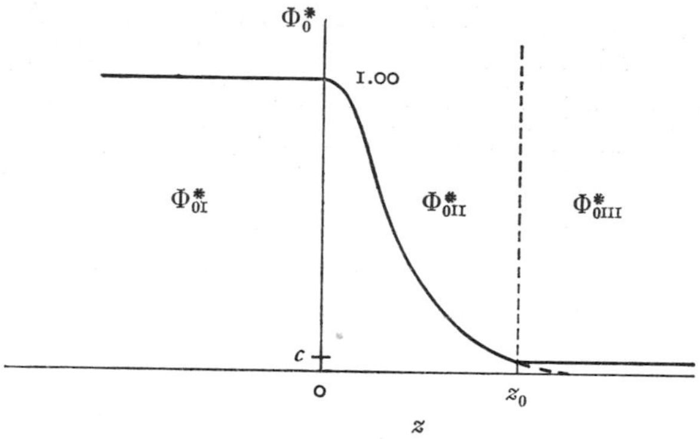

Fig. 4.3.1. The zero-order operator $\Phi_0^*(z)$. This operator, proposed by Croxton and Ferrier, applied to a homogeneous single-component fluid serves to establish a free surface in the xy-plane by 'switching off' the interactions between atoms as a function of the normal coordinate, z.

approximated by a linear combination of the liquid and gas bulk values of $g(r)$:

$$g_{(2)}(r_{ij}, \rho(z_i)) = \frac{\rho(z_i)-\rho_g}{\rho_1-\rho_g} g_{(2)}(r_{ij}, \rho_1) + \frac{\rho_1-\rho(z_i)}{\rho_1-\rho_g} g_{(2)}(r_{ij}, \rho_g). \quad (4.3.10)$$

Stable, self-consistent solutions to (4.3.8) using this model are obtained for one value of α only at any given temperature. This 'nested approximation' approach tends to obscure the physical processes responsible for the features of the transition zone.

Rather than attempt to modify the radial distribution function directly, Croxton and Ferrier[8] have presented a 'bootstrap' theory in which an operator $\Phi_0^*(z)$ is incorporated in the BGY equation whose effect is to provide an angle-dependent suppression of the pair interaction, rather like an angle-dependent coupling parameter. The extent of the suppression will depend upon the location of either of the atomic centres, and in particular their distance from the liquid surface. The spatial form of the operator is shown in fig. 4.3.1 and is taken to account for the collective inhomogeneity in which an atom finds itself. The operator is seen to have three distinct regions, two in which the operator assumes a constant value, and an intermediate region having an analytic form similar to, though not necessarily that of the attractive LJ component of the pair potential, $(\Phi^*_{0\,\mathrm{II}})$. The bulk liquid interactions remain unchanged, but interactions between atoms in the vicinity of the transition and beyond have their interactions largely suppressed. Thus, whilst the radial form of the bulk liquid pair density function is retained, angular dependence is impressed on the interaction by means of the operator. Clearly this is energetically indistinguishable from a true liquid surface with the appropriate modification of the distribution function.

This method avoids the reduction of the liquid surface to a step function (although this does represent a special case of the operator Φ^*). Neither is it necessary to resort to unrealistic potentials so as to accord with the various idealized equations of state on the basis of which the expressions for pressure and chemical potential depend. And finally, realistic forms of the radial distribution function may be retained without judicious analytical 'modification' or suppression of the computationally important long range oscillatory nature of this function. The precise form of the operator is open to discussion, but Croxton and Ferrier[(8)] provide a means of systematic improvement via a recurrence relation. They, however, work only to first order.

The operator $\Phi^*_0(z)$ is introduced in the BGY equation in the following way:

$$\rho_{\mathrm{L}} kT\nabla_1 g_{(1)}(z_1) + \rho^2_{\mathrm{L}} \int_2 \nabla_1 \{\Phi(r_{12})\,\Phi^*_0(z_2)\} g_{(2)}(r_{12})\,\mathrm{d}\mathbf{2} = 0. \quad (4.3.11)$$

It may be shown quite easily that this equation may be written

$$\frac{kT\nabla_1 g_{(1)}(z_1)}{g_{(1)}(z_1)} + \frac{\rho_{\mathrm{L}}}{2}\int_2 \{\Phi\nabla_1\Phi^*_0 + \Phi^*_0\nabla_1\Phi\} g_{(2)}(r)\,\mathrm{d}\mathbf{2} = 0, \quad (4.3.12)$$

where ρ_{L} is the bulk liquid number density. Integration of (4.3.12) subject to the boundary condition $\rho_{(1)}(-\infty) = \rho_{\mathrm{L}}$, yields

$$\rho_{(1)}(z_1) = \rho_{\mathrm{L}} \exp\left\{-\frac{\rho_{\mathrm{L}}}{2kT}\int_{-\infty}^{z_1}\int_2 \{\Phi\nabla_1\Phi^*_0 + \Phi^*_0\nabla_1\Phi\} g_{(2)}(r)\,\mathrm{d}\mathbf{2}\,\mathrm{d}z\right\}. \quad (4.3.13)$$

The integral separates into two distinct parts, S_1 and S_2, which are attributed to $\Phi\nabla\Phi^*_0$ representing the constraining effect of the operator at the liquid surface, and $\Phi^*_0\nabla\Phi$ which represents the modification of the nearest neighbour force, respectively (fig. 4.3.2).

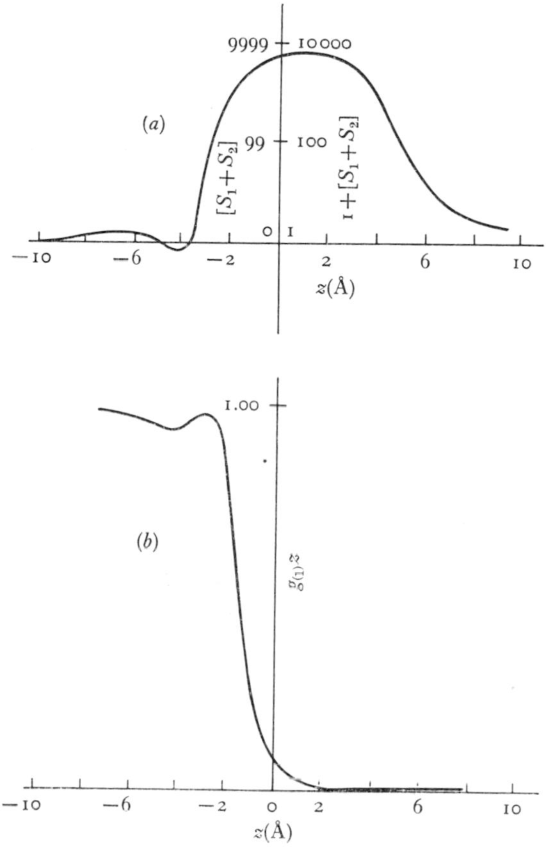

Fig. 4.3.2. (a) The form of the integral $\int_2 \{\,\}\,dz$ (4.3.13). The negative region at $z \sim -4$ Å is responsible for the oscillation in $g_{(1)}(z)$. (b) The profile $g_{(1)}(z)$ for liquid argon at the triple point. Instead of the usual monotonic form an oscillation is seen to develop at $z \sim -4$ Å. The transition is complete within two atomic diameters.

The component S_1 is responsible for the deviation from simple monotonic form of $g_{(1)}(z)$. The negative region in fig. 4.3.2(*a*) actually causes the development of a discrete peak in the density transition profile, and in fig. 4.3.2(*b*) we show the transition profile for liquid argon at the triple point. This spatial coupling operator method has been extended to include an *orientational coupling operator* in the case of long organic molecules at the surface of

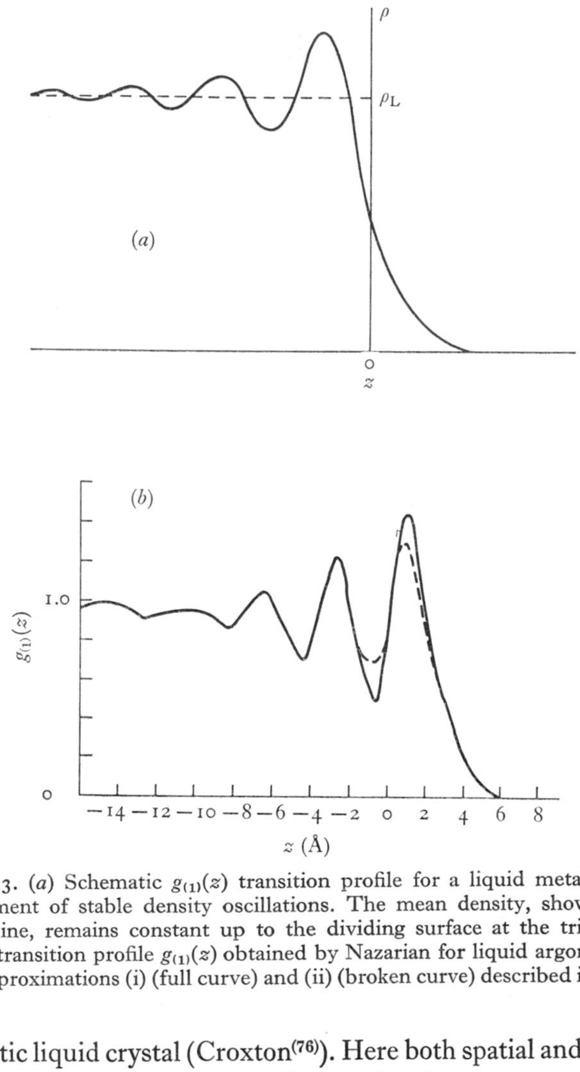

Fig. 4.3.3. (*a*) Schematic $g_{(1)}(z)$ transition profile for a liquid metal showing development of stable density oscillations. The mean density, shown by the broken line, remains constant up to the dividing surface at the triple point. (*b*) The transition profile $g_{(1)}(z)$ obtained by Nazarian for liquid argon at 90 °K using approximations (i) (full curve) and (ii) (broken curve) described in the text.

a nematic liquid crystal (Croxton[76]). Here both spatial and orientational contributions to the surface order develop and additional features arise in the statistical mechanics of the isotropic–nematic surface which are not encountered in simple liquid systems.

The principal feature of the transition profile is the oscillation located at about $z = -r_0$ (fig. 4.3.3)[74]. The development of stable density oscillations must be ascribed directly to the $g_{(2)}(r)\,\Phi(r)$

product which arises from the $\Phi\nabla\Phi_0^*$ term in the integrand. Clearly $g_{(2)}(r)\,\Phi(r)$ is highly sensitive to the precise form of the pair potential, and for the liquid metals, for example, we might expect the long range form of the pair potential together with the relatively large well depth to yield a stable oscillatory density transition[75] of the schematic form shown in fig. 4.3.3(*a*). Such a transition profile would have profound consequences for the thermodynamic functions of the liquid metal surface, as we shall see in §4.9.

Of course, as the temperature increases so the profile relaxes and the structure disappears in accordance with (4.3.13), until at the critical temperature the monotonic variation of density across the transition zone is regained, in agreement with the quasi-thermodynamic analyses[9]. The development of an oscillatory transition profile is currently the subject of some controversy. Whilst the statistical mechanical developments are greatly to be preferred, especially in the vicinity of the triple point, Toxvaerd still obtains monotonic transition profiles using (4.3.8), (4.3.9) and (4.3.10). In many ways this is hardly surprising bearing in mind the several smoothing procedures implied in the approximations (4.3.9) and (4.3.10). Toxvaerd mentions[73] that in his iterative solution of (4.3.8) an oscillation does develop after a few iterations, but subsequently dies out. The Croxton–Ferrier profile did not have to be determined iteratively, and so the oscillation cannot be ascribed to incomplete iteration. It is not a foregone conclusion, of course, that Toxvaerd's statistical mechanical treatment is incapable of yielding oscillatory profiles – it may be that for systems other than the liquid inert gases oscillations would develop. The approximations (4.3.9) and (4.3.10), however, obscure the mechanisms governing the structure of the transition zone, and to this extent the coupling operator $\Phi_0^*(z)$ has intuitive appeal.

Nazarian[74] emphasizes the importance of the statistical mechanical approach in the discussion of the density transition, and confirms the development of an oscillatory transition profile in liquid argon for two specific cases:

(i) for a pair correlation varying as

$$g_{(2)}^{\mathrm{L}}(r_{12}) \text{ if } z_1+z_2 < 0 \quad \text{and} \quad g_{(2)}^{\mathrm{V}}(r_{12}) \text{ if } z_1+z_2 > 0,$$

and (ii) a linear variation

$$g_{(2)}(z_1, z_2, r_{12}) = g^{\mathrm{L}}_{(2)}(r_{12}) + \left[\left(\frac{z_2}{z_{12}}\right) A(z_2) - \left(\frac{z_1}{z_{12}}\right) A(z_1)\right] [g^{\mathrm{V}}_{(2)}(r_{12}) - g^{\mathrm{L}}_{(2)}(r_{12})],$$

where $A(z)$ is the unit step function. Nazarian's profiles are shown in fig. 4.3.3 (*b*) and are seen to be of extreme oscillatory form. Undoubtedly such pronounced oscillations are unlikely to develop in liquid argon – indeed, as we shall see (§ 4.9), such a transition would have profound consequences for the temperature dependence of the thermodynamic functions of the liquid surface. Nevertheless, one feels that under certain circumstances the development of stable density oscillations as a response to the collective constraining field at the liquid surface is feasible, and that any integral equation describing the transition should admit to this possibility. Certainly, the extreme constraint imposed by an ideal wall induces density oscillations normal to the boundary(11, 26) – this, moreover, may be readily simulated and has been discussed by Bernal(27).

Recent optical scattering experiments in the vicinity of the critical point have shown that the density transition is a monotonic decreasing function from the liquid to the vapour phase. The measurements appear to be in good agreement with the semi-macroscopic theories of the transition profile to which the present analysis tends as $T \to T_c$. However, it must be realized that optical measurements only provide an assessment of the 'dielectric profile'(21). Near the critical point this may be reasonably interpreted as being proportional to the density; at lower temperatures a more subtle interpretation is required. Drude's analysis(22) incorrectly assumes that the dielectric tensor is isotropic throughout the transition zone. The experimental results are usually interpreted in terms of a 'minimum optical thickness' of the transition zone. Buff and Lovett(23) have reformulated Drude's theory, in a way such that it is valid at low temperatures. However, the low temperature result is not applicable to the critical region. Nevertheless, it should be stressed that in this type of experiment one sees only the long-wavelength Fourier components of the density profile, whereas the more detailed structure which may be present is completely washed out.

A recent investigation of the reflection spectrum of liquid mercury in the range 4500–10000 Å[63,71] shows that the normal reflectivity is very close to that predicted from a Drude dispersion relation based on the valence electron density and the d.c. conductivity of the liquid. However, ellipsometric studies seem to yield optical constants for liquid mercury considerably in excess of those predicted by the Drude theory. From an examination of the solutions of Maxwell's equations describing the interaction of electromagnetic radiation with an inhomogeneous conductor, Bloch and Rice[64] show that such anomalous optical behaviour may be attributed to a liquid metal whose boundary is not geometric, but a continuous density transition. (Even if the *ionic* boundary were geometric there is no reason why the distribution of conduction electron density should be so – § 4.7.) Bloch and Rice assume a simple form for the conductivity profile and find that excellent agreement between theory and experiment is found for a variety of parametric forms, *provided the conductivity passes through a maximum as the surface zone is traversed.* Clearly this requires an ionic density transition profile of the form shown in fig. 4.3.3, and we might speculatively conclude that there exists a close connection between anomalous Drude behaviour and the anomalous surface tension characteristics observed by White[18] and others, discussed in § 4.6.

Some recent preliminary high energy reflection electron diffraction experiments have been performed on certain liquid metal and alloy systems[10] in an attempt to determine the interphasal structure. Some evidence was obtained for the existence of a structural interphase in the case of liquid lead. Tin on the other hand showed no indication of surface layering. The lead–tin alloys exhibited intermediate characteristics. Further diffraction investigations are currently in progress and should provide a definitive determination of the density profile.

Disregarding the influence of vapour particles on the structure of the boundary layer of the liquid, Fisher[11] discusses the corresponding one-dimensional model of a transition layer next to an ideal infinitely high potential wall for which a full and rigorous solution is possible. The one-dimensional microdensity has been calculated by Fisher and Bokut'[12], and by Felderhof[26], for infinitely hard spheres with a square-well potential: such configurations have also been observed by simulation[13, 27]. In the case of hard spheres against an ideal wall the problem becomes tractable as a fortuitous consequence of the gradients being everywhere $\pm\infty$ or zero. Nevertheless, this model demonstrates the development of stable density oscillations initiated by the boundary wall.

Croxton and Ferrier[13] have also performed computer simulations of the argon liquid–vapour interface. Their results are discussed in detail in § 5.11.

4.4 The distribution function $g_{(2)}(z, r)$

In the absence of an accurate and explicit theory of the anisotropic pair distribution at the liquid surface, various approximate prescriptions have been suggested which allow a statistical mechanical calculation of the stress-tensor (4.2.1) to be concluded. The exact distribution is

$$\rho_{(2)}(z_1, \boldsymbol{r}_{12}) = \rho_{(1)}(z_1)\,\rho_{(1)}(z_2)\,g_{(2)}(z_1, \boldsymbol{r}_{12}) \tag{4.4.1}$$

and assuming the single-particle distribution has already been determined, attention is generally focused on the approximate representation of $g_{(2)}(z_1, \boldsymbol{r}_{12})$. The usual hierachical relation between adjacent orders of distribution still holds of course (see § 2.2), and we may test our prescription for $g_{(2)}$ against the condition

$$-1 = \int \rho_{(1)}(\mathbf{2})\,[g_{(2)}(\mathbf{1}, \mathbf{2}) - 1]\,\mathrm{d}\mathbf{2} \tag{4.4.2}$$

which it should satisfy. None of them do, however, and to that extent the one- and two-particle approximations are inconsistent.

Retaining the surface of density discontinuity for $\rho_{(1)}(z)$, KB set

$$\begin{aligned} \rho_{(1)}(z) &= \rho_{\mathrm{L}} \quad (z \leqslant 0), \qquad \rho_{(2)}(z_1, \boldsymbol{r}) = \rho_{\mathrm{L}}^2\, g_{(2)}(r) \quad (z \leqslant 0),\\ &= 0 \quad (z > 0), \end{aligned}$$

neglecting the vapour. More realistic density transitions involve some sort of attempt at the self-consistent relation of the anisotropic distribution $\rho_{(2)}(z_1, \boldsymbol{r}_{12})$ to $\rho_{(1)}(z_1)$. A realistic representation, first proposed by Green[83] and used by Berry and Reznek[84], is

$$\rho_{(2)}(z_1, \boldsymbol{r}_{12}) = \rho_{(1)}(z_1)\,\rho_{(1)}(z_2)\,g_{(2)}^{\mathrm{L}}(r_{12}),$$

where $g^{L}_{(2)}(r_{12})$ is the RDF appropriate to the bulk liquid phase. Certainly any prescription which retains the bulk liquid distribution throughout the transition cannot accurately describe the gas phase distribution, although this is not likely to be serious[2]. It can, however, lead to incorrect values of the gas density $\rho_{(1)}(+\infty)$. Nevertheless, there is much to be said for not introducing further parametrized interpolations between the two coexistent bulk distributions, $g^{L}_{(2)}$ and $g^{V}_{(2)}$. Toxvaerd's prescription, (4.3.9) and (4.3.10), has already been discussed and criticized on the grounds of its difficult assessment as a nested approximation. Nazarian's prescription, whilst as *ad hoc* as the others, is somewhat simpler to interpret, and as pointed out earlier, results in a strongly oscillatory density profile.

Croxton and Ferrier[16] give a variational determination of the pair distribution function in the vicinity of the transition zone. Since they assume no radial distortion of the distribution, their development amounts to a harmonic mode analysis of $g_{(1)}(z_1)g_{(1)}(z_2)g_{(2)}(r_{12})$. However, it is found that the thermodynamic functions of the liquid surface may be rather directly expressed in terms of the harmonic components of the anisotropic pair distribution function. The pair distribution is first expressed in terms of the separable product[70]

$$\rho_{(2)} = \rho_{L}^{2}\, g_{(2)}(r) \sum_{l \geqslant m=0}^{\infty}\sum^{\infty} A_{lm}(z, T) P_{l}^{m}(\cos\theta)\,\Phi(m\phi). \qquad (4.4.3)$$

The angular dependence of the density function is expressed in terms of a complete set of spherical harmonics whose variable parameters $\{A_{lm}\}$ are to be determined. Invariance to arbitrary translation and rotation in the xOy-plane (parallel to the surface) eliminates the ϕ-dependence for non-hydrodynamic equilibrium systems of spherically symmetric, pairwise-interacting molecules at a planar interface. From symmetry arguments many of the $\{A_{l0}\} = 0$, and a reasonable description of the θ-dependence of the distribution is given by a linear combination of the first and second unassociated harmonics, squared to increase the flexibility of the trial function a little:

$$\begin{aligned}\rho_{(2)}(z_1, \mathbf{r}) &= \rho_{L}^{2} g_{(2)}(r)\,\{A_{00}(z_1)\,P_{0}^{0}(\cos\theta) + A_{10}(z_1)\,P_{1}^{0}(\cos\theta)\}^2 \\ &= \rho_{L}^{2} g_{(2)}(r)\, A_{00}^{2}(z_1)\,\{P_{0}^{0} + \lambda(z_1)\,P_{1}^{0}\}^2. \qquad (4.4.4)\end{aligned}$$

$\lambda(z, T)$ has the significance of a hybridizing coefficient between the spherically symmetric bulk modes and the surface angular modes of the appropriate symmetry. It is clear that in this sp_z approximation the coefficient $\lambda \to 0$ as $z \to \pm\infty$. Hybridization of bulk liquid and vapour modes with the specifically interphasal configurations will diminish as the spatial and dynamical correlations decouple far from the interphase, the density distribution becoming spherically symmetric. It is also self-evident that λ is temperature dependent: at the critical point $\lambda \sim 0^+$ since the single-particle distribution is spatially extended and little anisotropy develops. The action of the operator Φ_0^* in creating an equilibrium free surface is to establish those harmonics having the appropriate symmetry. Fig. 4.4.1 shows the spatial form of $\lambda(z)$ determined from the variational minimization of the interfacial free energy. To obtain absolute values of the coefficients $A_{00}(z)$, $A_{10}(z)$ we make contact with the single-particle distribution and write

$$\int g_{(1)}(z_1)\, g_{(1)}(z_2)\, g_{(2)}(r_{12})\, \mathrm{d}\mathbf{2} = A_{00}^2(z_1) \int \{1 + \lambda(z_1)\cos\theta\}^2 g_{(2)}(r_{12})\, \mathrm{d}\mathbf{2},$$

giving,
$$A_{00}^2(z_1) = \frac{g_{(1)}(z_1) \int g_{(1)}(z_2)\, g_{(2)}(r_{12})\, \mathrm{d}\mathbf{2}}{2\{1 + \lambda^2(z_1)/3\} \int_0^\infty g_{(2)}(r)\, r^2\, \mathrm{d}r}, \tag{4.4.5}$$

and the amplitude of the first harmonic is given directly by

$$A_{10}(z) = \lambda(z)\, A_{00}(z). \tag{4.4.6}$$

As we might have anticipated (fig. 4.4.1), the first unassociated harmonic $P_1^0(\cos\theta)$ shows extensive hybridization with the spherically symmetric $P_0^0(\cos\theta)$ (bulk) modes only in the vicinity of the liquid surface.

As the temperature is raised, in particular, as $T \to T_c$, the singlet density profile $g_{(1)}(z)$ becomes a monotonic function of distance, and the transition zone broadens. It would be expected intuitively that hybridization with the P_1^0 mode decreases as a function of increasing temperature, $A_{10}(z)$ relaxing to zero as $T \to T_c$. At the

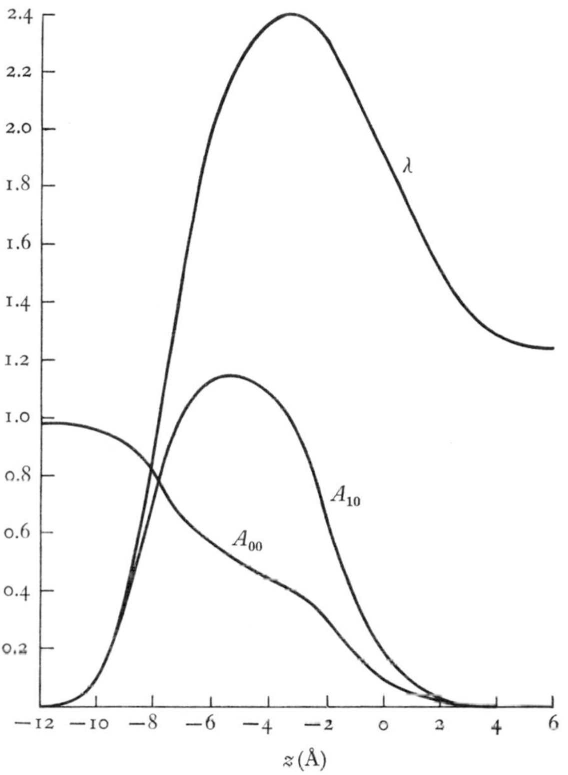

Fig. 4.4.1. The coefficients of the first and second unassociated zonal harmonics, $A_{00}(z)$, $A_{10}(z)$, and the hybridization ratio $\lambda(z) = A_{10}(z)/A_{00}(z)$ in the vicinity of the liquid argon surface at the triple point.

critical point $A_{10}(z, T_c) \sim 0$ over its entire range since there can be virtually no structural anisotropy under these conditions.

It is clear that $A_{10}(z)$ represents a measure of surface thickness since the coefficient is non-zero only in regions of anisotropy at the liquid–vapour interface (provided there is no hydrodynamic flow, nor that there are any thermal gradients). Fine structure related to $g_{(1)}(z)$ and $A_{00}(z)$ will be impressed on $A_{(10)}(z)$ in accordance with the relation $A_{10} = \lambda A_{00}$.

4.5 Thermodynamic functions of the liquid surface[17]

In the two previous sections the singlet and doublet molecular distribution functions were determined under conditions of transitional anisotropy at a planar liquid–vapour interface. The determinations were essentially of a statistical mechanical nature and in no way depended upon the quasi-thermodynamic arguments mentioned earlier. In the case of the pair distribution, however, the *radial* as distinct from the angular, modification was left undetermined, and does not therefore constitute a complete theory of the distribution function[70]. Nevertheless, this does represent a considerable improvement on previous analyses in that it takes explicit account of the interphasal contributions to the thermodynamic quantities associated with the interface.

The general expression for the surface tension in the Kirkwood–Buff stress tensor formalism is

$$\gamma = \int_{-\infty}^{\infty} \{P - P(z_1)\}\, dz_1,$$

where the tangential stress is given by

$$P(z_1) = kT\rho_{(1)}(z_1) - \frac{1}{2}\int_\tau \rho_{(2)}(z_1, \boldsymbol{r})\, \nabla\Phi r^3 \sin^2\theta \cos^2\phi \, d\theta \, d\phi \, dr$$

(cf. (4.2.7)). In the present analysis this may be written

$$P(z_1) = kT\rho_{(1)}(z_1) - \tfrac{1}{2}\rho_L^2 \int_\tau g_{(1)}(z_1)\, g_{(1)}(z_2)\, g_{(2)}(r)\, \nabla\Phi r^3 \sin^3\theta \cos^2\phi \, d\theta \, d\phi \, dr \tag{4.5.1}$$

or, in terms of the hybridization ratio

$$P(z_1) = kT\rho_L\, g_{(1)}(z_1) - \tfrac{1}{2}\rho_L^2 \int_\tau A_{00}^2(z_1) \times \{1 + \lambda(z_1)\cos\theta\}^2 g_{(2)}(r)\, \nabla\Phi r^3 \sin^3\theta \cos^2\phi \, d\theta \, d\phi \, dr. \tag{4.5.2}$$

In either bulk phase as $\lambda \to 0$, or as $\rho^2 g_{(1)}(-\infty)\, g_{(1)}(-\infty) \to \rho_L^2$, and $\rho_L^2\, g_{(1)}(+\infty)\, g_{(1)}(+\infty) \to \rho_V^2$ the vapour phase number density, so the above integrals reduce to the bulk expression

$$P = kT\rho - \tfrac{1}{6}\rho^2 \int g_{(2)}(r)\, \nabla\Phi(r)\, r \, d\tau. \tag{4.5.3}$$

Alternatively, we may take the z-component of the pressure, which for an equilibrium planar interface will be constant across the transition zone and equal to the bulk expression (4.5.3)

$$P = kT\rho_L g_{(1)}(z) - \tfrac{1}{2}\rho_L^2 \int_\tau A_{00}^2(z_1)$$
$$\times \{1 + \lambda(z_1)\cos\theta\}^2 g_{(2)}(r)\,\nabla\Phi(r)\,r^3 \sin\theta\cos^2\theta\,d\theta\,d\phi\,dr. \quad (4.5.4)$$

Again, in the limit $z_1 \to \pm\infty$, $\lambda \to 0$, (4.5.4) is seen to reduce to the preceding expression for the bulk pressure. At intermediate z_1, when $\lambda \neq 0$, and $g_{(1)}(z) \neq 1.00$, it is not explicitly clear that (4.5.4) remains constant across the transition; however, inserting (4.5.2) and (4.5.4) into the stress-tensor expression for γ we obtain directly

$$\gamma = -\pi\rho_L^2 \int_0^\infty g_{(2)}(r)\,\nabla\Phi(r)\,r^3 \int_{-\infty}^\infty A_{00}^2(z_1)\,\lambda(z_1)$$
$$\times \{1 + \tfrac{4}{15}\lambda(z_1)\}\,dz_1\,dr. \quad (4.5.5)$$

This expression of the surface tension in terms of the hybridization ratio is physically appealing, and would be expected to show the correct limiting thermal dependence since $\lambda(z) \to 0$ at the critical point. Equation (4.5.5) is of particular importance since it does not depend explicitly upon the sensitive difference between the potential and kinetic components of P and $P(z)$ in (4.5.3) and (4.5.4). The function $\{P - P(z)\}$ may also be used to locate the plane known as the *surface of tension* located at an origin $z_{ST} = 0$ such that

$$0 = \int_{-\infty}^\infty z\{P - P(z)\}\,dz. \quad (4.5.6)$$

The dividing surface at which the superficial density of matter vanishes, located at z_Γ is given by

$$0 = \rho_L \int_{-\infty}^{z_\Gamma} \{1 - g_{(1)}(z)\}\,dz + \rho_L \int_{z_\Gamma}^\infty \{\rho_V/\rho_L - g_{(1)}(z)\}\,dz. \quad (4.5.7)$$

The distance δ between the surface of tension and the Gibbs dividing surface is

$$\delta = z_{ST} - z_\Gamma \quad (4.5.8)$$

and is of importance since it determines the curvature dependence of the surface tension according to the rigorous thermodynamic Gibbs–Tolman equation[14],

$$\frac{d\ln\gamma}{d\ln r} = \frac{2(\delta/r)\{1 + \delta/r + \frac{1}{3}(\delta^2/r^2)\}}{1 + 2\delta/r\{1 + \delta/r + \frac{1}{3}(\delta^2/r^2)\}}, \quad (4.5.9)$$

where r is the radius of the spherical surface of tension and γ is the surface tension referred to the surface of tension.

The excess of the real two-phase system over a hypothetical reference system in which the bulk properties are constant up to the dividing surface provides a precise meaning for the various thermodynamic functions of the liquid surface. Of γ, u_S, s_S and Γ, the surface excess per unit area of energy, entropy and matter respectively, only γ is independent of the choice of origin. The thermodynamic relations applicable to a one-component planar interface are the free energy and Gibbs adsorption equations

$$u_S = \gamma - Ts_S - \Gamma\mu, \tag{4.5.10}$$

$$0 = s_S\,dT + \Gamma\,d\mu + d\gamma, \tag{4.5.11}$$

where μ is the chemical potential. Clearly, the location of the dividing surface on that plane at which the superficial density of matter vanishes is of particular importance, and only then is u_S related to γ by the thermodynamic relationship

$$\frac{d(\gamma/T)}{dT} = -\frac{u_S}{T^2} \tag{4.5.12}$$

and the surface excess entropy is given by

$$\frac{d\gamma(T)}{dT} = -s_S. \tag{4.5.13}$$

Under these circumstances the Helmholtz free energy per unit area and the surface tension become identical.

Kirkwood and Buff[(2)] have determined the surface excess energy to be

$$_{(KB)}u_S = -\frac{\rho_L^2}{2}\int_0^\infty \Phi(r)\,g_{(2)}(r)\,r^3\,dr$$

in the approximation of a density discontinuity at the liquid surface, and with no modification of the pair distribution. In the present analysis it may be easily shown that a more general expression of the surface excess energy is (cf. (4.2.10))

$$\left.\begin{aligned} u_S = 2\pi\rho_L^2\int_{-\infty}^{\infty}\int_{-\infty}^{\infty}\{g_{(1)}(z_1)\,g_{(1)}(z_2)-1\}\,dz_1\,dz_2 \\ \times\int_0^\infty \Phi(r)g_{(2)}(r)\,r^2\,dr, \\ z_2 = z_1 + r\cos\theta, \end{aligned}\right\} \tag{4.5.14}$$

where the origin is located at z_Γ.

The above expression for the surface excess energy may, of course, be written in terms of the hybridization ratio, λ;

$$u_S = \pi\rho_L^2 \int_0^\infty \Phi(r) g_{(2)}(r) r^2 \,dr \int_{-\infty}^0 \int_0^\pi \{A_{00}^2(z)(1+\lambda\cos\theta)^2 - 1\}$$

$$\times \sin\theta \,d\theta \,dz + \pi\rho_L^2 \int_0^\infty \Phi(r) g_{(2)}(r) r^2 \,dr$$

$$\times \int_0^\infty \int_0^\pi A_{00}^2(z)(1+\lambda\cos\theta)^2 \sin\theta \,d\theta \,dz, \quad (4.5.15)$$

$$= 2\pi\rho_L^2 \int_0^\infty \Phi(r) g_{(2)}(r) r^2 \,dr$$

$$\times \left[\int_{-\infty}^0 \{A_{00}^2(z)(1+\tfrac{1}{3}\lambda^2) - 1\} \,dz \right.$$

$$\left. + \int_0^\infty A_{00}^2(z)(1+\tfrac{1}{3}\lambda^2) \,dz \right]. \quad (4.5.16)$$

Again, it is clear from the above equation that the excess energy contributions originate in the interphase region where $\lambda \neq 0$. Further, as $T \to T_c$ and $\lambda \to 0$ for all z, the integral in the square brackets in (4.5.16) reduces to

$$\int_{-\infty}^0 \{A_{00}^2(z) - 1\} \,dz + \int_0^\infty A_{00}^2(z) \,dz. \quad (4.5.17)$$

Under these circumstances (4.4.5) reduces to

$$A_{00}^2(z) \to g_{(1)}(z) = 0.25 \quad (4.5.18)$$

at the critical point, where all distances are measured relative to the plane at which the superficial density of matter vanishes, in the approximation $\rho_V = 0$. In other words, the surface excess energy vanishes as $T \to T_c$ as indeed it should.

The resulting theoretical estimates on the basis of the above theory are compared with earlier estimates in table 4.5.1.

If we apply the low of corresponding states at the triple point we can obtain preliminary estimates of the surface properties of several other systems on the basis of the argon values. The atomic diameters for all systems except argon have been taken directly from Shoemaker *et al.* The results are shown in tables 4.5.2 and 4.5.3. The results for krypton and xenon are seen to be in substantially good agreement with experiment whilst the agreement

TABLE 4.5.1. *Comparison of results for surface properties of liquid argon at the triple point* (84.3 °K)

Argon	H	H†	PP	SPC	KB*	FM [88]	CF[17]	Experiment[15]
γ (dyn cm^{-1})	6.91	21.6	16.55	15.6	16.84	13.7	13.48	13.45
u_S (erg cm^{-2})	19.43	60.59	50.55	27.08	44.3	27.6	35.35	35.01
δ (Å)	2.67	2.65	2.01				3.84	

* Estimated from their 90 °K values by $\gamma = \gamma_0(1 - T/T_c))^{1.28}$, $u_S = \gamma - T(\partial\gamma/\partial T)$.
† Hill's original calculations corrected by PP.

TABLE 4.5.2. *Comparison of surface tensions for various systems on the basis of the law of corresponding states* (γ, dyn cm^{-1})

	σ (Å)	Corresponding states (CF)	SPC	Experiment
Kr	3.599	16.66	17.09	16.1
Xe	3.750	21.18	24.47	19.3
Ne	2.761	8.01	4.49	2.7
N_2	3.341	10.08	12.73	12.0
O_2	3.026	12.89	18.64	16.5
CH_4	3.579	13.82	15.83	16.0

TABLE 4.5.3. *Comparison of surface energies for various systems on the basis of the law of corresponding states* (u_S erg cm^{-2})

	ϵ/k (°K)	Corresponding states (CF)	SPC	Experiment
Kr	168.51	41.23	33.44	40.1
Xe	296.40	50.31	41.93	50.0
Ne	34.44	25.84	8.05	14.3
N_2	140.43	28.19	22.93	27.5
O_2	197.72	37.95	33.91	37.1
CH_4	181.25	34.40	30.56	35.8

for neon is poor[42, 43]. The homonuclear diatomics nitrogen and oxygen are not in good agreement either: it is questionable, however, whether the law of corresponding states can be legitimately extended to include these systems.

A surface tension relation has been derived from significant structure theory by Lu, Jhon, Ree and Eyring[65]. These authors introduce the approximation that the surface of a liquid consists of a monomolecular layer in which a molecule has a sublimation energy different from that of a bulk liquid molecule. The theory has been applied to simple liquids, polar liquids and liquid metals at various temperatures with reasonable success. This model whilst instructive, lacks much of the *a priori* objectivity of the truly molecular theories however.

4.6 Entropy–temperature relations at the liquid surface

The possibility of the development of stable density oscillations in the equilibrium $g_{(1)}(z)$ profile for certain liquids has been explicitly demonstrated in §4.3. The consequence of such a structured interphase is considerable, and a qualitative examination of the rate of production of excess entropy with temperature at the liquid surface suggests a markedly different $\gamma(T)$-characteristic to the monotonic decreasing function widely reported in the literature.

It was indicated in §4.3 that certain liquid systems, in particular the liquid metals, might develop a density transition of the schematic form shown in fig. 4.3.3. That is, one whose density transition is complete within a molecular diameter, and which possesses well-developed density oscillations. Provided the mean atomic density shown by a broken line in fig. 4.3.3 remains constant up to the dividing surface, it is clear that the entropy per unit area at the surface S_σ is considerably less than that of the bulk, S_β, and in consequence the excess entropy per unit area at the surface $s_S = (S_\sigma - S_\beta)$ is negative. This is supported by the recent semi-empirical relation of Egelstaff and Widom[44]. These authors note that the product $\gamma\chi_T$ is virtually independent of the nature of the liquid near the triple point, and represents a characteristic length proportional to the surface thickness. A comparison of liquid metals with non-metallic liquids is significant because interionic forces in metals are density dependent. Thus, had the density gradient at the interface not been sharp, it might have been expected to show up through a difference in the value of $\gamma\chi_T$ for metals and insulators. No such effect is found. Egelstaff and Widom conclude that the liquid–vapour density transition near the triple point is effectively complete within an atomic diameter, and there is not a diffuse surface layer. Of course, if such a situation is to develop at all, it can only be in the vicinity of the triple point. Equation (4.5.13) related the surface excess entropy to the temperature gradient of the $\gamma(T)$ characteristic thus,

$$\frac{d\gamma}{dT} = -s_S. \tag{4.6.1}$$

As mentioned above, for a system possessing a layered or structured interphase there will be an entropy deficiency at the liquid surface $-s$ will be negative yielding a positive gradient for $d\gamma/dT$. Eventually the surface will delocalize, and so the excess entropy will pass through zero to positive values. Such a $\gamma(T)$-curve exhibiting an inversion is contrary to the majority of experimental evidence, and probably all theoretical models.

However, White(18) in a series of *equilibrium* measurements of the $\gamma(T)$-characteristics of various liquid metals obtains inversions in a number of systems. It must be emphasized that, contrary to most experimental determinations of the surface tension, especially of liquid metals, the experiments of White were performed with the liquid in equilibrium with its saturated vapour so that there was no net flux of atoms across the liquid–vapour interface. The majority of $\gamma(T)$-determinations have neglected this outstandingly important point, and in consequence a non-equilibrium measurement of an equilibrium parameter is obtained(72). In fig. 4.6.1 we show the effect on the $\gamma(T)$-characteristic as measured by White under conditions of progressive vapour transport. The equilibrium curve shows a strong inversion. Under progessively non-equilibrium conditions the familiar monotonic decreasing characteristic is regained.

It is instructive to construct schematic plots for the 'classical' entropy-temperature relations, and compare them to the present

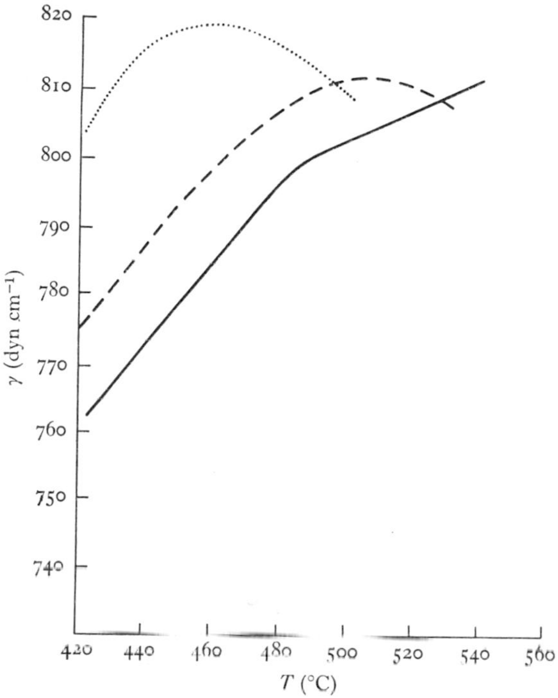

Fig. 4.6.1. The effect of vapour transport on the surface tension of 99.999+ per cent zinc (after White[18]). Full curve, no vapour transport; broken curve, moderate vapour transport; dotted curve, high vapour transport. (By permission from White, *Trans. Metall. Soc. AIME*, **236**, 796 (1966).)

model and its S–T characteristics. The classical curve shows the convergence of the bulk and surface entropies; initially $(S_\sigma - S_\beta) > 0$ at the triple point and finally $(S_\sigma - S_\beta) = 0$ at the critical point when the liquid–vapour boundary disappears. Such an $S(T)$-relation would generate a monotonic decreasing $\gamma(T)$-curve as is generally reported in the literature for most systems (fig. 4.6.2(*a*)).

However, in the case of inversion of the surface tension at a temperature T_i, and where the bulk and surface entropies are inverted at the triple point by virtue of density oscillations in the transition profile, then for the inversion to occur, and for $\gamma(T) \to 0$ as $T \to T_c$, the S–T curve must have some such form as shown in fig. 4.6.2(*b*), provided the rate of entropy production at the surface

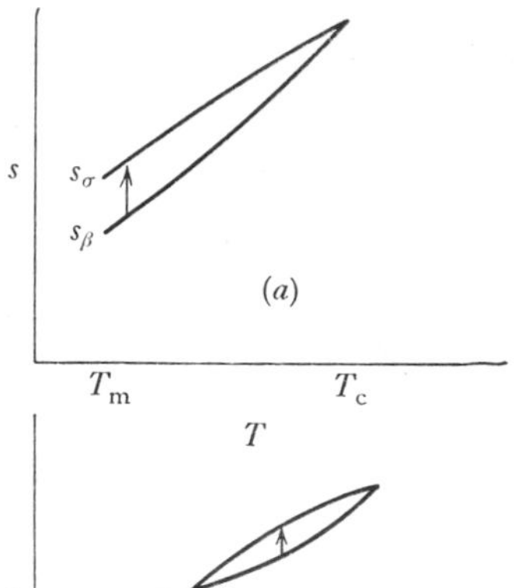

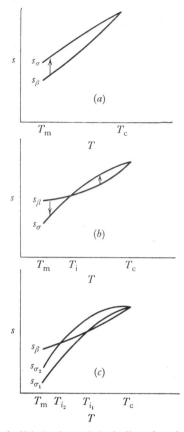

Fig. 4.6.2. (*a*) 'Classical' behaviour of the bulk and surface entropies as a function of temperature. The difference $S_\beta - S_\sigma = \partial\gamma/\partial T$ shows a monotonic decrease in γ with increasing T. (*b*) $S_\sigma < S_\beta$ inversion at T_m. Differing rates of entropy production produce inversion at T_i. $\gamma(T)$ is almost linear. (*c*) Increasing rate of vapour transport shifts the S_σ curve up the entropy axis, and T_i to lower values as observed by White[18] for zinc (cf. fig. 4.6.1.).

is greater than that of the bulk. Thus the gradient of the $\gamma(T)$-curve inverts, and remains largely constant over quite a wide temperature range subsequently. Systems showing almost coincidental bulk and surface entropy curves are presumably just those which have almost horizontal $\gamma(T)$-characteristics.

Under conditions of enhanced vaporization as White investigated little change would occur to the $S_\beta(T)$-curve, but presumably $S_\sigma(T)$

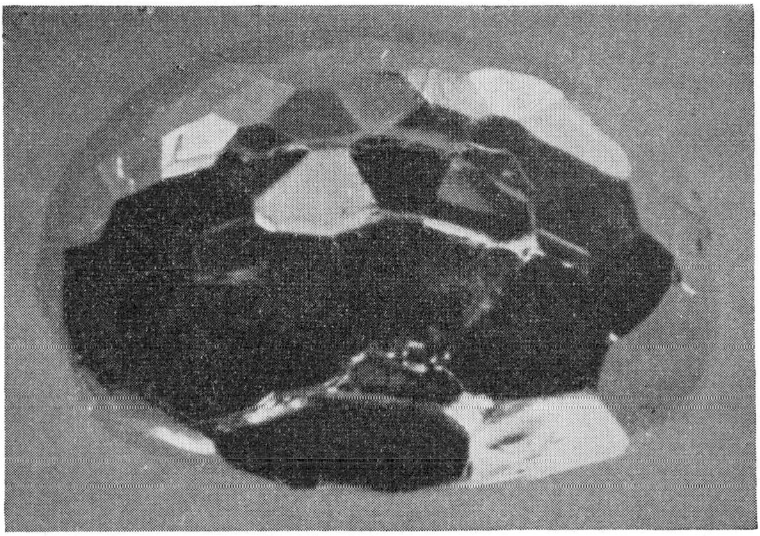

Fig. 4.6.3. 'Geodesic' faceting at the surface of a solidified sessile drop of liquid zinc. Each facet is a basal plane. (Reproduced by permission of White, *J. Inst. Metals*, **99**, 287 (1971).)

is shifted to a higher value at each point along the temperature axis. This can only have the effect of lowering the inversion temperature, and if streaming of the vapour becomes excessive, S_β and S_σ will reinvert, to give the classical behaviour (fig. 4.6.2(*c*)). This would seem to explain the behavioural characteristics of the $\gamma(T)$-curves in the deliberately non-equilibrium experiments of White on liquid zinc, and is possibly the mechanism responsible for the apparent classical behaviour of all liquid metals when full precautions for ensuring equilibrium have not been taken.

A particularly interesting observation of White[19] is that of surface faceting. It was found that the liquid metal sessile droplets used in the $\gamma(T)$-investigations developed surface faceting to an extent and perfection directly related to the system's tendency to show $\gamma(T)$-inversion. Thus, zinc and cadmium, for example, solidified into beautifully faceted droplets (fig. 4.6.3), whilst other metal systems (aluminium and tin) solidified into apparently smooth sessile drops. These latter metals showed no tendency towards

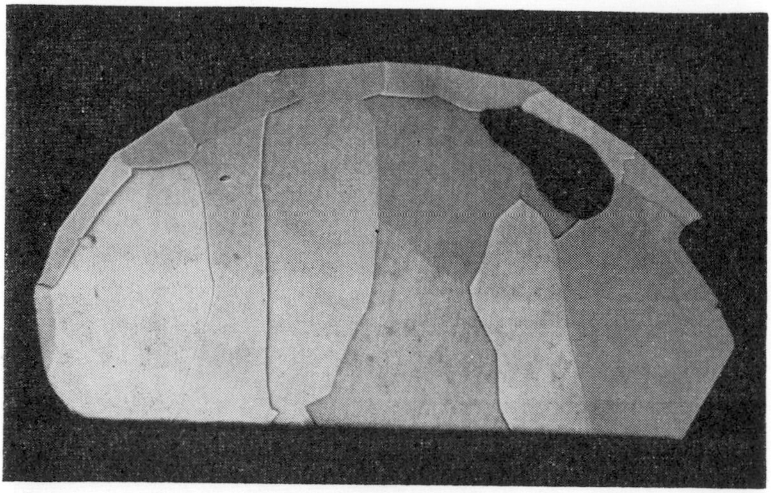

Fig. 4.6.4. Cross-section of a zinc droplet showing the development of grains both at the base and, independently, at the surface of the drop. The dark area is the shrinkage cavity arising from the volume decrease on solidification. (Reproduced by permission of White, *J. Inst. Metals*, **99**, 287 (1971).)

inversion. Laue back-reflection X-ray investigations on solidified drops of cadmium and zinc indicated that all the facets were basal planes, i.e. planes of densest packing.

That faceting initiates at the surface of the droplet (as well as at the base) is evident from the cross-section of a zinc droplet shown in fig. 4.6.4. The configuration of grains is consistent with solidification starting from several locations at the base and growing upwards, having begun independently at several locations on the surface. Very relevant is White's observation that the shrinkage cavity (dark region in fig. 4.6.4) in more than thirty drops sectioned, though usually located just below the surface, in no case broke out into the surface. Moreover, in one case the thin grain above the cavity could be seen to have 'dished in' due to the 'negative pressure' created when the shrinkage cavity was formed(19).

Bearing in mind the mechanism for γ-inversion proposed in the earlier sections, it is difficult to escape the conclusion that surface faceting is a direct consequence of the development of stable density oscillations at the liquid metal surface just above the triple point.

Goodman and Samorjai[24] have made low energy reflection electron diffraction investigations of the possibility of surface pre- or post-melting. One might expect the liquid surface to show indications of atomic rearrangement just prior to solidification, particularly in the case of the inverting systems. However, on the basis of the measurement of the mean square atomic displacement of the atoms *normal* to the liquid surface, the authors conclude that the atomic amplitudes of vibration are up to twice as great as the bulk amplitudes. Goodman and Samorjai conclude that this precludes the development of low entropy states at the liquid surface. However, corrections have to be applied to the diffraction results to take account of liquid metal surface potential effects, and the final result appears somewhat inconclusive. Certainly structural investigations must precede investigations of atomic dynamics at the liquid–vapour interface.

Bohdansky[25] has proposed a model for the temperature dependence of surface tension for liquid metals. The surface tension is expressed in terms of the free energy developed in bringing an atom from the bulk to the surface at absolute zero, less the change in free energy associated with the ratio of the bulk and surface characteristic frequencies of vibration. Very closely related, and historically earlier is Frenkel's[55] modification of Born and von Kármán's[56] treatment which treated the surface capillary waves as the ordinary elastic waves used to describe the thermal motion in the volume of the body. The approaches of Bohdansky and Frenkel are expressed largely in terms of physical unknowns, and whilst they make an interesting qualitative contribution, they cannot be usefully incorporated into any modern concept of the statistical thermodynamics of the liquid surface.

4.7 The surface tension of liquid metals[53, 57]

The high surface tensions of the liquid metals is very often, and mistakenly, ascribed to the presence of the conduction electrons whose presence, in actual fact, generally serves to *lower* the surface tension rather than increase it. The electronic contribution is almost entirely kinetic except for the development of an electrostatic double

TABLE 4.7.1 *Comparison of experimental liquid metal surface tensions with various theoretical estimates*

Metal	Expt	Samoilovich[62]	Breger and Zhukovitskii[60]	Huang and Wyllie[58]	Stratton[61]
Li	398			894	400
Na	191	400	777	440	190
K	101	224	333	180	70
Rb				140	50
Cs				110	40
Cu	1066	1128	3720	1820	740
Ag	927	748	2170	1140	450
Au	1134		2170	1250	450
Mg					
Zn	769	1397	2330		
Cd		1156	1670		
Hg	465	644	1550		
Al	505				
Ga	538		1330		
Sn	537		1040		
Pb	456		1150		
Sb	367		980		
Bi	389				
Fe	1409		2480		
Pt	1819		2120		

layer at the liquid metal surface where the electrons 'leak out' beyond the ionic boundary of the metal. Frenkel[59] once ascribed the entire surface tension of a liquid metal to this effect, and indeed he obtained $\gamma = 472$ dyn cm^{-1} for mercury which is in excellent agreement with modern determinations. An approach which embodies many of the principal features is that of Huang and Wyllie[58] which considers the electrons to be constrained in a potential 'trough'. The imposition of this boundary condition establishes one kinetic contribution to the internal energy which is proportional to the volume and a second contribution proportional to the area. It is in this that we are interested here. Breger and Zhukhovitskii[60] considered the constraint of the electrons in an infinitely deep trough, but this does not permit the development of the double layer which does undoubtedly occur. Finally, these authors account for the change in the vibrational characteristics of the ions as they

TABLE 4.8.1. *The index μ in the vicinity of the critical point*

	Two-dimensional lattice gas[32,35]	Three-dimensional lattice gas	Three-dimensional continuum fluid	Classical critical point[33,34]
Cahn–Hilliard	9/8	1.28	1.24	3/2
Widom	1	1.29	1.20	

approach the surface rather similarly to Bohdansky. No account is taken of the *structural* contributions at the surface. Stratton[61] has extended the Huang–Wyllie theory to the case where the electronic and ionic boundaries at the surface are not coincident. This, of course, slightly modifies the double layer, and a different expression for the surface entropy term due to ionic vibrations is adopted. Good agreement is obtained for the monovalent metals lithium, sodium and potassium. Generally, however, the agreement with experiment is very bad indeed (table 4.7.1).

4.8 Surface tension near the critical point

A theory of the thermal variation of surface tension in the vicinity of the critical point has been given by Cahn and Hilliard[28] (CH) as a redevelopment of an original theory of van der Waals[29]. On the CH theory there are two contributions to the free energy at the liquid surface attributed to density fluctuations and the density gradient. The equilibrium surface free energy and surface tension are then obtained as

$$F \propto V(\Delta\rho)^2/\rho^2\chi_T, \tag{4.8.1}$$

$$\gamma \propto L(\Delta\rho)^2/\rho^2\chi_T, \tag{4.8.2}$$

where $\Delta\rho$ is a number density fluctuation in a substantial subvolume of an effectively infinite volume. L is the thickness of the transition zone, and χ_T the isothermal compressibility. At the critical point the product $\rho^2\chi_T$ is more or less constant across the transition zone[30,31] and we fortunately do not have to be more specific. Whilst we cannot make an absolute determination of the surface tension, we can determine its temperature dependence. Writing

$\gamma \sim (\Delta T)^{\mu}$, Cahn and Hilliard obtain $\mu = 1.24$ for a continuum fluid which is in almost perfect agreement with Guggenheim's empirical relation $\mu = 11/9 = 1.22$. However, in spite of the successes of the CH theory it must be judged fundamentally incorrect: for a two-dimensional lattice gas it may be shown rigorously(36, 38) that $\mu = 1$ instead of the CH value of 9/8.

Widom(39) has modified the CH theory in dispensing with the second contribution to the surface free energy, and obtains identical expressions to (4.8.1), (4.8.2), but with a different proportionality constant. The index μ is redetermined and is now found to be an explicit function of the dimensionality of the system – a feature absent in the CH theory. A comparison of the values of μ for various systems on the basis of the two theories is given in table 4.8.1.

4.9 Experimental determinations of surface tension

The empirical relation

$$\gamma(T) = \gamma_0(1 - T/T_c)^{\mu} \tag{4.9.1}$$

suggested by van der Waals(29) appears to describe simple fluids remarkably well. Equation (4.9.1) has been subsequently shown by Guggenheim(37) to arise as a natural product of the law of corresponding states. $\gamma(T)$ has been measured over a very large range for argon, nitrogen and xenon, and these data appear to fit (4.9.1) to within 1.6 per cent for μ in the range 1.25–1.29. The value $\mu = 1.5$ originally predicted by van der Waals(29) is thus too high, whilst the value suggested by Guggenheim(37) (1.22) and Widom(39) (1.20) are much too low. In fact the original Cahn–Hilliard value of $\mu = 1.24$ appears to be most satisfactory. It is interesting to note that the three-dimensional lattice gas value on either the CH or Widom theory agrees most closely with experiment. In table 4.9.1 we give the parameters T_c, γ_0 and μ appearing in (4.9.1) for various determinations of the surface tension of argon, nitrogen and xenon. A good, though not recent, review of the surface tensions of argon and nitrogen has been given by Stansfield(40), and a more general review of the surface tension of simple liquids by Buff and Lovett(21), and by Croxton(85).

TABLE 4.9.1. *Surface tension parameters appearing in* (4.9.1)

System	T_c (°K)	γ_0 (dyn cm^{-1})	μ	Reference/Year
Ar	150.7	38.07	1.281	(40)/1958
Ar	150.72	37.78	1.277	(41)/1966
N_2	126.0	29.09	1.247	(40)/1958
N_2	126.23	28.42	1.232	(41)/1966
Xe	289.74	53.9	1.290	(42)/1965
Xe	289.74	54.5	1.287	(43)/1967

It is inappropriate here to review the experimental methods employed in the determination of the thermodynamic functions of the liquid surface: for these the reader is referred to the excellent review of White[45]. There is quite an extensive number of reviews of the surface tension of liquid metals[53].

It will be appreciated that there is no guarantee that the parametric representation (4.9.1) will hold right down to the triple point. Indeed, the relation may be expected to hold only for molecules of high symmetry and low acentric factor in the vicinity of the critical point. When the density transition at the liquid surface is complete within a few atomic diameters a quasi-thermodynamic approach is inapplicable. Clearly, such inversion of the surface tension as has been observed by White, and certain Russian workers, cannot be described by such an expression. For if

$$\gamma = \gamma_0\left(1-\frac{T}{T_0}\right)^{\mu}$$

then

$$\frac{\partial\gamma}{\partial T} = -\frac{\mu\gamma}{T_c - T}. \qquad (4.9.2)$$

Whereupon, at the inversion temperature T_i, $\delta\gamma/\partial T = 0$, and either μ or γ must be zero. Further, if $\partial\gamma/\partial T$ is to be positive over part of the temperature range, then μ must be negative. On the other hand, as $T \to T_c$, so we should expect the quasi-thermodynamic analysis to become appropriate, and $\mu \to 1.2^+$. We might therefore incorporate the deviation from the semi-macroscopic treatment in terms of a temperature-dependent μ. T_m, T_i and T_c represent the triple (melting), inversion and critical temperatures respectively, whilst μ_c is the critical limit to which $\mu(T)$ asymptotically approaches as $T \to T_c$. Writing

$$\gamma = \gamma_0 \left(1 - \frac{T}{T_c}\right)^{\mu(T)}, \tag{4.9.3}$$

$$\frac{\gamma'}{\gamma} = \ln\left(1 - \frac{T}{T_c}\right)\mu'(T) - \frac{\mu(T)}{T_c - T}, \tag{4.9.4}$$

where the primes indicate derivatives with respect to T. In the vicinity of the inversion $\gamma'/\gamma \to 0$, and we may simplify (4.9.4) to read

$$\mu' \ln\left(1 - \frac{T}{T_c}\right) = \frac{\mu}{T_c - T} \tag{4.9.5}$$

which may be numerically integrated to yield $\mu(T)$.

White's observation of a positive temperature coefficient for zinc is the first that has appeared in the literature, but it is not the first time a positive coefficient has been reported for a liquid metal. Positive coefficients have been observed for mercury[46], cadmium[47], copper[48, 49], for iron–nickel alloys[50] and for some other iron alloys[48]. The results of Pugachevich and Yashkichev[49] for copper are shown in fig. 4.9.1 (*a*).

White[18] points out that, particularly in the case of zinc, the $\gamma(T)$-characteristic for many liquid metals consists of two or more straight sections, with the consequence that

$$-T\frac{\partial^2 \gamma}{\partial T^2} = \frac{\partial u}{\partial T} = 0. \tag{4.9.6}$$

Thus, the superficial specific heat is zero, and this Einstein[51] has interpreted to mean that the entire energy of formation of the surface, u_S, is configurational.

Since all White's $\gamma(T)$-determinations were carried out under the same equilibrium conditions by the same experimental procedure we shall briefly consider the results for lead, tin and the lead–tin alloys. In fig. 4.9.1 (*b*) we show the $\gamma(T)$-characteristics for lead and tin. Both are linear functions of temperature, but the lead characteristic is almost temperature-independent which suggests that the rates of entropy production in the bulk and at the surface are virtually identical. Tin, on the other hand, shows γ to be a linear decreasing function of temperature, and this may be understood on the 'classical' model of a delocalized liquid–vapour density

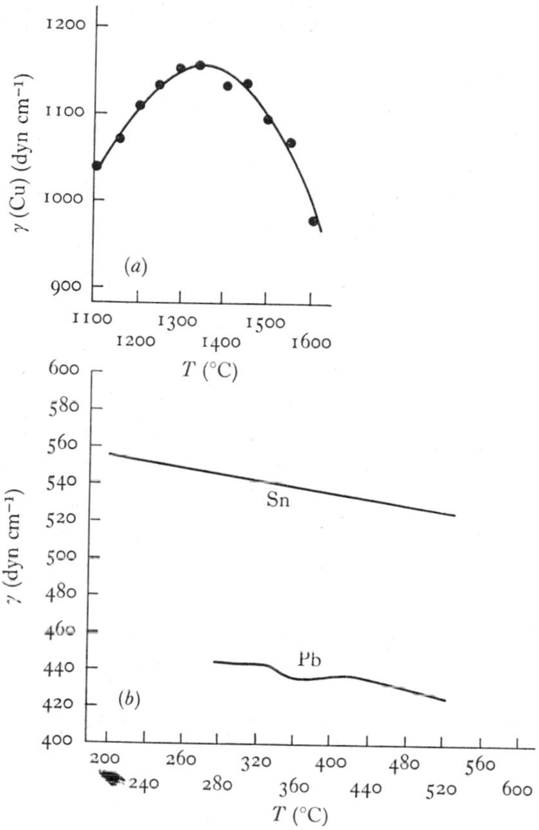

Fig. 4.9.1. (*a*) The surface tension of copper as determined by Pugachevich and Yashkichev[49], showing a pronounced inversion at ~ 1350 °C. (*b*) The surface tensions of 99.9999% lead and 99.99% tin, as determined by White[53].

transition. Intermediate characteristics are exhibited by the lead–tin alloys. The curves shown in fig. 4.9.1 are, of course, the equilibrium characteristics (liquid in equilibrium with its saturated vapour). Non-equilibrium characteristics obtained by establishing a net vapour transport from the surface depresses the γ-characteristic over the whole range. Thus it appears that we have both inverting and non-inverting characteristics under equilibrium conditions.

It will be recalled (§4.3) that the development of (low entropy) density oscillations at the liquid surface was contingent upon the form of the pair potential, $\Phi(r)$, the collective surface field parametrized by the operator $\Phi^*(z)$, and the absolute temperature T. No mention, either explicit or implicit, is made of the triple point T_m, other than if a stable oscillatory density profile is to develop at all in the liquid, then it will be at and just beyond T_m. The possibility exists then that if we could supercool a liquid enough, *all* liquid metals would eventually show the γ-inversion. As it is, however, only that region beyond T_m corresponding to the stable liquid phase is actually observed. This is evidently an over-simplification, for if we were to construct a *reduced* γ^*-characteristic, complete with an inversion, then clearly the whole family either would or would not show inversion. However, it is found that liquid metals in the same column of the periodic table do tend to exhibit the same $\gamma(T)$-characteristics. Thus, zinc, cadmium and mercury have all been reported to show inversion. Likewise, copper, silver and gold. On the other hand germanium, tin and lead apparently do not show such inversions. Several liquid metal systems, although never achieving a positive slope to its γ-characteristic, do show a tendency towards a zero gradient as the triple point is approached. One such example is indium, whose γ characteristic is well described by the quadratic[52]:

$$\gamma(t) = 568.0 - 0.04t - 7.08 \times 10^{-5}t^2 \quad (t\,°\mathrm{C}). \qquad (4.9.7)$$

Again it appears that if the liquid could be sufficiently supercooled the inversion might well be observed. No correlation has as yet been made between the surface tension of a metal on either side of the triple point – that is, in the solid and liquid phases.

4.10 The quantum liquid surface

The deviation of the quantum liquids from the classical estimates of the superficial thermodynamic functions is already apparent in the case of neon (cf. tables 4.5.2 and 4.5.3). According to the classical principle of corresponding states we should expect the reduced surface tension $\gamma^* = \gamma\sigma^2/\epsilon$ to be a universal function of the reduced temperature $T^* = kT/\epsilon$, for those substances which are described adequately by the Lennard-Jones potential. The experimental data available indicate that γ^* depends parametrically on the quantum

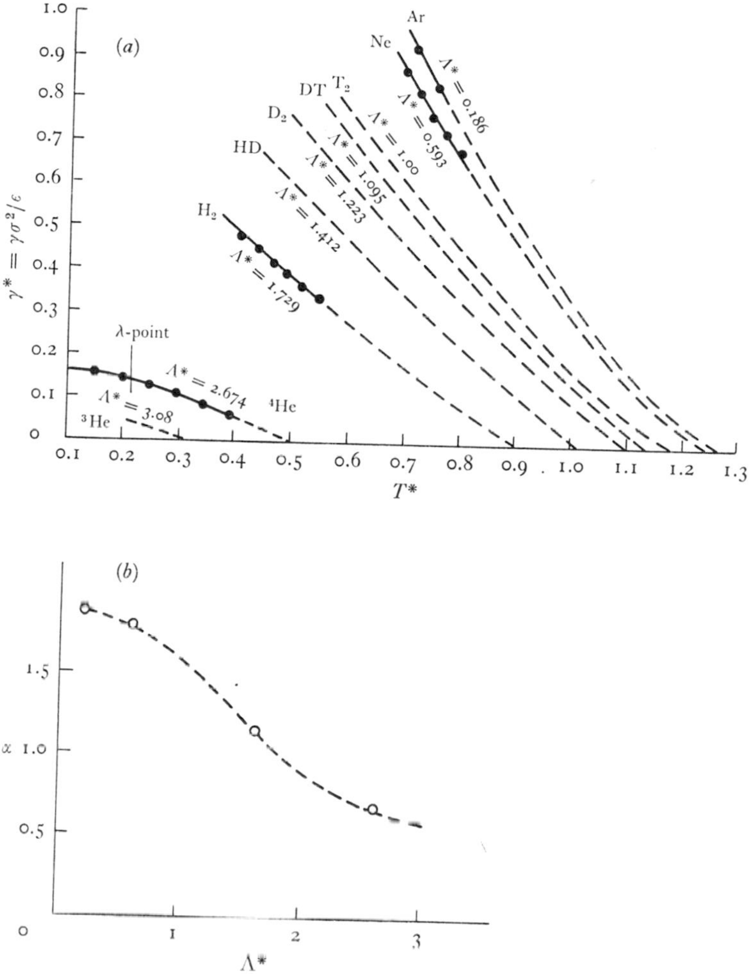

Fig. 4.10.1. (*a*) The surface tensions of the quantal fluids depart from the universal reduced curve appropriate to the classical fluids, such as argon. The numbers refer to the de Boer parameter Λ^*, upon which the characteristics depend parametrically via the function $\alpha(\Lambda^*)$ (cf. equation (4.10.1). (*b*) The coefficient $\alpha(\Lambda^*)$ which plays the role of γ_0 in the phenomenological $\gamma(T)$ relation. (Redrawn by permission from Hirschfelder, Curtis and Bird, *Molecular Theory of Gases and Liquids* (Wiley).)

mechanical de Boer parameter Λ^*. The reduced phenomenological relation is now written

$$\gamma^* = \alpha(\Lambda^*)\left(1-\frac{T^*}{T^*_c}\right)^{\mu} \tag{4.10.1}$$

where the coefficient α introduces the Λ^*-dependence. Fig. 4.10.1 (*a*) shows the concurrence of the experimental and 'corresponding state' characteristics. Fig. 4.10.1 (*b*) shows the explicit dependence of α upon Λ^*: this is for $\mu = 11/9$ in (4.10.1).

The discrepancy between the classical predictions and the experimental observation of the surface tension and energy of quantum liquids may be understood in terms of the pair momentum current density tensor at the liquid surface, or what we might loosely call the 'kinetic temperature tensor'. We have seen in § 2.21 that for classical systems in thermodynamic equilibrium the temperature tensor is isotropic and identical to the thermodynamic temperature. But as soon as the kinetic component of the Hamiltonian

$$(-\hbar^2 \sum_i \nabla_i^2/2m)$$

becomes operationally dependent upon the distribution $(\prod_{i<j} f_{ij})$, then clearly the *kinetic* contribution to the pressure will be both a function of position and of the kinetic temperature τ which is generally $\geqslant T$, the thermodynamic temperature. From (2.21.11) we may propose that $P(z)_{\text{quantum}} \geqslant P(z)_{\text{classical}}$ so that in the stress-tensor formalism:

$$\gamma = \int_{-\infty}^{\infty} \{P - P(z)\}\, dz.$$

We may conclude that the quantum γ is depressed below its classical estimate, and this is indeed found to be the case.

There have been several attempts to establish a quantum-statistical counterpart to the Kirkwood–Buff statistical treatment of classical liquids. A proposition of Atkins[(54)] enables us to determine the limiting law of temperature dependence of the surface tension of liquid ^{4}He. This is for sufficiently low temperatures that the whole liquid may be regarded as superfluid. Analysis is in terms of hydrodynamic capillary waves[55], as follows. The surface part of the free energy per unit area may be written

$$F = \gamma_0 + \frac{2\pi}{\beta}\int \ln\left(1 - e^{-\hbar\omega/kT}\right)\sigma(k)\,dk; \quad \beta = (kT)^{-1}, \tag{4.10.2}$$

where $\sigma(k)\,dk$ is the distribution of capillary states per unit area in the range k to $k+dk$,

$$\sigma(k)\,dk = \frac{k\,dk}{(2\pi)^2} \tag{4.10.3}$$

assuming a continuous distribution of k-states at the liquid surface. γ_0 represents the residuum of free energy at absolute zero which may be attributed to change in zero point oscillation, plus the usual potential energy excess, and may be identified as the surface tension at $T = 0$. Equations (4.10.2) and (4.10.3) may be combined to give

$$F = \gamma_0 - \frac{\hbar}{4\pi}\int \frac{k^2}{e^{\hbar\omega/kT} - 1}\,d\omega. \tag{4.10.4}$$

At sufficiently low temperatures only the oscillations at low frequencies are important, i.e. those with small wave numbers (long wavelengths). Such oscillations are hydrodynamic capillary waves for which

$$\omega^2 = \frac{\gamma k^3}{\rho} \sim \frac{\gamma_0 k^3}{\rho}, \tag{4.10.5}$$

where ρ is the density of the liquid. Hence

$$\gamma = \gamma_0 - \frac{\hbar}{4\pi}\left(\frac{\rho}{\gamma_0}\right)^{\frac{2}{3}}\int_0^\infty \frac{\omega^{\frac{4}{3}}}{e^{\hbar\omega/kT} - 1}\,d\omega. \tag{4.10.6}$$

Since the integral converges rapidly the upper limit may be taken as infinity. Integration gives

$$\gamma = \gamma_0 - \frac{(kT)^{\frac{7}{3}}}{4\pi\hbar^{\frac{4}{3}}}\left(\frac{\rho}{\gamma}\right)^{\frac{2}{3}}\Gamma(\tfrac{7}{3})\,\zeta(\tfrac{7}{3}) \tag{4.10.7}$$

$$= \gamma_0 - \frac{0.13(kT)^{\frac{7}{3}}\rho^{\frac{2}{3}}}{h^{\frac{4}{3}}\gamma^{\frac{2}{3}}}, \tag{4.10.8}$$

where Γ and ζ are the gamma and zeta functions.

It must be pointed out that this derivation applies only to a superfluid boson system (^4He). Capillary waves of this kind do not exist in a Fermi liquid (^3He) since the viscosity increases indefinitely as $T \to 0$. Further, the analysis requires that only relatively long surface waves be considered, having a frequency which is smaller than the relaxation time of the corresponding liquid at the temperature in question. Fortunately at very low temperatures the integrand of (4.10.4) contributes only at long wavelengths. If we were to extend this treatment to higher temperatures the capillary waves would be gradually replaced by non-dispersive waves of the Rayleigh type: the surface would behave essentially as a solid. The frequency spectrum of the surface waves constituting the heat motion of the free surface of a liquid body must thus be separated into two components: (i) the low frequency of capillary part $\nu < 1/\tau_0$, and (ii) the high frequency or Rayleigh part $\nu > 1/\tau_0$, where τ_0 is the relaxation time of the liquid.

We may alternatively approach the problem along the same lines as that proposed in §§4.3, 4.4, 4.5 by Croxton and Ferrier, and by Croxton[69]. Again the singlet density profile across the quantum liquid–vapour interface may be determined from the one-dimensional BGY equation in the Φ^* operator formalism (equation (4.3.11)). Experimental forms of the ^{4}He and ^{3}He radial distribution function lead us to expect a relatively structureless transition profile, assuming a Lennard-Jones interaction. This aspect of the problem is indifferent to the particle statistics: such information is contained implicitly in the form of $g_{(2)}(r)$.

However, when we come to make a variational determination of the anisotropic pair density distribution function $\rho_{(2)}(z_1, \boldsymbol{r}_{12})$ by minimization of the expectation energy, we are confronted with the full quantum-mechanical problem of the variational minimization of $\langle \mathscr{H} \rangle$ with respect to the parametric coefficients of the total wave function.

The expectation value of the Hamiltonian may be written

$$\langle \mathscr{H} \rangle = \frac{\begin{aligned} N(N-1)\int \cdots \int \prod_{i<j=1}^{N} f^{\dagger}(\boldsymbol{r}_{ij}) \\ \times \left\{ \frac{\hbar^2}{2m} \frac{\nabla_1^2 f(\boldsymbol{r}_{12})}{f(\boldsymbol{r}_{12})} - \tfrac{1}{2}\Phi(r_{12}) \right\} \prod_{i<j=1}^{N} f(\boldsymbol{r}_{ij})\, \mathrm{d}\mathbf{3} \ldots \mathrm{d}\boldsymbol{N} \end{aligned}}{\int \cdots \int \left| \prod_{i<j=1}^{N} f(\boldsymbol{r}_{ij}) \right|^2 \mathrm{d}\mathbf{1} \ldots \mathrm{d}\boldsymbol{N}}, \tag{4.10.9}$$

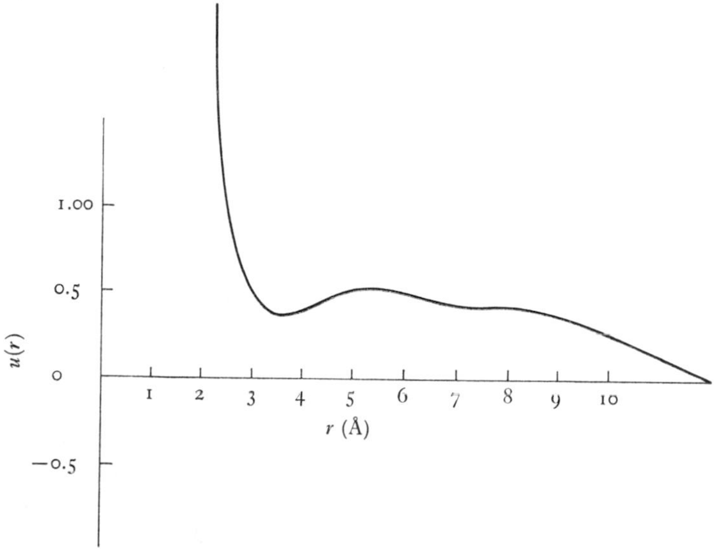

Fig. 4.10.2. The radial Bijl–Dingle–Jastrow u-function for a ground-state boson fluid determined from Abe's iterative scheme (equations (4.10.10), (4.10.11)).

where the dagger † signifies the complex conjugate. We take a symmetrical, correlated ground-state wave function of a modified Bijl Dingle–Jastrow type (BJD′) to describe the ground state of the N-body boson fluid

$$\Psi(1 \dots N)$$

$$= \prod_{i<j=1}^{N} f(\mathbf{r}_{ij}) = \prod_{i<j=1}^{N} \exp\{\tfrac{1}{2}u(r_{ij})\, \Gamma(\theta, \phi)\}$$

$$= \prod_{i<j=1}^{N} \exp\left\{\tfrac{1}{2}u(r_{ij}) \sum_{l=0}^{\infty} \sum_{m=0}^{l} A_{lm}(z_i)\, P_l^m(\cos\theta_{ij})\, \Phi(\pm im\phi_{ij})\right\}. \tag{4.10.10}$$

Here the z axis is directed from the liquid to the vapour phase normal to the liquid surface, and the angles θ_{ij}, ϕ_{ij} define the orientation of the r_{ij}-vector in these coordinates. The radial function,

$u(\boldsymbol{r}_{ij})$, assumed separable, must be determined from some relation as that of Abe[67] given below (cf. § 2.11)

$$-\nabla_1 \ln g_{(2)}(12) = \nabla_1 u(12) + \sum_{n=1}^{\infty} (-1)^n \rho^n$$
$$\times \int_3 \cdots \int_{n+2} \{\nabla_1 g_{(2)}(1, n+2)\} g_{(2)}(n+1, n+2)$$
$$\times \prod_{i=3}^{n+1} \{g_{(2)}(1,i) g_{(2)}(i-1,i)\} \mathrm{d}\mathbf{3} \ldots \mathrm{d}(\boldsymbol{n}+\mathbf{2}). \quad (4.10.11)$$

Using the RDF data of Gordon[68] an iterative solution of Abe's expansion yields the u-function shown in fig. 4.10.2[69].

The set of associated Legendre coefficients $\{A_{lm}(z)\}$ are to be variationally determined as in the classical case (§ 4.4). Full account however must now be taken of the kinetic component of the Hamiltonian operator. Again on symmetry grounds invariance to arbitrary rotation and translation in the xy-plane requires that all $m \equiv 0$, reducing the general tesseral set to the zonal harmonic set in (4.10.10). The potential component of (4.10.9) may now be written in terms of the operator $\Phi^*(z)$:

$$\langle \mathscr{H} \rangle = \frac{N(N-1) \int \cdots \int \prod_{i>j=1}^{N} f^{\dagger}(\boldsymbol{r}_{ij}) \times \left\{ -\frac{\hbar^2}{2m} \frac{\nabla_1^2 f(\boldsymbol{r}_{12})}{f(\boldsymbol{r}_{12})} + \tfrac{1}{2}\Phi(r)\,\Phi^*(z) \right\} \prod_{i>j=1}^{N} f(\boldsymbol{r}_{ij}) \,\mathrm{d}\mathbf{3} \ldots \mathrm{d}\boldsymbol{N}}{\int \cdots \int \prod_{i>j=1}^{N} |f(\boldsymbol{r}_{ij})|^2 \,\mathrm{d}\mathbf{1} \ldots \mathrm{d}\boldsymbol{N}}. \quad (4.10.12)$$

The expectation kinetic energy is

$$\frac{\langle \mathscr{T} \rangle}{N} = -\frac{\rho \hbar^2}{2m} \int \left[\frac{\nabla_1^2 f(\boldsymbol{r}_{12})}{f(\boldsymbol{r}_{12})} \right] g_{(2)}(\boldsymbol{r}) \,\mathrm{d}\boldsymbol{r}$$
$$= -\frac{\rho \hbar^2}{2m} \int \left\{ \frac{\Gamma u'}{r} + \frac{\Gamma}{2} \left[u'' + \frac{u'^2 \Gamma}{2} \right] \right.$$
$$\left. + \frac{u}{2r^2} \left[\Gamma'' + \Gamma'^2 \frac{u}{2} \right] + \frac{u \Gamma' \cos\theta}{2r^2 \sin\theta} \right\} g_{(2)}(\boldsymbol{r}) \,\mathrm{d}\boldsymbol{r}, \quad (4.10.13)$$

where the primes on u and Γ represent the r and θ derivatives, respectively. In so far as the BDJ′ wave function is an adequate representation of the total N-body ground state function $\Psi(1 \ldots N)$,

(4.10.13) is exact. However, we now assume that we may describe the pair distribution function as

$$g_{(2)}(\mathbf{r}) = g_{(2)}(r)\,\Gamma^2(\theta), \tag{4.10.14}$$

whereupon

$$\frac{\langle \mathscr{T} \rangle}{N} = -\frac{\rho\hbar^2}{2m}\int_\tau \left\{\Gamma^3\left[\frac{u'}{r}+\frac{u''}{2}\right]+\Gamma^4\frac{u'^2}{4}+\Gamma^2\Gamma'\frac{u\cos\theta}{2r^2\sin\theta}\right.$$
$$\left.+\Gamma^2\Gamma'^2\frac{u^2}{4r^2}+\Gamma^2\Gamma''\frac{u}{2r^2}\right\}g_{(2)}(r)\,d\tau. \tag{4.10.15}$$

which is convergent since u, $u' \to 0$ as $r \to \infty$. Similarly, the expectation potential energy:

$$\frac{\langle \mathscr{V} \rangle}{N} = \tfrac{1}{2}\rho\int \Phi(r)\,\Phi^*(z)\,g_{(2)}(r)\,\Gamma^2(\theta)\,d\tau. \tag{4.10.16}$$

This latter expression is identical to the classical expectation energy, §4.5. Here, however, we have to variationally minimize the sum $\langle \mathscr{T} \rangle + \langle \mathscr{V} \rangle$ with respect to the parametric set $\{A_{l0}(z)\}$. Obviously practical interest is in a trial function containing a finite number of components; the computational difficulties involved in solving the secular determinant for the $\{A_{l0}\}$ become considerable for anything other than a few components. The secular problem in the vicinity of the liquid surface may be reduced on symmetry grounds leaving only the zonal harmonics of the general tesseral set, and only the first two terms $l = 0, 1$ are retained in the angular trial function:

$$\Gamma(\theta) = [A_{00}P_0^0(\cos\theta) + A_{10}P_1^0(\cos\theta)]. \tag{4.10.17}$$

We now set

$$B = -\left[\frac{u'}{r}+\frac{u''}{2}\right]\frac{\rho\hbar^2}{2m}g_{(2)}(r), \qquad C = -\frac{u'^2\rho\hbar^2}{4\cdot 2m}g_{(2)}(r),$$

$$D = -\frac{u\cos\theta}{2r^2}\frac{\rho\hbar^2}{2m}g_{(2)}(r), \qquad E = -\frac{u^2\sin^2\theta}{4r^2}\frac{\rho\hbar^2}{2m}g_{(2)}(r),$$

$$F = -\frac{u\cos\theta}{2r^2}\frac{\rho\hbar^2}{2m}g_{(2)}(r), \qquad V = \Phi\Phi^*\frac{\rho^2}{2}g_{(2)}(r),$$

where the terms Γ', Γ'^2 and Γ'' appearing in (4.10.15) have been incorporated. $\langle\mathscr{H}\rangle$ is minimized with respect to A_{00}, A_{10}, as in the classical case:

$$\begin{aligned} 0 = {} & 4A_{00}^3 C_{0000} + 3A_{00}^2 B_{000} + A_{00} A_{10}(6B_{001} - 2D_{00} - 2F_{00}) \\ & + A_{10}^2(3B_{011} - 2D_{01} - 2F_{01}) + A_{00} A_{10}^2(12C_{0011} + 2E_{00}) \\ & + 12A_{00}^2 A_{10} C_{0001} + A_{10}^3(4C_{0111} + 2E_{01}) \\ & + 2A_{00} V_{00} + 2A_{10} V_{10}. \end{aligned} \tag{4.10.18}$$

Similarly, minimization with respect to A_{10} yields

$$\begin{aligned} 0 = {} & 4A_{00}^3 C_{0001} + 4A_{10}^3(E_{11} + C_{1111}) + A_{00}^2(3B_{001} - D_{00} - F_{00}) \\ & + A_{00} A_{10}(6B_{011} - 4D_{01} - 4F_{01}) + 3A_{10}^2(B_{111} - F_{11} - D_{11}) \\ & + A_{00}^2 A_{10}(12C_{0011} + 2E_{00}) + A_{00} A_{10}^2(12C_{0111} + 6E_{01}) \\ & + 2A_{00} V_{01} + 2A_{10} V_{11}. \end{aligned} \tag{4.10.19}$$

Under these circumstances it may be directly shown that the surface tension becomes

$$\begin{aligned} \gamma = {} & -\frac{\rho_L \hbar^2 \pi}{3m} \int_{-\infty}^{\infty} \int_0^{\infty} \{A_{00}^3(z)\,\lambda(z)\,(3 + \tfrac{1}{2}\lambda^2(z))\,B(r) \\ & + 2A_{00}^4(z)\,(2 + \lambda^2(z))\lambda(z)\,C(r) \\ & + 2(2D(r) - E(r))\,A_{00}^2(z)\,\lambda(z)\}\,g_{(2)}(r)\,r^2\,\mathrm{d}r\,\mathrm{d}z \\ & - \pi\rho_L^2 \int_{-\infty}^{\infty} \int_0^{\infty} A_{00}^2(z)\,\lambda(z)\,(1 + \tfrac{4}{15}\lambda(z))\,g_{(2)}(r)\,\nabla\Phi(r)\,r^3\,\mathrm{d}r\,\mathrm{d}z, \end{aligned} \tag{4.10.20}$$

where $\lambda(z) = A_{10}(z)/A_{00}(z)$ is the hybridization ratio between the first and second unassociated harmonics. λ maximizes in the vicinity of the liquid–vapour transition zone. The second integral in (4.10.20) is identical to its classical counterpart, and represents the configurational contribution to the stress tensor arising from the anisotropic structure of the liquid surface. The first integral is peculiar to quantal systems, and represents the kinetic residuum present even at absolute zero. We see that the first term represents a quantum kinetic temperature tensor $\boldsymbol{\tau}$ in regions of atomic anisotropy, and, moreover, $\boldsymbol{\tau} > T$, the thermodynamic temperature.

The first integral in (4.10.20) serves to lower the surface tension below its classical estimate. The surface excess energy u_S is given as

$$u_S = -\frac{\rho_L \hbar^2 \pi N}{2m} \int_{-\infty}^{\infty} \int_0^{\infty} \{A_{00}^3(z)\,\lambda(z)\,(3+\tfrac{1}{2}\lambda^2(z))\,B(r)$$
$$+2A_{00}^4(z)(2+\lambda^2(z))\,\lambda(z)\,C(r)$$
$$+2(2D(r)+E(r))\,A_{00}^2(z)\,\lambda(z)\}g_{(2)}(r)\,r^2\,dr\,dz$$
$$+2\pi\rho_L^2 \int_0^{\infty} \Phi(r)\,g_{(2)}(r)\,r^2\,dr \left[\int_{-\infty}^{0} \{A_{00}^2(z)\,(1+\tfrac{1}{3}\lambda^2(z))-1\}\,dz\right.$$
$$\left.+\int_0^{\infty} A_{00}^2(z)\,(1+\tfrac{1}{3}\lambda^2(z))\,dz\right]. \tag{4.10.21}$$

Again, the second term is identical to the classical expression for the excess configurational energy arising at the liquid surface. For a classical system there would be no kinetic contribution to u_S, but here the tensor in the transition zone serves to lower the surface energy. We may understand this kinetic deficiency at the quantum liquid surface by comparing the energies associated with the bulk and surface excitations. In the vicinity of the zero point the bulk and surface dispersion curves become quite different. Assuming a classical dispersion relation for surface capillary waves on the surface of liquid ^{4}He in a gravitational field, Bowley(70) estimates the lowering of the surface energy in replacing the bulk excitations by their surface counterpart as $\Delta u_S = -0.18$ erg cm^{-2}.

Singh(86) has pointed out that we may regard the ^{4}He(II) system as an ideal, degenerate assembly of non-interacting phonons. Using (4.10.4) in conjunction with the dispersion relation $\omega = ck$ (c = sound velocity in ^{4}He(II)) the excess free energy developed by the introduction of bounding surfaces turns out to be

$$\gamma = \gamma_0 - \pi m \zeta(2)\,k_B T^2/2h^2$$
$$= \gamma_0 - 7{\cdot}5 \times 10^{-3} T^2 \text{ erg cm}^{-2}. \tag{4.10.22}$$

Of course, in a liquid the normal modes are not purely harmonic, so in addition to the phonon contributions we might also expect other elementary excitations in the form of turbulence and vortex flow. Equations (4.10.8), (4.10.22) are, as yet, experimentally indistinguishable(87).

CHAPTER 5

NUMERICAL METHODS IN THE THEORY OF LIQUIDS

5.1 Introduction

In the previous chapters approximate integral equations were developed for the pair distribution. From this function we were able to establish virtually all the thermodynamic functions of interest, although it was clear that certain approximations had, of necessity, to be invoked either on the grounds of mathematical expediency as in the KBGY class of equations, or on more physical grounds in the case of the PY–HNC cluster expansions. The development of *approximate* integral equations was, of course, enforced by the impossibility of obtaining a direct evaluation of the N-body partition function for dense fluids, and as yet there appears to be no general method of calculating this quantity. It will be recalled that the difficulty was mathematical rather than conceptual, originating in the collective nature of the total potential Φ_N even in the pairwise additive approximation. Powerful numerical techniques have been brought to bear upon the various integral equations, and these are generally evaluated by iterative techniques. The question arises as to whether these complex calculational techniques should not be used directly in rigorous variants of the theory. An ideal case would, of course, be the possibility of calculating directly the configuration integral of a system of a large number of particles.

It appears that with the advent of large electronic computers we have at our disposal a means of calculating an ensemble average in terms of the accessibility of states of the system. The 'states' of the system may be purely configurational and determinate as in the molecular dynamics approach, where the representative point evolves in accordance with a classical Liouville equation, subject to constraints of microscopic reversibility and ergodicity. The distribution is here established as a time-average integration over the evolution. Alternatively the states may be purely probablistic and indeterminate, as in the Monte Carlo approach where, after many trial configurations of a relatively small number of particles, a distribution of states is established whose occupational probability closely resembles that implied in the partition function. It should be noted that the equivalence of statistical mechanical averages and time averages of dynamical systems depends essentially on passage to the macroscopic limit $N \to \infty$, $V \to \infty$, $N/V = \text{constant}$. Farquhar[1] emphasizes that for small finite systems the various alternative statistical mechanical ensembles give results which are different by at least $\mathcal{O}(N^{-1})$.

We have already encountered some results of the machine calculations in chapter 2 where their role in providing 'experimental' data for idealized systems of particles is quite clear. The wealth of microscopic data generated by such an 'experiment' provides an unambiguous basis for the testing and refinement of the existing theories. Indeed, the machine calculations have now reached a level of sophistication such that we may temporarily abandon comparison with experiment and instead study idealized systems which are simple enough to aid theoretical developments. The simplification can only involve the choice of a particular form of potential which, nevertheless, contains the essential features of a real potential. As far as the theory of liquids is concerned these ideal abstractions may be used as bases for expansions, just as the ideal gas serves as a basis for the virial coefficient expansion.

We shall begin this chapter by introducing the two principal methods of machine calculation. The molecular dynamics approach will be discussed first, being perhaps conceptually the simpler. This technique does have the great advantage over the Monte Carlo method in that it enables us to deal with non-equilibrium transport phenomena provided the relaxation time for the process is significantly smaller than the computation time. Since the maximum temporal extent for which a dynamic event can be followed without inordinate expenditure of computer time is $\sim 10^{-11}$ second, we are clearly limited to microscopic non-hydrodynamic events. We shall then consider the Monte Carlo method which has the important advantage that it may relatively easily be extended to quantum-mechanical systems in which the exchange symmetry of the single particle wave function must be preserved.

The remainder of the chapter will be devoted to a review of a number of investigations based on either the molecular dynamics or Monte Carlo technique. These reviews are unrelated within themselves, other than by technique of investigation, but serve both to supplement the analytical investigations reported in earlier chapters, and to demonstrate the technique in its own right. When a problem is of particular methodological interest, for example the machine calculations for liquid water, the ground state of liquid ^{4}He and the molecular dynamic calculations for the liquid surface, we shall not hesitate to devote ourselves equally to this aspect of the computation.

5.2 Molecular dynamics

The net force on an atom i may be determined classically by taking the vector sum

$$-\sum_{j}^{N-1} \nabla\Phi(ij),$$

presupposing a knowledge of the distribution of atomic centres and the interaction potential, $\Phi(ij)$. If the configurational and momentum coordinates of each atom at some time t are stored, then the classical trajectory of atom i over a period of time Δt may be determined by solution of the Newtonian equations. The new position coordinate in the absence of any external fields will be

$$\boldsymbol{q}_i(t+\Delta t) = \boldsymbol{v}_i(t)\,\Delta t - \frac{1}{2m_i}\sum_{j}^{N-1} \nabla\Phi(ij)\,(\Delta t)^2 + \boldsymbol{q}_i(t) \qquad (5.2.1)$$

whilst the new velocity coordinate will be

$$\boldsymbol{v}_i(t+\Delta t) = \boldsymbol{v}_i(t) - \frac{1}{m_i}\sum_{j}^{N-1} \nabla\Phi(ij)\,(\Delta t). \qquad (5.2.2)$$

The distribution of velocities should be Maxwellian for an equilibrium system, of course, and if this is the case then the temperature of the system of N particles is given as

$$T = \frac{m_i}{3k}\sum_{i=1}^{N} \boldsymbol{v}_i^2 \qquad (5.2.3)$$

and this is generally monitored in the course of an 'experiment'. If T falls outside a specified range, $\sim \pm 1$ degree, then *all* the velocities are scaled to bring the system temperature back into range. Since fluctuations in the temperature are to be minimized the molecular dynamics computations are generally performed with a somewhat larger number of centres than the corresponding Monte Carlo calculation.

The calculations are generally based on the assumption that classical dynamics with a two-body central force interaction such as that described above will give a reasonable description of the motion of the constituent atoms. A number of practical considerations enforce further assumptions, however, and the essential limitations of the computer schemes have to be kept in mind. Pairwise additivity of the potential function, although not a necessary restriction, is generally employed because of the great simplification that ensues. Another great restriction is the relatively small number of degrees of freedom that can be dealt with on even the largest conceivable computer. In actual fact, this does not prove to be a serious problem, and can in fact be turned to advantage by studying how various properties depend upon the number of degrees of freedom. For a system of classical structureless particles, $N \sim 1000$ represents the current limit to the number of particles which may be dealt with in a reasonable amount of computer time. This of course represents a cubical array of side ~ 10 atomic diameters at liquid densities, and whilst number fluctuations within that volume are of necessity zero, a statistical mechanical error is engendered by the fact that only a finite number of particles are used. An estimate of the statistical mechanical error can be obtained by comparing calculations based on the same parameters using different numbers of particles[20]. To a certain extent 'surface effects' arising from the cell boundaries of the array may be overcome by imposing the so-called periodic boundary condition: the basic cubical array is replicated by a further 26 'images' whose atomic coordinates are obtained by adding or subtracting L from each Cartesian coordinate. Thus when a particle moves out of, say, the right hand face of the cell it reappears through the left hand face – rather like the k-space Brillouin zone scheme for electrons.

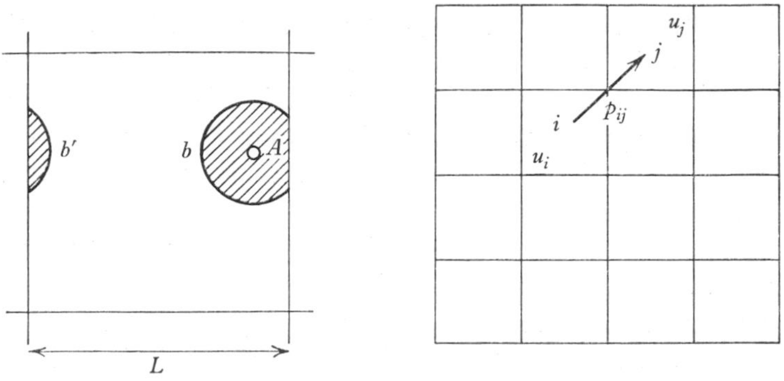

(a) (b)

Fig. 5.2.1. (*a*) Schematic two-dimensional array showing periodic boundary conditions. The range of the interaction of particle *A* is indicated by the shaded region. Clearly the dimension of the array must be such that no *bb′*-interaction can result. (*b*) The transition between configurational states *i*, *j*. represents the transition probability for the transition $i \to j$. u_i, u_j represent the absolute occupation probabilities of the states *i*, *j*.

Thus density and energy is conserved. If for convenience we take a two-dimensional array (fig. 5.2.1 (*a*)) particle *A* will take account of the shaded environment. Clearly the cell dimension must be considerably greater than the range of the pair interaction, i.e. there must be no *bb′*-correlation. Partly for this reason and partly for computational expediency the interaction is truncated beyond a certain range ~ 10 Å. It is important to realize that the application of periodic boundary conditions to simulate an infinite system cannot, in itself, diminish the statistical mechanical error: this is strictly dependent upon the number of particles in the system. Precisely the same communal entropy problems must arise in the molecular dynamics and Monte Carlo method as in the cell models of the liquid state. The entropy of the system will, in general, therefore be too low, whilst the free energy will be too high. However these problems are not likely to be important for $N \sim 100$ or more.

Nelson[110] has suggested the use of stochastic boundary conditions according to which a particle crossing the cell wall is lost from the system. However, new particles are introduced across the walls at random intervals, in random positions and with random velocities. The rationale behind such a boundary condition is to admit the possibility of density fluctuations, thus making the system a member of the grand canonical ensemble. It will also remove all traces of wall effects. Control would, of course, have to be exercised to ensure that there was no long range change in density, or accumulation of momentum or energy. Short term fluctuations would, however, be allowed as they occurred.

It is also advantageous, particularly at high densities, to choose N and the shape of the cell in such a way that the periodic boundary condition generates a perfect lattice appropriate to the system under study when the particles in the fundamental cell are arranged in a suitably ordered manner. Argon, for example, crystallizes in a face-centred cubic lattice, and for this system it is therefore convenient to use a cubic cell and choose $N = 4n^3$, where n is an even integer, i.e. $N = 4, 32, 108, 256, 500, 864, \ldots$.

The total potential of the system Φ_N may be calculated as a sum of pair terms $\Phi(r)$; the corresponding pair virial function $p_v(r)$ for potentials without discontinuities, is given by $p_v(r) = r(\mathrm{d}\Phi(r)/\mathrm{d}r)$ yielding the total virial P_v. Interactions between the particles in the fundamental cell, plus those in adjacent cells within the range of interaction, are included. Contributions from particles beyond the truncation of the range are not calculated explicitly, but are usually accounted for by integration over a uniform particle density. Clearly this procedure is in error unless the range is greater than that for which $g_{(2)}(r) \to 1.00$. Equilibrium properties are calculated as time averages. Data on equilibrium properties such as specific heat, compressibility, and thermal pressure coefficient, is obtained from the magnitude of fluctuations in the potential energy and virial, and it is sometimes difficult to obtain high accuracy. No particular difficulty is encountered in calculating the pressure or total energy of the system at liquid densities, except in the vicinity of a phase change, when large fluctuations in the calculated properties occur as the system separates into two phases. In the melting region the isotherm generally has two branches, one corresponding to the solid phase and the other to the liquid (see fig. 3.4.1). Near the critical point there is a further difficulty. The relatively small size of the fundamental cell suppresses the large density fluctuations in density which characterize the critical region in macroscopic systems. One effect of this is to increase the critical temperature of the 12–6 fluid by ~ 7 per cent[2], and another is that the specific heat is underestimated[3].

The trajectory of a given atom will evolve as a series of discrete linear steps whose length, for a given velocity, will depend directly upon the time increment Δt. The conflicting interest of expediency and significance of the calculated result lead to $\Delta t \sim 10^{-14}$ s. If coarser time-graining is used overlapping configurations of atoms develop which subsequently separate with infinite, or near-infinite, velocities depending upon the functional form of the repulsive interaction. The total energy of the system is constant as mentioned above, except for small fluctuations caused by the use of a finite time interval Δt, in the numerical integrations. But with the computation time varying approximately as N^2 it is not feasible to use a finer discretization than $\Delta t \sim 10^{-14}$.

In the case of hard spheres the development of overlapping configurations is inevitable, but provided the time step is small enough the many-body interaction can be resolved as a series of two-body encounters. When a contact or overlap configuration occurs the centres separate classically with a conservation of momentum, and infinite velocities do not develop.

No probablistic elements enter into the molecular dynamics computations at all other than in the choice of the initial coordinates and velocities of the particles. The former are chosen pseudo-randomly, i.e. so as to preclude overlap configurations, although they may be started from a periodic lattice arrangement. The velocity coordinates may have random directions and identical velocities such that the total kinetic energy is appropriate to the associated temperature. The atoms may alternatively be released from rest and the configuration allowed to relax. The velocities may be subsequently scaled so as to attain the experimental temperature. The velocity distribution should become Maxwellian, and Rahman[4] shows that for a system of 864 argon atoms at 94.4 °K the system rapidly attains equilibrium (table 5.2.1).

Alder and Wainwright[5] have determined the Boltzmann H-

TABLE 5.2.1. *Equilibration of a molecular-dynamic system at 864 argon atoms at 98.4 °K*

Number of collisions, ν	$\bar{T}$ (°K) for steps 1 to ν	$(\langle T^2\rangle - \bar{T}^2)^{\frac{1}{2}}/\bar{T}$
100	94.64	0.0167
200	94.47	0.0161
300	94.55	0.0158
400	94.55	0.0155
500	94.67	0.0160
600	94.51	0.0170
700	94.43	0.0170
780	94.45	0.0165

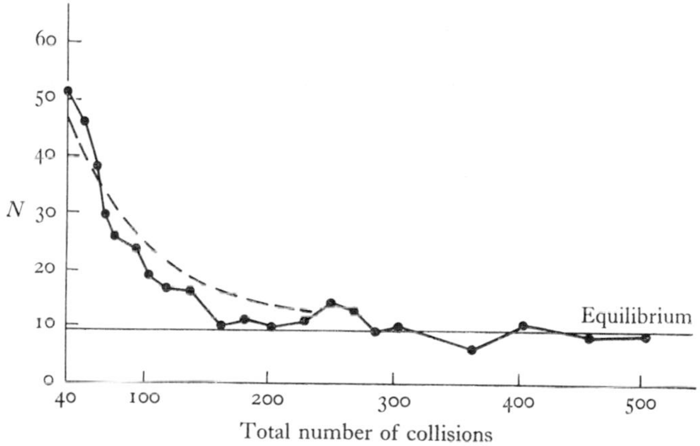

Fig. 5.2.2. Progression towards equilibrium of a 100-particle system of hard spheres as a function of number of collisions. N represents number of particles in an interval about the initial non-equilibrium velocity from which all particles were started. The broken line represents the evolution according to the Boltzmann transport equation.

function: this again is found to rapidly attain its equilibrium value. The attainment of the equilibrium velocity distribution is strikingly shown in fig. 5.2.2. Fig. 5.2.2 makes a direct comparison between the convergence of a 100-particle hard sphere system and the solution of the Boltzmann equation towards equilibrium. Various tests may be applied to establish that the distribution is Maxwellian. For example, the ratio of the widths w_1, w_2 and w_3 of Rahman's(4) velocity distribution at heights $e^{\frac{1}{2}}$, e^{-1} and e^{-2} of the maximum (argon, 98.4 °K) should be 1.78:2.51:3.55 whereas in fact they are 1.77:2.52:3.52.

5.3 The Monte Carlo method

The term Monte Carlo has come into use to designate numerical methods in which specifically stochastic elements are introduced, in contrast to the completely deterministic algebraic expressions of the molecular dynamic approach. The particular form of approach used in liquid state physics is that devised by Metropolis *et al.*[6]. The method and the results have been extensively reviewed in the literature[7,8]. Its essential feature is that it produces a Markov chain in which the individual Markov states are points in the usual configuration space of statistical mechanics for a system of N molecules confined to a volume V at temperature T. Thus if in fig. 5.2.1 (*b*) the cells (not to be confused with the molecular dynamic array, or cell) represent different configurational states then the system will populate the ith configurational state, or the ith cell with an absolute occupation probability u_i. If the system is in a state j, and $u_i > u_j$, then clearly there will be a strong transition probability p_{ji} for the transition $j \rightarrow i$. The occupation probabilities may be defined as

$$u_j = \operatorname*{Lt}_{T\rightarrow\infty} (T_j/T), \tag{5.3.1}$$

where T_j is the time spent in the jth configuration, and T is the total time. We have the normalization condition immediately that

$$\sum^{B} u_j = 1.00, \tag{5.3.2}$$

where B is the number of accessible states. Further, we have the principle of macroscopic reversibility which ensures

$$p_{ij}u_i = p_{ji}u_j \quad \text{for all } ij \tag{5.3.3}$$

and since a transition $i \rightarrow j$ must occur, even if it is $i \rightarrow i$, then

$$\sum_{j=1}^{B} p_{ij} = 1; \quad p_{ij} \geqslant 0. \tag{5.3.4}$$

The ergodicity condition now follows, for if i and j are any two admissible states (i.e. states for which u_i and u_j do not vanish), then the n-step transition probability $p_{ij}^{(n)}$ is non-zero[19]. This means that all states of the system are accessible from all other states: the transition probability may be vanishingly small, but it is not zero. Notice that we may define u_j in terms of $p_{ij}^{(n)}$ as

$$p_{ij}^{(n\to\infty)} \to u_j \tag{5.3.5}$$

and

$$\sum_{i=1}^{B} u_i p_{ij} = u_j \quad \text{for all } j. \tag{5.3.6}$$

Now the average of any classical quantity $\langle f\rangle$ is given by

$$\langle f\rangle = \frac{\int_\Omega f(x)\,P(x)\,\mathrm{d}x}{\int_\Omega P(x)\,\mathrm{d}x}, \tag{5.3.7}$$

where f is any sufficiently well-behaved function of x, and $P(x)$ is an unnormalized probability density of an M-dimensional vector in a region Ω. For the statistical mechanical averages with which we shall be principally concerned

$$P(x) = \exp\left\{-\frac{\Phi_N(1, \ldots, N)}{kT}\right\}, \tag{5.3.8}$$

and provided the volume element Δx is small enough we may approximate (5.3.7) by

$$\langle f\rangle = \sum_{i=1}^{B} f(x_i)\,u_i, \tag{5.3.9}$$

where

$$u_i = \frac{P(x_i)}{\sum_j P(x_j)}. \tag{5.3.10}$$

This latter 'space' average is identical to the time average (5.3.1). Thus, the Monte Carlo method consists of generating a set of molecular configurations by random displacements of the particles in the model. The configurations are accepted or rejected according to a criterion which ensures that a given configuration occurs with a probability u_j proportional to the Boltzmann factor $\exp(-\beta\Phi_{N_j})$ for that configuration. We note that pairwise additivity of the molecular forces of the system is not essential. The Monte Carlo method is valid for any law of interaction between the particles, provided the ergodicity is maintained (see below). This amounts, of course, to deciding upon p_{ij}. This p_{ij} prescription must satisfy the steady-state and the ergodicity conditions (5.3.4), (5.3.6). In fact there is a certain latitude of choice open to us regarding the transition matrix p_{ij}. Ideally one would wish to choose it in such a way as to minimize the variance σ^2/n of $\langle f \rangle$ where n is the number of steps in the Markov chain, corresponding to a fixed investment of computing time. Wood[8] observes that a complicated choice of p_{ij} producing a small σ^2 might very well require more computing time to reach some fixed value of σ^2/n than a simpler transition matrix with a larger σ^2. In practice the dependence of σ^2 upon p_{ij} is so opaque as to make this approach of use in an *a posteriori* fashion only.

A prescription for the transition matrix which is commonly used is the following: a particle of the system is selected, either serially or at random, and given a random displacement from state i to state j. Let us suppose that the increase in the total configurational energy is $\Delta\Phi_N^{ij}$, then if $\Delta\Phi_N^{ij}$ is negative the move is accepted and the new configuration replaces the old one. This corresponds to $u_j > u_i$. If $\Delta\Phi_N^{ij}$ is positive however, the move is accepted only with the probability $p_{ij} = \exp(-\Delta\Phi_N^{ij}/kT)$: the machine then selects a random decimal number in the range 0 to 1, and compares it to $\exp(-\Delta\Phi_N^{ij}/kT)$. If the exponential is the greater the move is allowed, otherwise the move is rejected. If a move is rejected the previous configuration is necessarily counted again. The occupation density of cells of configuration space (fig. 5.2.1 (*b*)) thus develops as the Boltzmann factor, $\exp(-\Phi_N/kT)$. The overall chain average of any function therefore converges to the canonical ensemble average of the same quantity as $n \to \infty$. Thus, instead of selecting configurations at random and weighting them with a factor

$$\exp(-\Phi_N/kT),$$

the procedure is to select configurations with a frequency proportional to $\exp(-\Phi_N/kT)$ and weight them equally. It is this modification which prevents, incidentally, a direct calculation of the partition function.

The Monte Carlo approach is particularly useful for strictly isothermal processes since the temperature is a fixed parameter, unlike the molecular dynamics case. The molecular dynamics method is best applied to isochoric (NVT) ensembles: changes in density engender changes in temperature, and a PV-isotherm is consequently somewhat difficult to follow. The Monte Carlo computations may, however, be applied equally to both isothermal–isochoric, and isothermal–isobaric (NPT)[8–11] ensembles.

The mean potential energy $\langle\Phi\rangle$ so obtained determines the molar configurational internal energy $U^\dagger$. The mean virial

$$\langle P_\mathrm{v}\rangle = \langle \sum_i \boldsymbol{r}_i . \nabla_i \Phi_N\rangle \tag{5.3.11}$$

determines the pressure according to

$$PV = RT - (\tilde{N}/3N)\langle P_\mathrm{v}\rangle, \tag{5.3.12}$$

where $\sim$ signifies a molal quantity, and the dagger † a configurational quantity. Isothermal changes of the Helmholtz free energy A can be calculated from $\langle P_\mathrm{v}\rangle$ values, and isochoric changes of the configurational entropy from $\langle\Phi\rangle$-values:

$$\Delta A = -\frac{\tilde{N}}{N}\int_{v_1}^{v_2} P\,\mathrm{d}v, \tag{5.3.13}$$

$$\Delta S^\dagger = \int_{T_1}^{T_2} \frac{C_v^\dagger\,\mathrm{d}T}{T}, \quad C_v = \frac{\tilde{N}}{N}\left(\frac{\partial\langle\Phi\rangle}{\partial T}\right)_v, \tag{5.3.14}$$

where v is the volume of the Monte Carlo system.

To reduce computing labour Wood and Parker[12] have suggested that the sum of all pair energies is replaced by one in which the potential due to a uniform density is substituted for the effect of distant particles:

$$\Phi_N = \sum_{r_{ij}\leqslant r_m} \Phi(r_{ij}) + \frac{4\pi N}{2v}\int_{r_m}^{\infty} \Phi(r)\,r^2\,\mathrm{d}r \equiv \Phi_N^*(r_\mathrm{m}). \tag{5.3.15}$$

McDonald and Singer[13] in their computations take $r_\mathrm{m} = 6.5$ Å, and thus determine $\langle\Phi_N^*(r_\mathrm{m} = 6.5\,\text{Å})\rangle$. They further correct this by computing $\Phi_N^*(r_\mathrm{m} = 6.5\,\text{Å})$ and $\Phi_N^*(r_\mathrm{m} = 8.4\,\text{Å})$ for a small number of configurations at each density and adding the mean difference of $\{\Phi_N^*(r_\mathrm{m} = 8.4\,\text{Å}) - \Phi_N^*(r_\mathrm{m} = 6.5\,\text{Å})\}$ to $\langle\Phi^*(r_\mathrm{m} = 6.5\,\text{Å})\rangle$ in all cases. They apply the same procedure to the virials. For the potential energy the 'effective' long range potential amounts to 15–20 per cent of the total. For the virial, the long range contribution is of opposite sign and of the same order of magnitude as the exactly calculated short range contribution. It is estimated that the error in Φ_N does not exceed 0.5 per cent and 5 per cent for P_v by this method.

Many of the computational details such as the effect of the number of degrees of freedom and the application of periodic boundary conditions are similar to those of the molecular dynamic method and are discussed in the preceding section. Both schemes apply to small finite systems, usually utilize the same periodic boundary conditions and generate essentially 'exact' solutions. In the molecular dynamics case asymptotically exact solutions are obtained for time averages of phase space functions by integrating over the Newtonian phase trajectories after any initial transients associated with the initial conditions have died down. In the Monte Carlo approach, however, ensemble averages are established by taking a random walk in configuration space, giving the configurational projection of the distribution in phase. The molecular dynamics trajectory, of course, is in the full position–momentum space, but both the Monte Carlo and the molecular dynamics integration over phase trajectory leads to the same asymptotic result. In the former case the states are weighted equally but the transition matrix p establishes the distribution in phase, whilst the molecular dynamics approach introduces no probablistic elements. The transition is entirely deterministic but the time average of the phase trajectory establishes a phase distribution which, it is contended, is identical to the Monte Carlo and the true equilibrium ensemble average in the limit of the time average $T \to \infty$, and $n/N \to \infty$. Thus, for the two approaches to be asymptotically equivalent we require $\langle f \rangle \equiv \bar{f}$ where

$$\langle f \rangle = \frac{\int \dots \int f(x)\, P(x)\, \mathrm{d}x}{\int \dots \int P(x)\, \mathrm{d}x} \tag{5.3.16}$$

$$\simeq \sum_{i=1}^{B} f(x_i)\, u_i \quad \text{from (5.3.10),}$$

and $$\bar{f} = \underset{T\to\infty}{\mathrm{Lt}} \frac{1}{T}\int_0^T f(t)\,dt \tag{5.3.17}$$

$$\simeq \sum_{i=1}^{B} f(T_i)\frac{T_i}{T} \tag{5.3.18}$$

which from (5.3.1) enables us to make a tentative identification of the space and time averages $\langle f\rangle$ and $\bar{f}$. The equivalence or otherwise of these two types of average will be found to be intimately related with considerations respectively of equilibrium or non-equilibrium.

We have not as yet mentioned the problem of ergodicity. The ergodicity requirement (5.3.4) that the u-step transition element $p_{ij}^{(n)}$ between two admissible states i, j be $\geqslant 0$ must essentially be preserved if the finite Markov chains are ultimately to converge on the ensemble average. Breakdown of ergodicity then, arising from *inaccessibility* of states, presents a serious problem to the Monte Carlo method, and one which must be effectively overcome. If we consider a system whose interaction potential becomes nowhere infinite, then the values of the transition element $p_{ij} = (-\Delta\Phi_N^{ij}/kT)$ is always non-zero, and between any two states i, j the n-step transition probability is likewise non-zero. All the states are therefore accessible one from another, and their totality forms one ergodic class. In this case any Markov chain with transition probabilities defined by (5.3.3) and (5.3.4) converges to the canonical ensemble in the sense mentioned above.

For realistic pair potentials $\Phi(r)$ only becomes infinite when the centres coincide, and for these transitions $p_{ij} = 0$ violating the ergodicity condition. However, these states may be specifically excluded and the remainder of configuration space will possess the property of ergodicity. Provided the cells of configuration space are fine enough the inaccessible volume is negligible. Ergodic problems develop when the regions of inaccessibility become finite, as in the case of hard spheres or discs. At high densities situations will develop when transitions become impossible and the system separates into isolated ergodic classes without transitions between them. The same problem arises in the molecular dynamic approach, and under these conditions the Markov chain or the time average of the phase trajectory does not converge to the Gibbs ensemble average. Notice that this is a problem of the statistical mechanics, not of the theory of Markovian processes: provided the attainability of states is preserved and the ergodicity condition holds the Markov chain will converge to the canonical ensemble, but for small statistical errors. Ergodic difficulties arise only for certain forms of the pair potential, and at very high packing densities: these problems require further study and will not be discussed here. However, for the hard sphere and hard disc systems, particularly at the change of phase, it is as well to be aware of the difficulties that may be associated.

It is clear that only a small number of the possible configurations that might develop in an infinite system are actually realizable in the restricted number of states accessible to the Monte Carlo system. These excluded states fall roughly into one of two classes depending as to whether the configuration of the particles is close to or far from the attainable configurations of the array. The statistical weights associated with unattainable configurations *near* to those attainable within the basic cell have correspondingly close statistical weightings, and are therefore approximately taken into account. Large scale fluctuations, however, are precluded by the very constraint of the basic cell dimension, and these states cannot contribute to the statistical sum. Such fluctuations are, however, of negligible importance and their exclusion has little effect upon systems far from the critical point.

One further point concerns the development of *quasi-ergodic* states[12]. It is sometimes found that whilst a system is formally in one ergodic class, the system in fact separates into several isolated groups with vanishingly small transition probabilities between them. Furthermore, the development of these quasi-ergodic configurations seems to be associated with the dimension of the basic array, for whilst a larger system might be perfectly ergodic, a smaller fundamental ensemble *at the same density* nevertheless fails to satisfy the ergodicity criterion which would normally be ensured by thermal rearrangement. This problem can only be effectively overcome by taking sufficiently long Markov chains such that all states are eventually attained, or alternatively to run several Markov chains with different initial conditions.

Oppenheim and Mazur[20] have investigated the effect of periodic boundary conditions on the properties of a gaseous system, and find that the first-order correction to the first and second virial coefficients is $\mathcal{O}(N^{-1})$. Thus, they obtain

$$\left.\begin{aligned} B_2 &= B_{2\infty}\left(1-\frac{1}{N}\right), \\ B_3 &= B_{3\infty}\left(1+\frac{1}{5N}-\frac{6}{5N^2}\right), \end{aligned}\right\} \tag{5.3.19}$$

where ∞ refers to the infinite system and N is the number of molecules in the basic array. The effect of periodic boundary conditions upon the pressure and internal energy of a one-dimensional system has also been shown to require a first-order correction $\mathcal{O}(N^{-1})$.

5.4 Hard discs

A system of hard discs was first investigated by Metropolis *et al.*[6], and has been subsequently re-examined by Wood[14] and Rotenberg[15] for the Monte Carlo *NVT* ensemble. Wood[10] has more recently reported hard disc calculations for the isothermal–isobaric (*NPT*) ensemble. For hard discs and hard spheres the elements of the transition matrix are either unity or zero. Thus, provided the ergodicity requirement is satisfied the transitions are all equally probable, and the generation of Markov chains is now greatly simplified: for any random displacement which does not lead to overlapping of the particles the transition is accomplished. Otherwise the atom is returned to its original position. The simplification arising in this case allows the generation of much longer Markov chains for a given investment of computer time.

The Monte Carlo[6,14,15] points and those obtained by molecular dynamic numerical integration of the equations of motion[17,18] are shown in fig. 1.4.1. The results are seen to be in substantial agreement, including what might be identified as a first-order phase transition. Also shown is the Ree–Hoover–Padé 3–3 approximant to the equation of state using the virial coefficients B_2 to B_6. The Alder and Wainwright results show the van der Waals

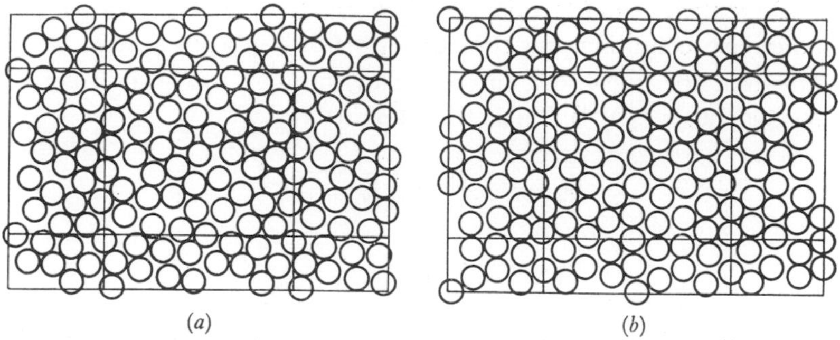

Fig. 5.4.1. Configurations from Wood's $N = 48$ Monte Carlo hard disc simulation[16]. Configuration (*a*) is associated with the upper (higher-pressure) 'fluid' *PV* locus of fig. 1.4.1 (*a*) whilst (*b*) is associated with the lower 'crystalline' branch. (Redrawn by permission from Wood, *Physics of Simple Liquids* (eds. Temperley, Rowlinson and Rushbrooke; North-Holland Publ. Co.).)

loop, and both the Monte Carlo and the molecular dynamic simulations show the solid and fluid branches. The scatter of the points about the curves is presumably a combination of an $\mathcal{O}(N^{-1})$ effect related to the variance, the possible $\mathcal{O}(N^{-1})$ ergodic effect, and an $\mathcal{O}(N^{-1})$ effect recently noticed by Hoover and Alder[18], and also by Lebowitz *et al.*[12], which seems to arise when momentum fluctuations in a molecular dynamic ensemble are suppressed.

Both Wood[16] and Alder and Wainwright[22] observe a definite correlation in the geometric arrangements of the particles and the fluctuations in pressure between the solid and fluid branches of the hard disc isotherm. In fig. 5.4.1 we show two of Wood's Monte Carlo configurations ($N = 48$), (*a*) being associated with the upper fluid locus and (*b*) with the lower crystalline locus. Transitions from one locus to the other and vice versa with corresponding variations in the atomic configuration were observed. Rotenberg[15] observed only solid–fluid transitions with none in the reverse direction: this may be due to the relatively short chains used. A particularly interesting, although unsuccessful attempt, was made by Rotenberg[15] to resolve the coexistent phases by applying a gravitational field to the configuration. His inability to resolve the phases lead him to question as to whether the phase transition was of first order. The molecular dynamics results, however, leave no doubt that the transition is of the melting–crystallization type:

identification of the type of transition follows from the coefficient of self-diffusion on both sides of the transition region.

Wood has recently reported results for an isobaric–isothermal (NPT) ensemble of hard discs[55]. The equation of state agrees within statistical error with the $P(3, 3)$ Padé approximant of Ree and Hoover based on the six-term virial expansion.

5.5 Hard spheres

The computational details involved in the study of a hard disc system are very simply extended to the case of a hard sphere ensemble in both the molecular dynamic and Monte Carlo approaches. However, the computational constraints regarding the number of atoms, the dimension of the basic cell and problems of ergodicity all become more severe in three dimensions. Fig. 3.4.1 shows the rigid sphere isotherm as calculated in the (NVT) Monte Carlo case by Rosenbluth and Rosenbluth[23], Wood and Jacobson[24], and Rotenberg[15, 25]. Molecular dynamics results have been obtained by Alder and Wainwright[26]. The behaviour of the hard sphere realizations is essentially the same as already described for hard discs: the isotherm splits into two branches, the fluid branch extending up to $V/V_0 = 0.667$, and the solid branch from densities $\gtrsim 0.625$. The usual 'indecisive' behaviour is found in the coexistence range of densities, and ergodic problems are apparently quite troublesome. The Padé (3, 3) approximant of Ree and Hoover[17] to the equation of state is also shown in fig. 1.4.1. Whilst it describes the fluid branch very well, there is no indication whatsoever of a phase transition, which is hardly surprising since the approximant cannot attempt to represent the essentially cooperative behaviour in terms of six-body integrals ($P(3, 3) = f(B_2, ..., B_6)$). The virial series up to B_5 and B_6 are also shown and are seen to underestimate the pressure: the next few virial coefficients should, presumably, be positive, and in fact Ree and Hoover[17] have shown that B_{20} is the first negative coefficient for hard spheres. The Padé approximant is not so transparently related to the virial expansion, but not until the approximant embodies collective behaviour can there be any hope of duplicating the Alder–Wainwright van der Waals loop. We might reasonably enquire why the loop develops at all, since for an infinite system the isotherm would have to be of positive, or at least zero slope in the transition region [113]. Alder and Wainwright[22] suggest that for infinite systems the loop probably derives from the fact that the constraint of constant density is imposed (by virtue of the fixed cell volume) at each density point, regardless of how many particles are dealt with. In consequence there is an upper limit on the size of crystallites which may develop: crystallites of average size greater than the number of particles dealt with being impossible to achieve. The effect of this constraint turns out to stabilize the predominant phase: thus the system was either all solid, or when a rare fluctuation disordered enough of the system, it became entirely fluid. This stabilization of the more abundant phase causes the pressure to be high on the fluid side and low on the crystal side. We must therefore conclude that a phase transition which might be complete in an infinite system is not complete in a finite system, this being so since a sizable portion of the system lies in the fluid–crystal boundary region of intermediate density, and consequently adopts more of the character of the predominant phase.

5.6 Square-well fluids

Monte Carlo calculations on the square-well fluid have been reported by Andews and Benson[108] in one dimension, and by Rotenberg[109] for three dimensions. Corresponding molecular dynamics investigations have been performed by Nelson[110]. Rotenberg also refers to some unpublished square-well results of Alder. The square-well potential is a three-parameter model, (5.6.1), the range of the attractive well being given as $g\sigma$, where g is a simple metric, its depth ϵ, and the collision diameter σ:

$$\Phi(r) = \begin{cases} \infty & r < \sigma, \\ -\epsilon & \sigma \leqslant r \leqslant g\sigma, \\ 0 & g\sigma < r < \infty. \end{cases} \tag{5.6.1}$$

In the limit $\epsilon/kT \to 0$ due either to ϵ tending to zero or T tending to infinity, we should regain the hard sphere characteristics.

The investigation of Andrews and Benson utilized the very long ranged potential $g = 150$. They demonstrated the development of dense physical clusters as the temperature was decreased, but do not report the equation of state. The development of these dense atomic clusters may be demonstrated even for much shorter ranged potentials ($g < 2$) at sufficiently low temperatures, $\epsilon/kT \ll 1$.

Nelson, using the method of molecular dynamics, and Rotenberg using the Monte Carlo approach, both make extensive investigations of the equation of state of a three-dimensional square-well fluid. Nelson also reports shear viscosity and self-diffusion data estimated from the velocity autocorrelation function (see also §§ 5.9, 6.13). Unfortunately the range parameters differ in the two computations – 1.50 and 1.85 for Rotenberg and Nelson respectively – and this makes the comparison of their results somewhat less conclusive. The virial equation of state is now given as

$$\frac{PV}{NkT} = 1 + \frac{2\pi\rho\sigma^3}{3}\left\{g_{(2)}(\sigma) - \left[1 - \exp\left(-\frac{\epsilon}{kT}\right)\right] g_{(2)}(g\sigma) g^3\right\}, \quad (5.6.2)$$

where $g_{(2)}(\sigma)$, $g_{(2)}(g\sigma)$ are understood to be functions of density. In the hard sphere case only the first term in the curly bracket is operative, and a monotonic isotherm is inevitable. The presence of the second term, however, permits the development of van der Waals loops, the density derivative of the square-well isotherm being given as

$$\frac{P'}{NkT} = 1 + \frac{2\pi\sigma^3}{3}\left\{[g_{(2)}(\sigma) + \rho g'_{(2)}(\sigma)] - g^3\left[1 - \exp\left(-\frac{\epsilon}{kT}\right)\right][g_{(2)}(g\sigma) + \rho g'_{(2)}(g\sigma)]\right\}, \quad (5.6.3)$$

where the prime represents the partial derivative with respect to density. We again see that the rigid sphere isotherm is recovered in the limit $\epsilon/kT \to 0$. In fig. 5.6.1 we plot the square-well Monte Carlo isotherms of Rotenberg for $\epsilon/kT = 0$, 0.33, 1 and 2, together with Nelson's molecular dynamic computations at $\epsilon/kT = 0.58$, 0.77. Unfortunately Nelson's results do not extend to sufficiently high densities so as to confirm the very pronounced van der Waals loops obtained by Rotenberg for $\epsilon/kT = 1, 2$. What is curious about

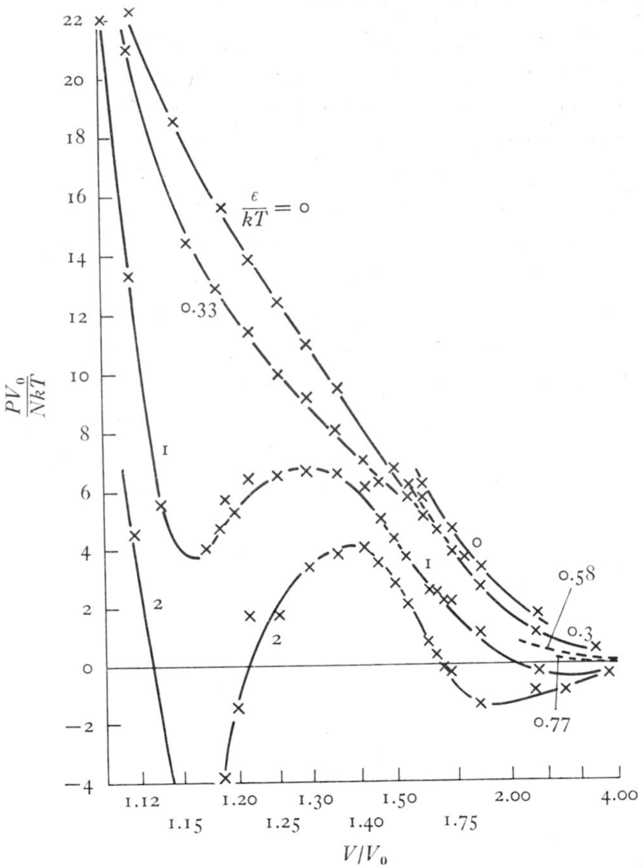

Fig. 5.6.1. Equation of state for a periodic system of square-well molecules. Broken curves represent the Monte Carlo ($N = 256$, $g = 1.85$) results of Nelson[110]. (Redrawn by permission from Wood, *Physics of Simple Liquids* (eds. Temperley, Rowlinson and Rushbrooke; North-Holland Publ. Co.).)

these loops is the second low density set occurring at $V/V_0 \sim 3$. Alder[111] has observed qualitatively similar double sets of loops over the same density range in determining the cell-theory isotherms. In that case, however, the high density loops were of much greater amplitude, and persisted up to a critical point $T_c \sim 5\epsilon/k$. Rotenberg's have vanished at $T_c \sim 3\epsilon/k$. Alder, moreover identifies the low density set with the liquid–gas transition, and the high density set with the liquid–solid transition: this, however, implies a solid–liquid critical point which is generally believed not to exist, and certainly has never been experimentally detected. Clearly this point warrants further intensive investigation.

Rotenberg points out that the development of significantly negative pressures for $\epsilon/kT = 1, 2$ must be attributed to the effect of the periodic boundary conditions. This might be a case for the implementation of Nelson's 'stochastic boundary conditions' mentioned in § 5.2. The development of unstable portions of the square-well isotherm for which $(\partial P/\partial \rho)_{N,T}$ is negative is attributed to the finiteness of the sample. These unstable branches would not establish in a larger sample wherein spontaneous mechanical fluctuations could develop.

Davis *et al.*(112) have calculated various transport properties for the square-well fluid and Nelson makes some comparison of these results with his simulations.

5.7 Lennard-Jones fluids

Two-dimensional Monte Carlo(23) and molecular dynamics(27) computations have been made for Lennard-Jones discs, but here we shall consider only the three-dimensional calculations.

Verlet and Levesque's(29) computations are carried out in the reduced quantities $r^* = r/\sigma$ and $T^* = kT/\epsilon$, otherwise the calculations are based on the experimental parameters of Michels *et al.*(28): $\sigma = 3.405$ Å, $\epsilon/k = 119.76$ °K. The two main objectives of the Lennard-Jones computations, in contrast to the essentially theoretical role of the hard sphere and square-well systems, are to simulate as closely as possible the extensive thermodynamic data, and to provide reference isotherms for the testing of the various second-order theories of the distribution function. In the former case, for non-critical states, discrepancy can be in part attributed to inadequacy of the 12–6 potential, whilst the latter provides a very severe test of the theory, the pair potential being explicitly defined. We shall begin by considering the calculated values of the thermodynamic properties of argon over a range of temperatures and densities.

In fig. 5.7.1 we show the 126 °K Monte Carlo isotherm taken

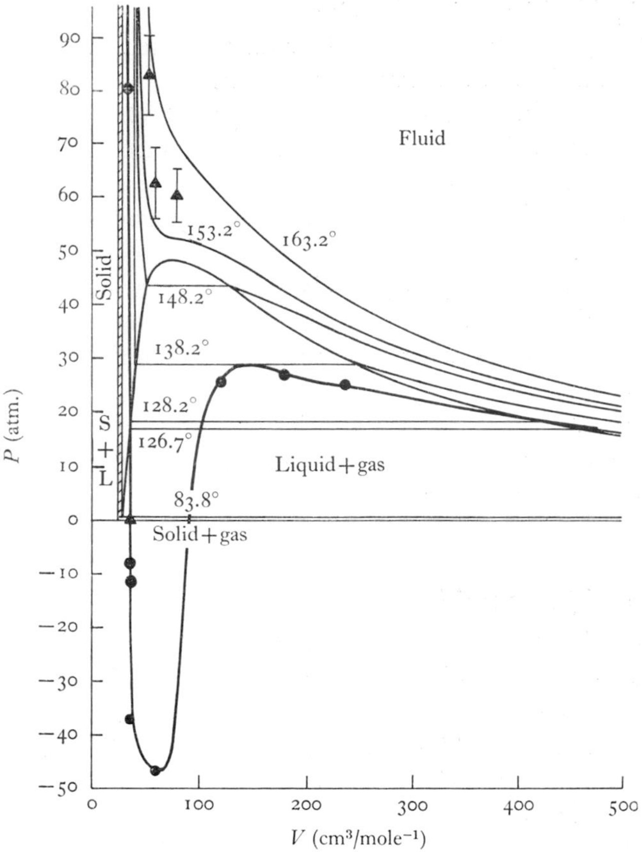

Fig. 5.7.1. Comparison of the 126 °K Monte Carlo isotherm with the experimental equations of state data on argon. (After Wood[8].) (Redrawn by permission from Wood, *Physics of Simple Liquids* (eds. Temperley, Rowlinson and Rushbrooke, North-Holland Publ. Co.).)

from table 4 of Wood's review[8]. The experimental isotherms and liquid vapour transition curves are those of Michels *et al.*[30]. The van der Waals loop is of typically asymmetrical shape having a broad positive peak approximately 10 atmospheres above the equilibrium vapour pressure, and a deep, narrow, negative phase less well defined. As Wood[8] observes, there is an inherent difficulty in calculating PV/RT for liquid states since this quantity is the difference between the statistical estimate $\langle P_v \rangle / 3NkT$ and unity. At

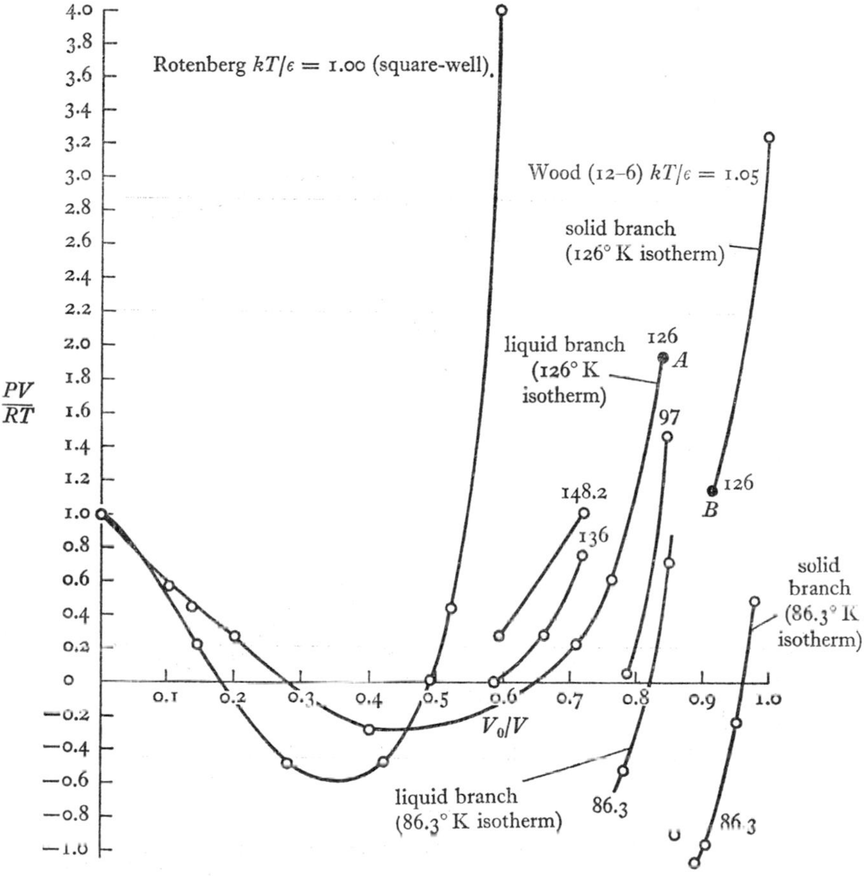

Fig. 5.7.2. Comparison of the Monte Carlo equation of state for argon (Wood[8], McDonald and Singer[13]) with Rotenberg's square-well simulation[109]. Wood takes kT/ϵ = 1.05, whilst Rotenberg takes kT/ϵ = 1.00.

liquid densities the difference is of the order of the statistical error. It seems that the negative loop is a bona fide consequence of the periodic boundary conditions, as suggested by Alder and Wainwright[26] (see § 5.5), rather than due to chance fluctuations from the 'true' positive value, and the effect seems to show no clear-cut dependence upon N. Inadequate definition of the van der Waals loop makes application of the Maxwell equal area rule difficult to

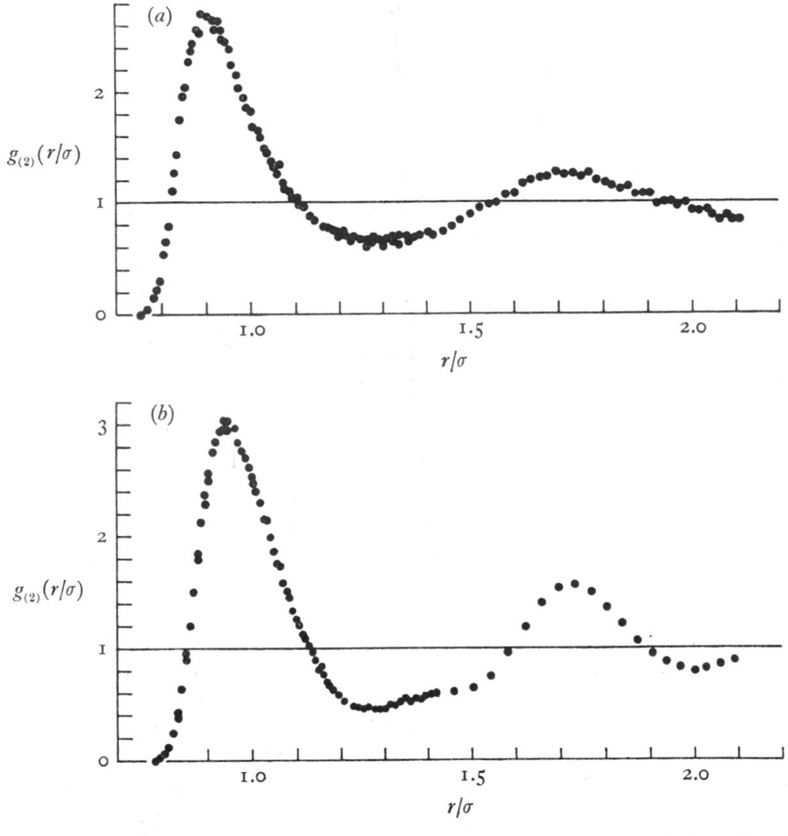

Fig. 5.7.3. The radial distribution function in the vicinity of the solid–liquid phase transition for liquid argon at 126 °K, as determined by Wood *et al.*[12] by the Monte Carlo technique. (*a*) The liquid phase, $\tau = 1.2$, $\rho = 0.84$; (*b*) the solid phase, $\tau = 1.1$, $\rho = 0.92$. (Redrawn by permission from Wood, *Physics of Simple Liquids* (eds. Temperley, Rowlinson and Rushbrooke; North-Holland Publ. Co.).)

apply, and the liquid–vapour tie line cannot be drawn in with any confidence.

Monte Carlo results taken from Wood[8], McDonald and Singer[13] for a liquid argon system, and Rotenberg's square-well computation are shown in fig. 5.7.2. The results are in substantially good agreement with experiment throughout the range of stability of the liquid. The 86.3 °K isotherm of McDonald and Singer[13], and the

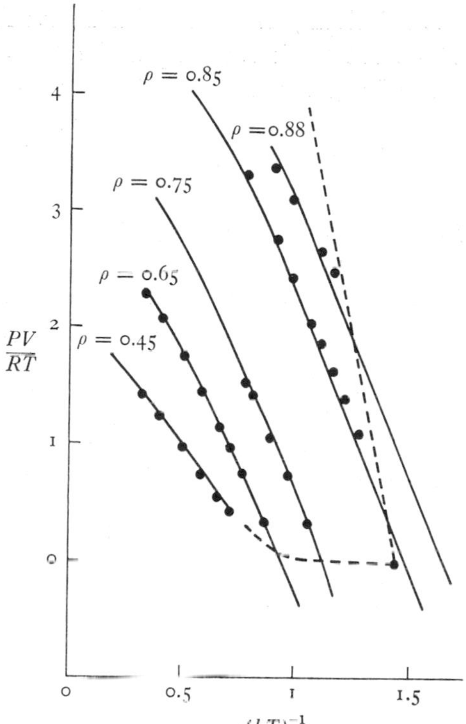

Fig. 5.7.4. Compressibility factor PV/RT as a function of temperature for argon. The points are the molecular dynamics results of Verlet[31]; the curves are the experimental results of Michels *et al.*[28]. The broken lines represent the gas–liquid, liquid–solid coexistence curves. (Redrawn by permission from Verlet, *Phys. Rev.* **159**, 98 (1967).)

126 °K isotherm of Wood[8], show solid and liquid branches. Wood and Parker[12] have examined the 'crystallography' of the points A and B on the 126 °K argon isotherm. The $\rho = 0.92$ configuration was recognized as being essentially fcc with small displacements, relative to the lattice spacing, of the molecules from their initial fcc lattice positions. There were no intermolecular interchanges. On the other hand, the $\rho = 0.84$ configuration was basically liquid-like with randomly large atomic displacements from their original positions. A comparison with the experimental solid–liquid co-existence boundary shows that the $\rho = 0.92$ state is almost certainly on a metastable extension of the solid phase isotherm. Wood,

Parker and Jacobson[34] display the radial distribution function at various densities. The distributions corresponding to $\rho = 0.92$ and $\rho = 0.84$ are shown in fig. 5.7.3 and bearing in mind the essentially fcc configuration on the solid branch isotherm, the $\rho = 0.92$ distribution function is disappointingly structureless. Wood[8] points out, however, that the determinations are for $N = 32$, so that beyond $r/\sigma = \sqrt{2}$ the structure is directly influenced by the periodicity imposed on the system.

Verlet's[31] molecular dynamics calculations again for liquid argon, show excellent agreement with experiment[28] (fig. 5.7.4). There are small systematic discrepancies between the experimental and calculated internal energies, but these are removed if the depth of the potential well is reduced by only 2 per cent[33]. In Rotenberg's[25] calculations the square-well interaction was given by

$$\left.\begin{aligned} \Phi(r) &= \infty & r &< \sigma, \\ \Phi(r) &= -\epsilon & \sigma &< r < 1.5\sigma, \\ \Phi(r) &= 0 & r &> 1.5\sigma \end{aligned}\right\} \qquad (5.7.1)$$

and consequently the equation of state is

$$\frac{PV}{NkT} = 1 + \frac{2\pi\rho\sigma^3}{3}\{g_{(2)}(\sigma) - [1 - \exp(-\epsilon/kT)](1.5)^3 g_{(2)}(1.5\sigma)\}, \qquad (5.7.2)$$

where $g_{(2)}(\sigma)$, $g_{(2)}(1.5\sigma)$ is understood to be a function of density. Clearly this system will exhibit ideal behaviour in the limit $\rho \to 0$ and when

$$g_{(2)}(\sigma) = [1 - \exp(-\epsilon/kT)](1.5)^3 g_{(2)}(1.5\sigma),$$

this identifying the Boyle temperature. Furthermore, the isotherm will cross the density axis when

$$\{(1.5)^3 [1 - \exp(-\epsilon/kT)] g_{(2)}(1.5\sigma) - g_{(2)}(\sigma)\} = 3/2\pi\rho\sigma^3.$$

Inasfar as the square-well virial is an adequate representation of the corrresponding Lennard-Jones function (fig. 5.7.5), the square-well fluid affords considerable insight into the thermodynamics of Lennard-Jones systems. It is evident that the positive component of $g_{(2)}(r)\,(\mathrm{d}\Phi(r)/\mathrm{d}r)$ is responsible for isotherms having $PV/RT < 1.0$.

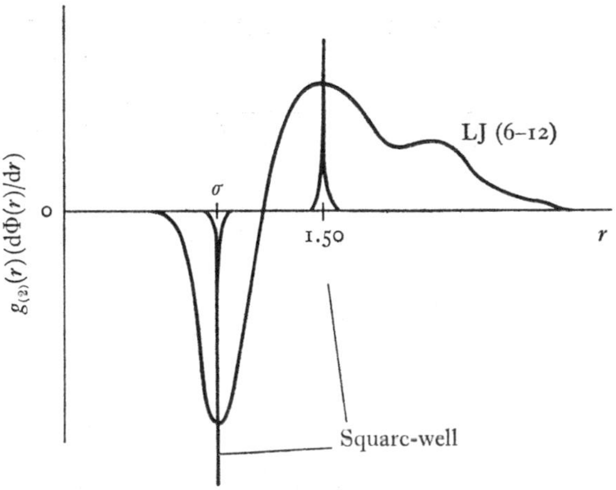

Fig. 5.7.5. Comparison of the Lennard-Jones and square-well virials $g_{(2)}(r)\,(d\Phi(r)/dr)$. For hard sphere interaction only the negative delta function at $r = \sigma$ is developed.

It is quite clear that Rotenberg's square-well isotherm in fig. 5.7.2 qualitatively reproduces the Lennard-Jones isotherm at the same reduced temperature ($T^* = kT/c \sim 1.0$). However, the temperature-independent collision diameter in the square-well case prevents the optimal location of the negative δ-function in fig. 5.7.5. The square-well fluid, unlike its hard sphere counterpart, separates into two fluid phases below its critical temperature.

Verlet and Levesque[29] have calculated the LJ (6–12) Monte Carlo isotherm at $T^* = 1.32$, 1.35 and 1.4, and they conclude from a combination of Monte Carlo calculations and calculations based on the PY 2 integral equation that T^*_{crit} should lie between the reduced limits 1.32 and 1.36 for a LJ (6–12) system. The observed reduced critical temperature of 1–26 is interpreted as an inadequacy of the Lennard-Jones potential, although suppression of the density fluctuations which characterise the critical point of macroscopic systems is undoubtedly partially responsible in the case of the small periodic assemblies investigated numerically. What is of interest, however, is the *increase* in the discrepancy when presumably better intermolecular potentials are used in the PY 2 calculations.

In table 5.7.1 we compare the thermodynamic properties of

TABLE 5.7.1. *Thermodynamic properties of argon (after McDonald and Singer*[13]*)*

T (°K)	V (cm³)	Mean p.e. $\left\langle \frac{-\Phi}{N} \right\rangle$ (erg × 10¹⁴)	Mean virial $\left\langle \frac{P_v}{N} \right\rangle$ (erg × 10¹⁴)	P (dyn cm⁻² × 10⁻⁸) MC	P expt	$\frac{-U}{RT}$ MC	$\frac{-U}{RT}$ expt	$\frac{C_v}{R}$ MC	$\frac{C_v}{R}$ expt	$-\frac{1}{V}\left(\frac{\partial V}{\partial P}\right)_T$ (cm² dyn⁻¹ × 10¹⁰) MC	expt	$\left(\frac{\partial P}{\partial T}\right)_V = \left(\frac{\partial S}{\partial V}\right)_T$ (dyn cm⁻² deg⁻¹ × 10⁻⁷) MC	expt	$\frac{1}{V}\left(\frac{\partial V}{\partial T}\right)_P$ (deg⁻¹ × 10³) MC	expt
86.3	24.78	11.90	1.64	1.57		9.98		0.91		1.60		1.74		2.78	
	25.60	11.51	4.12	−0.43		9.66		0.96		1.66		1.82		3.02	
	26.44	11.15	6.68	−2.36		9.36		1.27		1.71		2.44		4.17	
	26.90	10.93	7.32	−2.79		9.17		1.23							
	27.50	10.56	6.36	−2.03		8.85		1.28							
	27.50	10.30	0.66	2.13	2.10	8.64		1.35		1.87	1.73	2.30	1.94	4.30	3.36
	28.03	10.12	2.51	0.76	0.92	8.49		1.03		2.17	1.91	1.73	2.07	2.75	3.95
	28.48	9.95	3.34	0.17	0.09	8.35		1.09		2.34	2.06	1.77	2.15	4.14	4.43
	28.95	9.76	4.18	−0.42	−0.67	8.19		1.02		2.50	2.19	1.72	2.22	4.30	4.86
	29.68	9.50	5.55	−1.34		7.97		0.85		2.76		1.40		3.86	
	30.92	9.14	7.58	−2.60				0.68		3.10		1.05		3.32	
97.0	27.50	10.14	−2.21	4.55	4.24	7.57		0.88		1.88	1.94	1.52	1.21	2.86	2.35
	28.03	9.97	0.01	2.87	3.01	7.44		0.93		2.18	2.14	1.47	1.39	3.20	2.97
	28.48	9.81	0.85	2.23	2.16	7.32		0.83		2.31	2.29	1.42	1.50	3.28	3.44
	28.95	9.62	1.60	1.68	1.39	7.18		0.91		2.45	2.43	1.50	1.59	3.68	3.86
	29.68	9.38	3.43	0.40	0.35	7.01		0.71		2.71	2.64	1.18	1.69	3.20	4.46
108.2	28.48	9.59	−2.87	5.61		6.41		1.13							
	29.68	9.24	0.78	2.54	2.29	6.18		0.78							
136.0	33.51	7.97	1.42	2.52	2.37	4.24	4.15	0.56	0.58	4.73	4.53	0.83	1.21	3.91	5.26
	36.58	7.33	4.20	0.79	0.95	3.90	3.82	0.47	0.57	6.70	8.30	0.60	0.97	4.03	8.08
	40.00	6.72	5.88	−0.12	0.28	3.58		0.46	0.58	17.31	23.50	0.56	0.61	9.76	14.29
148.2	33.51	7.87	−0.35	3.89		3.85	3.75	0.61	0.61	2.89		0.88		2.55	
	35.00	7.58	1.68	2.49	2.79	3.69	3.60			4.27	4.72				
	36.58	7.25	2.80	1.93	2.01	3.54	3.46	0.50	0.57	6.50	6.58	0.64	0.89	4.16	5.86
	38.19	6.95	3.38	1.45	1.52	3.39	3.32			9.15	9.66				
	40.00	6.64	4.54	0.80	1.13	3.25	3.18	0.44	0.57	15.42	16.73	0.54	0.65	8.25	10.87

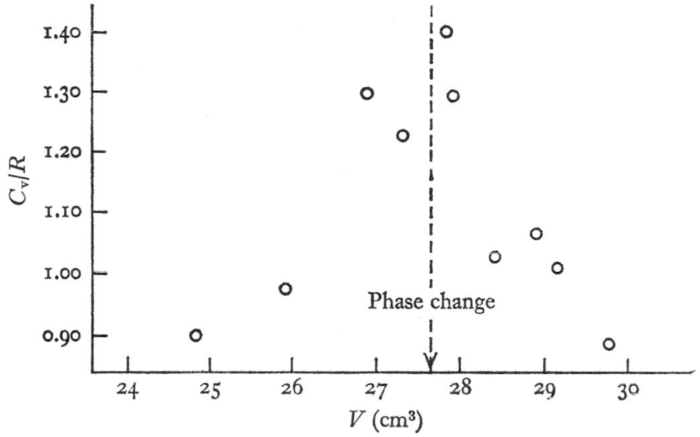

Fig. 5.7.6. Pronounced maximum in the isochoric specific heat of argon at the phase change, 86.3 °K taken from the Monte Carlo data of McDonald and Singer[13].

liquid argon with experiment. The data have been taken from McDonald and Singer[13]. McDonald and Singer[13] also estimate the latent heat of vaporization at 86.3 °K from the experimental vapour pressure[35] (0.91×10^6 dyn cm^{-2}) and their Monte Carlo data. They obtain a value of 1543 cal mole^{-1} compared with the measured value[35] of 1564 at 85.7 °K. For the latent heat of fusion they obtain 246 cal mole^{-1} compared with the experimental[35] value 284 cal mole^{-1} (melting pressure[36] at 84 °K, 1.03×10^8 dyn cm^{-2}). Since the molar specific heat is determined by fluctuations about the mean

$$C_v = \tfrac{3}{2}R + \frac{1}{kT^2}(\langle\Phi^2\rangle - \overline{\Phi}^2), \qquad (5.7.3)$$

where the overbar represents averaging along the Markov chain, it might be anticipated that a deterioration in agreement with experiment will occur as critical fluctuations become progressively more important. This is indeed the case as seen in table 5.7.1. The computed and observed values are in good agreement at 136 and 148.2 °K, but become significantly poorer as the critical volume is approached ($V_{\text{crit}} \sim 53.2$ cm^3) as the periodic boundary suppresses all large density fluctuations. The pronounced maximum in the

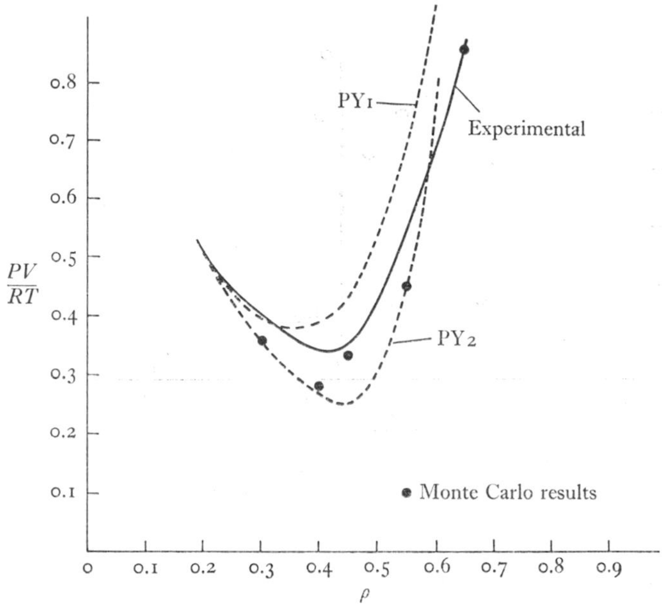

Fig. 5.7.7. Comparison of the Monte Carlo and experimental equations of state for liquid argon at the critical temperature, $T^* = 1.35$. Also shown are the PY1 and PY2 isotherms for argon (Verlet and Levesque[29]; Verlet[31]).

vicinity of the phase change shown in fig. 5.7.6 is indicative of large fluctuations in the molecular arrangement: a similar phenomenon at the critical point giving rise to an abnormally large specific heat is observed experimentally, but the constraints of periodicity prevent its machine simulation.

We now come to consider the role of machine calculations in evaluating the various integral equations for the distribution function. In this case there can be no ambiguity, and discrepancies cannot be attributed to inadequacy of the pair potential, although, as we have seen, the LJ 6–12 interaction with the parameters of Michels *et al.* appears to give a very satisfactory representation of the experimental data.

Verlet and Levesque[29] have calculated the reduced isotherm $T^* = 1.35$ over a range of densities using the Monte Carlo method. The equation of state on the basis of the PY1 and PY2 (Verlet) integral equations is compared with this isotherm (fig. 5.7.7).

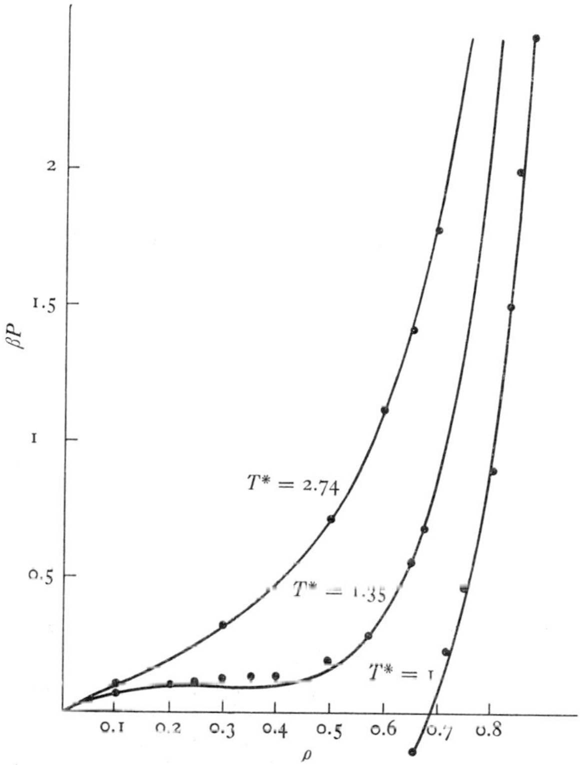

Fig. 5.7.8. Verlet's reduced molecular dynamics isotherms. The $T^* = 1.35$ isotherm is the most nearly critical, and it is this locus which is compared with the PY1 and PY2 estimates in fig. 5.7.7.

Using a reduced LJ 6–12 interaction both theoretical curves appear equally at variance with the experimental data. However, more significantly, the PY2 isotherm is seen to be in much better agreement with the Monte Carlo data than the PY1 curve, and it is this that enables us to conclude that the PY2 theory is a valid second approximation to the PY1 equation, yielding a consistent improvement over the first-order theory.

Verlet[31] concludes that the overall agreement between the machine calculations and experiment is surprisingly good: it appears that the Lennard-Jones potential is a quite satisfactory interaction as far as the equilibrium properties are concerned.

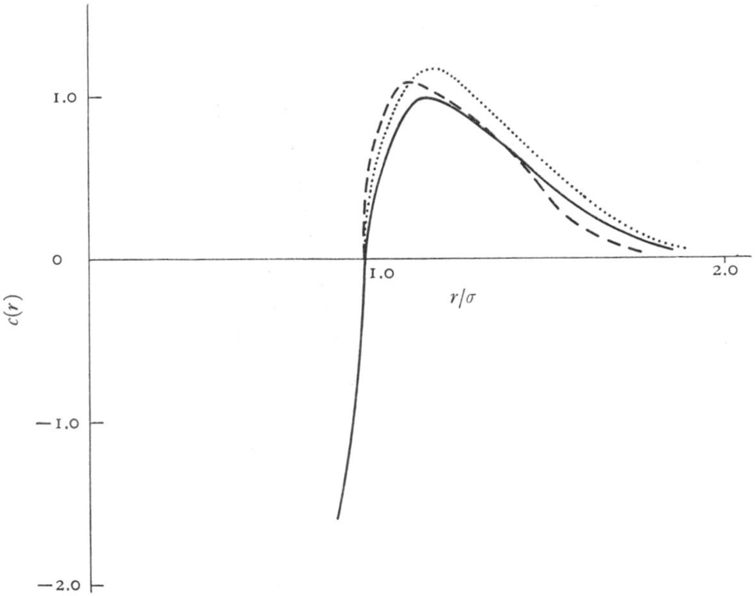

Fig. 5.7.9. The direct correlation function $c(r)$. The Mayer f-function

$$(\exp(-\Phi/kT)-1)$$

to which the direct correlation tends at large r is shown by the dotted line, whilst the full curve represents the results of Verlet's molecular dynamics calculations[46]. The broken curve represents $c(r) = g_{MD}(r)/[\exp(-\Phi(r)/kT)-1]$, where $g_{MD}(r)$ is the pair distribution function obtained from molecular dynamics.

Verlet observes, however that if xenon were chosen for the comparison instead of argon, the agreement would not be so good. For instance, for $T^* = 1.35$, $\rho = 0.75$ the value of PV/RT from molecular dynamics is 0.86; experimentally the same quantity is 0.86. For xenon (with the reduction parameters $\epsilon/k = 225.3$ °K, $\sigma = 4.07$ Å) the value 1.05 is obtained[51]. In fig. 5.7.8 we show Verlet's three molecular dynamic isotherms at $T^* = 2.74$, 1.35 and 1.0. Clearly the $T^* = 1.35$ isotherm is the most nearly critical and it is this one which is compared with the PY1, PY2 theories in fig. 5.7.7.

In a more recent paper, Verlet[46] investigates the direct correlation function, $c(r)$. First, the effect of truncating the interaction potential at some r_v is investigated. It is shown that the effect of the tail of the potential for $r > r_v$, which has been neglected in the molecular dynamics calculation, would not have changed $g_{(2)}(r)$ appreciably for $r < r_v$ had it been included. The results can thus be extended to an uncut potential.

Then, a procedure is developed to extrapolate $g_{(2)}(r)$ to any r since in determining the structure factor $S(k)$ as the Fourier transform of $h(r) = g_{(2)}(r) - 1$ it is inadvisable to use truncated distribution functions. Verlet finds some $r_c \leqslant r_v$ beyond which $h(r)$ is small. For $r < r_c$ we have the molecular dynamics results, whilst for $r > r_c$ we may presumably make an analytic extension using the PY relation

$$c(r) = g_{(2)}(r)\left\{\exp\left(\frac{\Phi(r)}{kT}\right) - 1\right\} \tag{5.7.4}$$

and the Ornstein–Zernike equation

$$h(r) = c(r) + \rho\int h(r')\, c(|\mathbf{r}-\mathbf{r}'|)\, d\mathbf{r}'. \tag{5.7.5}$$

The OZ relation enables us to continue $g_{(2)}(r)$ outwards and $c(r)$ inwards, and thus these two correlation functions are known for all r. The use of the PY equation to get the potential once $g(r)$ and $c(r)$ are known can be put to a direct test. The Lennard-Jones potential would be recovered if the PY equation were exact. The results show that at densities around the critical point ($\rho < 0.5$) the Lennard-Jones potential is recovered within a few per cent. An example of the direct correlation function calculated by the above procedure is shown in fig. 5.7.9 for a Lennard-Jones fluid.

The direct correlation shows the form expected intuitively an excluded volume for $r < 1$, and a maximum located on the potential minimum, within a few percent. For large r, $c(r)$ tends to the Mayer f-function, and this is shown by the dotted curve in fig. 5.7.9 however the f-function over-estimates the direct correlation at small r, and this is of course to be expected since the de-correlating effect of the third and subsequent neighbours is neglected in the Mayer approximation. The broken line represents the Percus–Yevick approximation to the direct correlation function. The radial distribution function used to estimate $c_{PY}(r)$ from (5.7.5) is that obtained by molecular dynamics.

As mentioned above, once $c(r)$ and $g_{(2)}(r)$ are known the validity of the PY equation can be tested by using it to calculate a potential

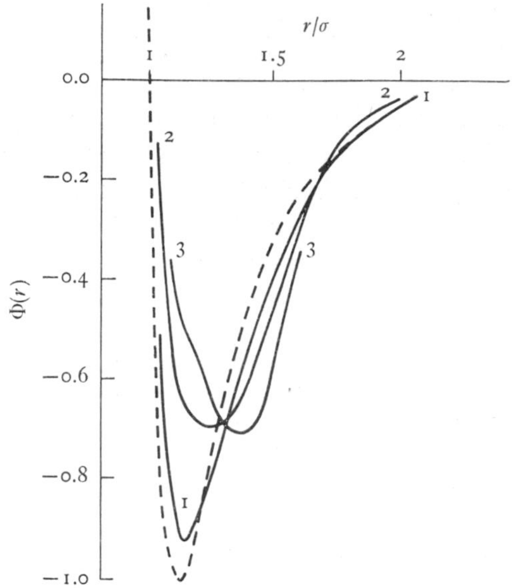

Fig. 5.7.10. Potential obtained from the correlation function using the PY equation. Curve 1: $T^* = 1.328$, $\rho = 0.5426$. Curve 2: $T^* = 1.05$, $\rho = 0.75$. Curve 3: $T^* = 1.127$, $\rho = 0.85$. The Lennard-Jones (6–12) potential which would be recovered if the PY equation were exact is shown by the broken curve. (Redrawn by permission from Verlet, *Phys. Rev.* **165**, 201 (1968).)

which should be strictly equal to the initial Lennard-Jones potential if that equation were exact. Thus one can know when the PY equation can be used to obtain the potential using experimental data.

For densities around the critical, Verlet[46] concludes that the potential can be recovered to within 1 per cent in the vicinity of the minimum, and never in greater error than 4 per cent for large r. The discrepancy becomes progressively more serious with increasing density as shown in fig. 5.7.10. This casts considerable doubt on the reliability of potentials determined by inversion of the PY equation, particularly in the case of liquid metals. Determination of pair potentials by the molecular dynamic method are considered in greater detail in the next section (§ 5.8).

With $g_{(2)}(r)$ analytically extended as described above, Verlet then examines the location of the first peak k_0 of the Fourier transform of the total correlation function $h(k)$. The location of the first peak appears to be an excluded volume effect whereupon we may expect k_0 to be of the order of $2\pi/\sigma$. The short range correlation which follows may, apparently, be described in terms of a damped sine wave of period $r_0 = 2\pi/k_0$. More generally, Verlet analyses the long range form of $h(r)$ in terms of the dominant poles of $h(k)$ and $c(k)$.

Recently Fehder[52] has investigated 'anomalies' which are often observed in the radial distribution function of simple liquids. These small subsidiary features on the experimental curves have been the source of considerable speculation and controversy. Unfortunately the situation is obscured by the complexity of the analytical procedures that must be amployed to obtain $g_{(2)}(r)$ from experimental data. Several authors[53] have attributed the subsidiary phenomena to the finite truncation of the Fourier inversion integral, whilst others[54] believe that these irregularities lie in the data themselves. Moreover, the subsidiary structure appears in some measurements whilst not in others.

Fehder's molecular dynamics simulation of a two-dimensional Lennard-Jones system over a range of temperature and density states enables any subsidiary phenomena associated with the principal form of the distribution function to be identified without the complications engendered by Fourier inversion. Wood[8] does not discuss the 'anomalous' structure of his distribution functions obtained by the Monte Carlo method, although such subsidiary phenomena are clearly visible. However, the small replication distance ($r/\sigma = \sqrt{2}$) and the small number of particles ($N = 32$) in his studies does not permit conclusive identification.

Fehder observes three *persistent* subsidiary features in addition to the principal structure of the radial distribution function. These are described below:

Feature I: A small shoulder at the base of the first principal maximum occurring at a reduced radius of 1.55–1.60. A slight shift to smaller radius is observed with decreasing temperature. The feature appears only very weakly in the functions for densities above 0.70, and at low densities is frequently obscured by broadening of the first peak.

Feature II: A small shoulder at the base of the second peak, generally centred at a radius of about 1.85 although a shift to larger radius is observed with decreasing temperature. This feature is not clear for reduced densities above 0.70, and at lower densities it is absorbed into the second maximum at low temperatures.

Feature III: A shoulder or small sub-peak on the large-radius side of the second principal maximum is apparent at a radius of about 2.4, but shifts to a larger radius with decreasing temperature. This feature is not seen at high densities and may disappear with temperature broadening of the second maximum at low densities.

The recurrent appearance of these features in the simulated distribution leads Fehder to conclude that they are not artefacts but bona fide subsidiary structure. Graphical displays suggest that Features I and II may be attributed to the presence of alternative configurations for local ordering within the model fluid. The two-dimensional hexagonal close-packed configuration is the most prevalent mode of ordering observed in the microstructure of the model. But at densities below 0.65, small local groups of particles are also frequently observed in the square close-packed configuration. If we assume that the nearest neighbour distance in both configurations would then be 1.59 and 1.94 respectively, these distances correspond closely to the radii at which Features I and II appear.

The disappearance of Feature I at high densities can be explained in terms of the difference in the packing densities for the ideal hexagonal and square configurations, but the similar disappearance of Feature II is not so easily understood. Fehder, however, suggests a Lennard-Jones–Devonshire (LJD) smeared-cell effect as a possible explanation. The mean potential within a cage of nearest neighbours may, over a limited range of relatively high density, develop a flat bottom. On the LJD model this occurs for reduced densities of about 0.7. The well is found to be essentially flat over a range of about 0.2σ; at densities higher and lower than this figure a different situation prevails. This kind of effective potential could lead to a lesser degree of local order than is present either at higher or lower densities. Fehder[52] points out that Feature II disappears from the radial distribution function at some reduced density between 0.63 and 0.70.

Thus, as the density of a simple fluid is increased three distinct stages are apparent in the detail of the atomic microstructure. At low densities the attractive component of the interparticle potential is the principal ordering agent, while at very high densities, geometric effects arising from the repulsive component of the potential become dominant. For a narrow range of moderately high densities, however, neither component may be effective over at least short distances and hence, in this narrow range of densities the fluid may show a lesser degree of short range order than for either higher or lower densities.

5.8 Pair potentials

The inversion of the various integral equations relating $\Phi(r)$ to $g_{(2)}(r)$ to yield the effective interatomic pair potential has been discussed in §§3.6 and 5.7. In particular, Johnson, Hutchinson and March (JHM)[37] have shown that inversion of experimental scattering data by means of the BGY and PY integral equations yields a long range oscillatory (LRO) pair potential for a number of metals, but gives the usual van der Waals potential for liquid argon. Now while LRO potentials have been derived theoretically[38] and observed experimentally[39] for solids, it is not obvious that these results can be extrapolated to the liquid state. Further, as we observed in §3.6, the analysis itself may not be capable of yielding accurate enough information to give the pair potentials quantitatively[40].

Paskin and Rahman[41] have investigated the dynamics and structure of liquid sodium using various LRO potentials by the method of molecular dynamics. The JHM procedure was to deduce the effective pair potential from a knowledge of the radial distribution function: Paskin and Rahman consists in going from an assumed form of pair potential to $g_{(2)}(r)$. An estimate of the self-diffusion coefficient is also made.

The molecular dynamic system consisted of 686 particles in a cubic array of side 30.449 Å at a packing density of 0.0243 atoms Å^{-3}. The parameters were chosen to fit sodium at 373 °K. In fig. 5.8.1 (*a*) we compare the various LRO potentials. The circles with error bars represent average values obtained by JHM using the BGY

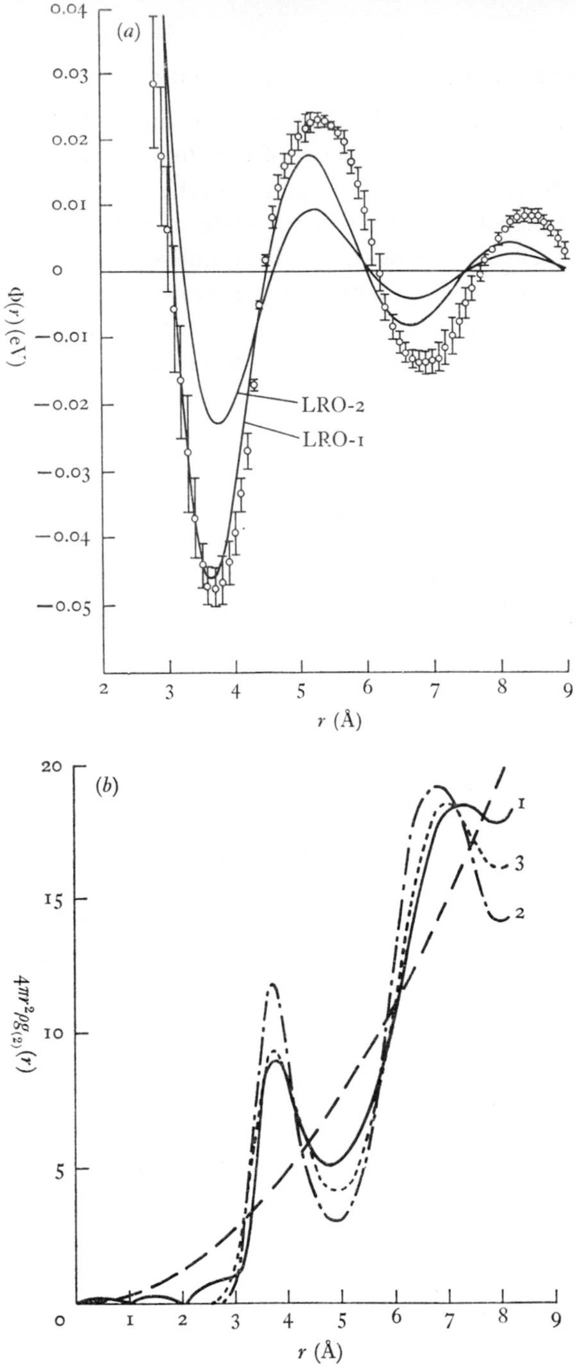

Fig. 5.8.1. (*a*) Comparison of pair potentials used for liquid sodium by Paskin and Rahman (LRO-1, LRO-2) with the BGY inversion of experimental data by Johnson, Hutchinson and March. (*b*) The function $4\pi r^2 \rho g_{(2)}(r)$ for liquid sodium. Curve 1 is derived from neutron scattering data. Curves 2 and 3 represent the LRO-1 and LRO-2 simulations, respectively. (Redrawn by permission from Paskin and Rahman, *Phys. Rev. Letters*, **16**, 300 (1966).)

TABLE 5.8.1. *Parameters appearing in equation (5.8.1)*

	A (eV)	β
LRO-1	0.048	0.5954
LRO-2	0.027	0.5689

equation on data obtained at 114 and 203 °C. Paskin and Rahman used two model potentials: LRO-1 was chosen so as to fit the JHM potential at the first minimum, whilst the LRO-2 potential was chosen to give a better fit to the observed radial distribution function[(42)]. The LRO potentials used were of the form

$$\Phi(r) = \Phi_{\text{LRO}}(r) + \Phi_{\text{R}}(r), \tag{5.8.1}$$

where
$$\Phi_{\text{LRO}}(r) = -A(r_0/r)^3 \cos\{7.812[(r/r_0)+\beta]\} \tag{5.8.2}$$

and
$$\Phi_{\text{R}}(r) = 0.78 \exp(5.0724 - 10.7863r/r_0)\,\text{eV}. \tag{5.8.3}$$

Here $r_0 = 3.72$ Å and A and β are taken as adjustable parameters for varying the LRO part of the potential. The values used are shown in table 5.8.1. The Born–Mayer repulsive part of the potential, Φ_{R}, was included to take care of the hard ionic core and ensure the desired behaviour at small distances. The parameters of Φ_{R} were those appropriate to sodium[(43)], and were the same in LRO-1 and LRO-2.

A comparison of theory (with potentials truncated at $r = 8.18$ Å) and experiment[(42)] is shown in fig. 5.8.1 (*b*). It is seen that the LRO-1 potential ($\sim \Phi_{\text{JHM}}$) yields a much sharper $\rho g_{(2)}(r)$, and in fact deviates by a factor of about 1.6 from the mean, than does the experimental curve. The LRO-2 curve, of course, was chosen to fit the curve better. Paskin and Rahman observe that keeping the hard core form of the pair potential constant, the attractive LRO branch of the interaction can have an appreciable effect near the melting point, which suggests that the hard sphere approximation may be somewhat less adequate than was hitherto supposed. Also, the LRO-2 potential, rather than the LRO-1 (or JHM) potential appears to yield the best $\rho_{(2)}(r)$ and in fact the LRO-2 form has a quantitative similarity to Cochran's potential[(43)] in the region of

TABLE 5.8.2. *Diffusion coefficients for liquid sodium*

	$cm^2\ s^{-1}$
D (LRO-1)	1.9×10^{-5}
D (LRO-2)	5.8×10^{-5}
D (expt)[44]	4.2×10^{-5}

the first minimum, deduced from the lattice dynamics of solid sodium.

However, agreement between LRO-2 and the observed structure is not a sufficient criterion for accepting the latter potential as a good approximation[43], and Paskin and Rahman go on to calculate the self-diffusion coefficient for liquid sodium interacting via the LRO-1 and LRO-2 potentials. Their results are summarized in table 5.8.2. It is seen that the experimental value lies between the two theoretical values: this does not enable us to decide finally upon the form of the pair potential. However, it does appear that use of the pair potential derived from the BGY and PY inversion of experimental data does not allow us to recover the radial distribution function. Further, oscillations in the diffusion coefficient at small times suggests that a LJ (6–12) form of interaction is inappropriate.

Ashcroft and Lekner[45], in the case of the liquid metals, and Verlet[46] in the case of the Lennard-Jones fluids, not to mention the various hard sphere perturbation theories[47], conclude that it is the repulsive part of the potential function which is primarily responsible for the essentially geometric, excluded volume effects which establish the liquid structure in the vicinity of the triple point. This is a conclusion nearly opposite to that of JHM. Schiff[47] has made an extensive investigation of the effect on the structure factor $S(k)$ and the velocity autocorrelation function $\psi(t)$ for a range of potential functions which may be regarded as caricatures of the true effective liquid metal pair potential. The hard core diameter σ and the well depth ϵ were chosen such that the density and temperature of the liquid metal, when expressed in units of σ^{-3} and ϵ, be of the order of the reduced density and temperature of liquid argon at its triple point. This ensures that the thermodynamic properties

TABLE 5.8.3. *Liquid metal model potentials as used by Schiff*[47]

Potential	Remarks
$\Phi_1(r) = [(\cos 2k_{F1}r)/r^3]$ $\times(A_1+B_1/r^2+C_1/r^4)$ $+[(\sin 2k_{F1}r)/r^4](D_1+E_1/r^2)$	Only potential used by Schiff having oscillations of 'realistic' amplitude. Oscillations fit those given theoretically by Pick[48]. Repulsive core varies as r^{-7} and is intermediate between the soft Born–Mayer core and the LJ hard core
$\Phi_2(r) = A_2(\cos 2k_{F2}r+B_2)/r^3$ $+E_2 \exp(F_2-G_2r/r_0)$	This is the LRO-2 function of Paskin and Rahman[41]. Oscillations are of wavelength $2k_F = k_0$, where k_0 is the position of the first peak of the structure factor. In fact the value of $2k_F$ appropriate to liquid sodium is $0.9k_0$. Core is of 'soft' Born–Mayer form
$\Phi_3(r) = [(\cos 2k_{F3}r)/r^3](A_3+B_3/r^2)$ $+[(\sin 2k_{F3})r/r^4](C_3+D_3/r^2)$ $+E_3 \exp(F_3-G_3r/r_0)$	Same Born–Mayer core as $\Phi_2(r)$, but with deeper bowl and larger amplitude oscillations
$\Phi_4(r) = [(\cos 2k_{F4}r)/r^3](A_4+B_4/r^2)$ $+[(\sin 2k_{F4}r)/r^4](C_4+D_4/r^2)$ $+E_4/r^{12}$	As $\Phi_3(r)$, but with a Lennard-Jones hard (r^{-12}) core
$\Phi_5(r) = [(\cos 2k_{F5}r)/r^3]$ $\times(A_5+B_5/r^2+C_5/r^4)$ $+[(\sin 2k_{F5}r)/r^4](D_5+E_5/r^2)$ $+F_5/r^{12}$	Similar to $\Phi_1(r)$, but with hard Lennard-Jones core, and used for aluminium

TABLE 5.8.4. *Parameters of the different two-body liquid metal model potentials* (Schiff[47])

		A	B	C	D	E	F	G	r_0	$2k_F$
Sodium	Φ_1	0.19	1.02	0.08	0.43	2.54				5.987
	Φ_2	−0.78	0.57			15.11	5.07	10.79	1.15	6.82
	Φ_3	−0.42	0.56	2.96	1.46	15.11	5.07	10.79	1.15	5.987
	Φ_4	0.32	5.24	5.39	5.76	4.06				5.987
Aluminium	Φ_5	0.66	4.22	2.61	0.54	0.67	1.49			8.97

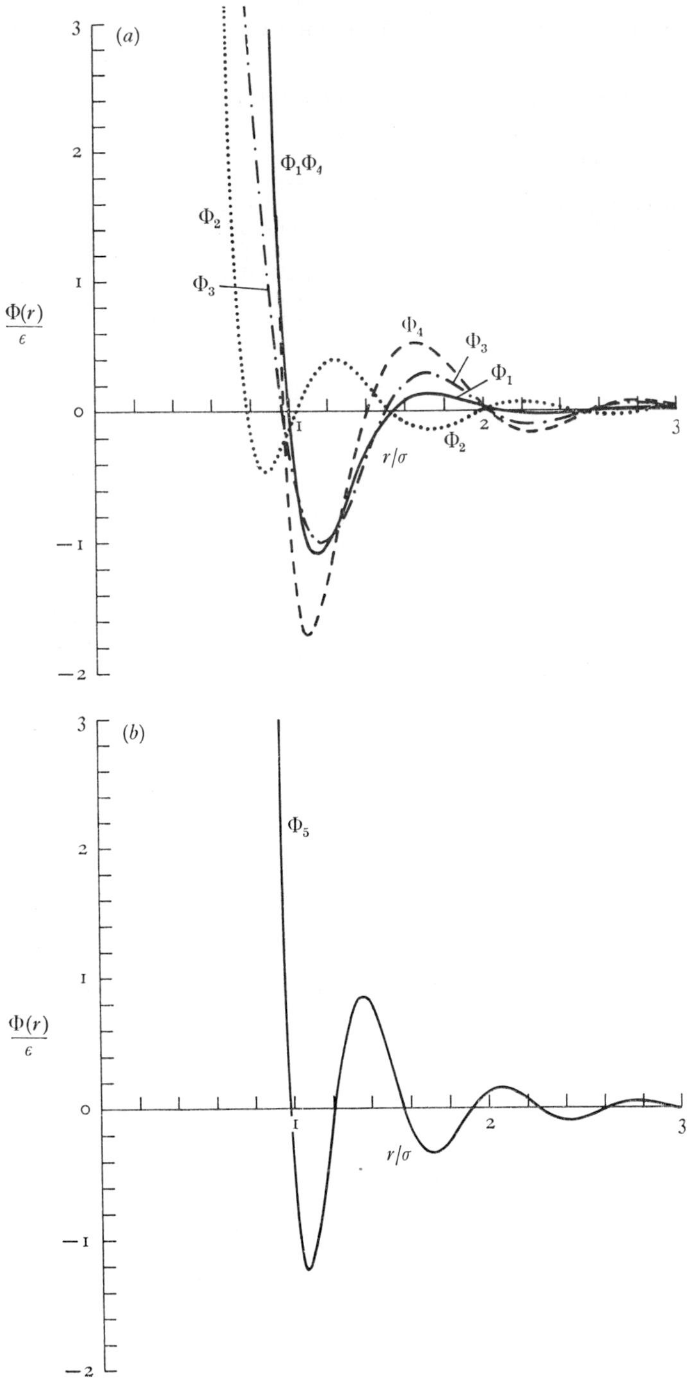

Fig. 5.8.2. (*a*) The liquid sodium potentials listed in table 5.8.3. (*b*) The liquid aluminium potential listed in table 5.8.3.

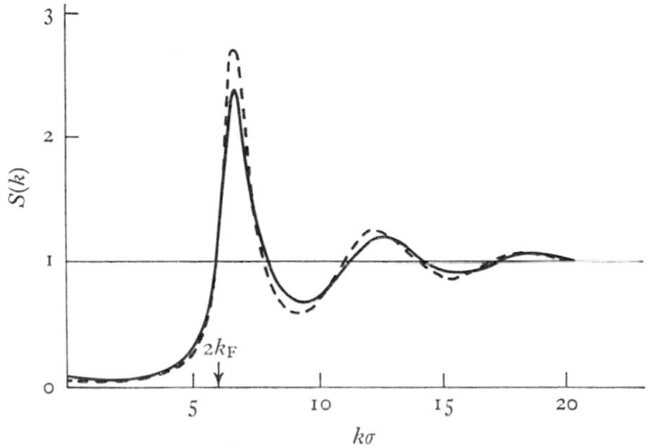

Fig. 5.8.3. The structure factors obtained by Schiff[47] for liquid sodium with potential $\Phi_1(r)$, $\rho = 0.83$, $T^* = 0.97$: dashed line, $\Phi_1(r)$ truncated at $r = 1.53$ (no oscillations); solid line $\Phi_1(r)$ truncated at $r = 3.20$ (oscillations).

of the artificial fluids are not too unreasonable. The potentials were then constructed, using these values of σ and ϵ, but with different core steepnesses and with or without long range oscillations. In the case of sodium Schiff takes $\sigma = 3.24$ Å, $\epsilon = 599$ °K. The potentials used are described below, tabulated in tables 5.8.3 and 5.8.4 and shown in fig. 5.8.2(*a*).

Potential functions $\Phi_2(r)$, $\Phi_3(r)$ and $\Phi_4(r)$ have oscillations up to three times as large as the eventual oscillations in liquid alkali metals.

For liquid aluminium, the other metal system investigated, Schiff takes $\sigma = 2.56$ Å, $\epsilon = 1198$ °K; the potential function is shown in fig. 5.8.2(*b*), and is analogous to the Pick[48] function $\Phi_1(r)$ used for sodium. Schiff also takes the potential calculated by Ashcroft and Langreth[50] (taken to be zero for $r > 2.24\sigma$ Å).

There appears to be no significant difference in the thermodynamic functions for the Born–Mayer and Lennard-Jones cores. The main difference in the radial distribution functions is the intuitively obvious softer leading edge to the first peak in the case of the Born–Mayer repulsion. Correspondingly, the structure factor associated with the soft potential is more damped, although the principal first peaks are virtually identical. The short range form

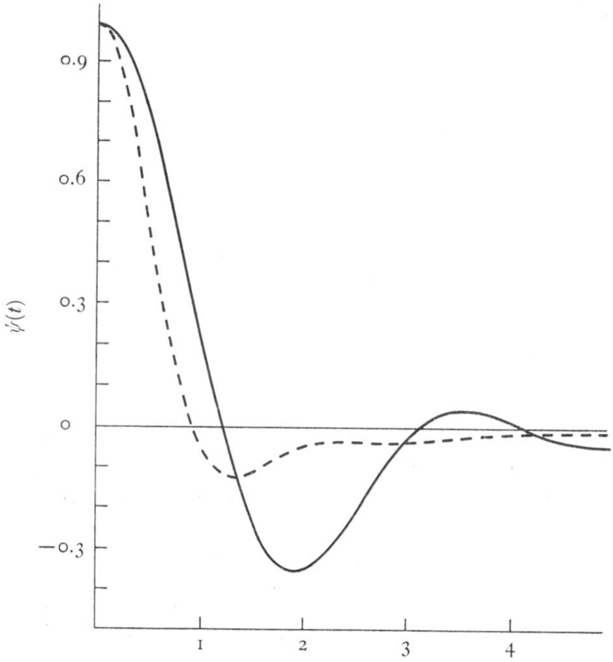

Fig. 5.8.4. Velocity autocorrelations obtained by Schiff[47] for liquid sodium with two potentials of different core steepnesses: solid line, $\Phi_3(r)$ (truncated at $r = 1.5$) $\rho = 0.83$, $T^* = 0.64$; dashed line, Lennard-Jones (6–12) (truncated at $r = 2.5$) $\rho = 0.84$, $T^* = 0.73$. (Redrawn by permission from Schiff, *Phys. Rev.* **186**, 151 (1969).)

of the pair potential engenders a progressive discrepancy in the large k form of the structure factor (see fig. 5.8.3).

The effect of a soft core on the velocity autocorrelation function seems to be to establish oscillations independent of the LRO form of the pair potential: $\psi(t)$ shown in fig. 5.8.4 is obtained using $\Phi_3(r)$ truncated at $r = 1.5\sigma$, that is *without* oscillations. The Lennard-Jones autocorrelation, on the other hand, shows no oscillatory tendencies, and presumably the difference must be attributed to the nature of the core. The corresponding spectral densities obtained from the Fourier inversion of $\psi(t)$ suggest that the soft potential is responsible for the better agreement with experiment (fig. 5.8.5), although the accuracy of the latter is not very high. As to the

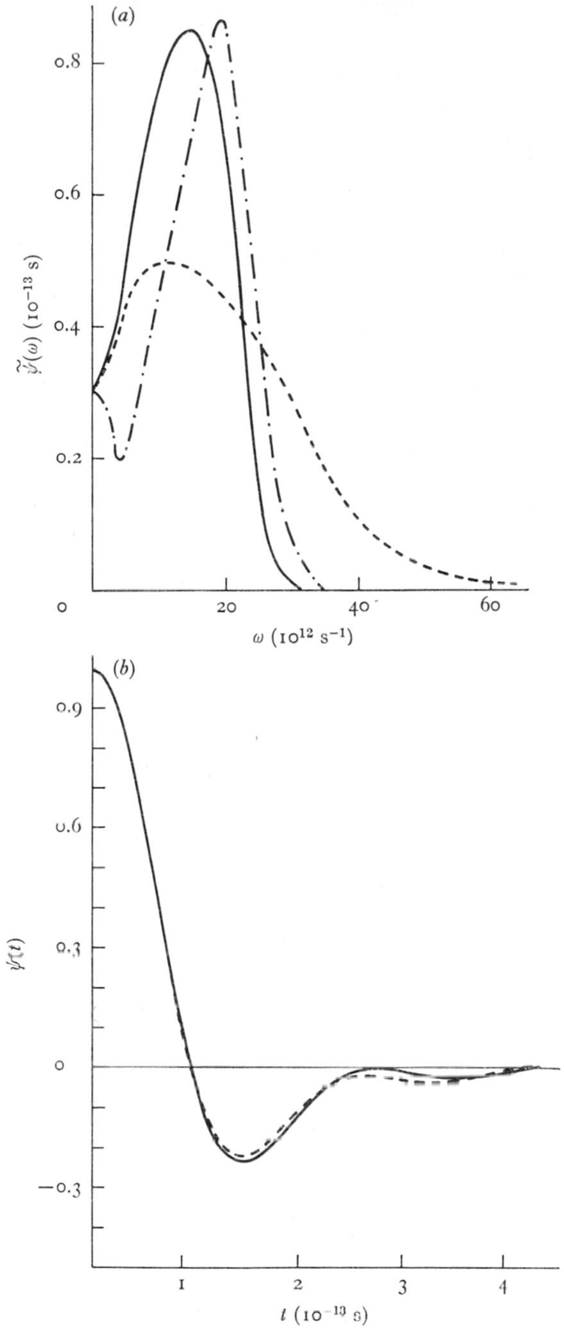

Fig. 5.8.5. (*a*) Fourier transforms of the velocity autocorrelations shown in (*b*): solid line, $\Phi_3(r)$ (truncated at $r = 1.5$) $\rho = 0.83$, $T^* = 0.64$; dashed line, Lennard-Jones (6–12) (truncated at $r = 2.5$) $\rho = 0.84$, $T^* = 0.73$; dot-dashed line, experimental curve for liquid sodium. (*b*) Velocity autocorrelation functions obtained with potential $\Phi_1(r)$, $\rho = 0.83$, $T^* = 0.97$: dashed line, truncated at $r = 1.53$ (no oscillations); solid line, truncated at $r = 3.20$ (oscillations). (Redrawn by permission from Schiff, *Phys. Rev.* **186**, 151 (1969).)

diffusion coefficient, it seems to be rather insensitive to the steepness of the core: the soft potential yields $D = 4.4 \times 10^{-5}\,\mathrm{cm^2\,s^{-1}}$, and the Lennard-Jones one $D = 4.9 \times 10^{-5}\,\mathrm{cm^2\,s^{-1}}$ whereas the experimental value for liquid sodium is $D = 4.2 \times 10^{-5}\,\mathrm{cm^2\,s^{-1}}$.

We now consider the effect of Friedel oscillations. Schiff computes the structure factor using $\Phi_1(r)$ truncated at $r = 1.53\sigma$ (no oscillations) and at $r = 3.20\sigma$ (oscillations). It is apparent from fig. 5.8.3 that the Friedel oscillations have little effect upon the structure factor, nor upon the velocity autocorrelation function. Since the effect of *realistic* oscillations seems to give no clear-cut effect, potentials with stronger oscillations were then used. The $\Phi_2(r)$ (LRO-2 of Paskin and Rahman) potential yields a first peak in the structure factor of 2.95 (with oscillations) in contrast to Paskin and Rahman's value of 2.4 for $S(k_0)$, which is in better agreement with the experimental value of 2.5. Schiff suggests that the discrepancy here is probably due to Paskin and Rahman's truncation of the radial distribution function at $r = 2.5\sigma$: Schiff extrapolates $g_{(2)}(r)$ beyond some cut-off radius by assuming that $g_{(2)}(r)$ satisfies the PY equation for r larger than this value.

It is found that Friedel oscillations in the long range form of the pair potential enhance the oscillations in $\psi(t)$ arising from the soft core. The potentials $\Phi_3(r)$ and $\Phi_4(r)$ in which the oscillations are still larger generate an even higher first peak in the structure factor, and enhance the structure of $\psi(t)$.

In the case of aluminium, Schiff has only computed the structure factor; the oscillations in the potential have the effect of lowering the first peak and deepening the first minimum.

Thus, generally, the effect of Friedel oscillations on $S(k)$ and $\psi(t)$ is small if their amplitude is of the order of the theoretical predictions. If the amplitude is two or three times larger, the Friedel oscillations increase the height of the first peak of $S(k)$ and the oscillations of $\psi(t)$. Schiff finally considers the dependence of Friedel oscillations of 'realistic' amplitude upon their wave vector $2k_F$ by means of a simple model. In that model the height of the first peak of the structure factor $S(k_0)$ is a maximum when $2k_F = k_0$. The possibility exists of observing such a resonance effect by neutron or X-ray scattering in a liquid lithium–magnesium alloy. A resonance occurs at $2k_F = k_0 = 6.8\,\text{Å}^{-1}$ which, on the basis of a free electron model, should occur around 60 per cent magnesium.

Levesque and Vieillard-Baron[74], have made a comparison of the commonly used two-body potentials for liquid argon. The equation of state for several potentials is determined by a combination of PY2 (at critical densities) and molecular dynamics computations (at high liquid densities). They find that the agreement with experiment, once the critical region has been extracted, is best for the LJ 6–12 potential which is therefore, of all the simple two-body potentials, the best effective interatomic potential for rare gases.

A very important rationalization of the intermolecular force in condensed argon has recently been made by Barker, Fisher and Watts[61]. They have calculated the thermodynamic properties of liquid argon by Monte Carlo and molecular dynamic techniques, using accurate pair potential functions determined from the properties of solid and gaseous argon together with the Axilrod–Teller[62] three-body interaction. The fair quantitative agreement reported in other sections of this chapter between the machine results and experiment must be regarded as largely fortuitous: the consistent observation has been made that the use of more realistic potentials leads to a worse agreement with experiment, although as we have seen from Levesque and Vieillard-Baron's work[74], the Lennard Jones 6–12 seems to be the best *purely two-body* (effective) potential interaction at liquid densities. The actual interaction may differ from the 6–12 model in two ways, firstly in the form of the two-body interaction, and secondly in the presence of many body interactions. The effect of neglecting *all* many-body interactions has been studied by Dymond and Alder[63], however, the simplest many-body approximation is to assume that the dominant contribution arises from the triple-dipole interaction[62]. Such a form of interaction has been widely considered[64–71]. Bobetic and Barker[66] (BB) found that with this approximation they were able to determine a pair potential which gave excellent agreement with experiment for the thermodynamic properties of liquid argon. However, the BB pair potential was found to give only fair agreement with experiment,

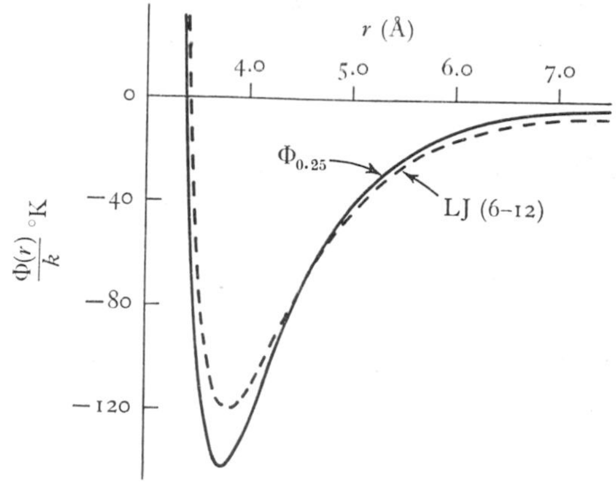

Fig. 5.8.6. Comparison of the Lennard-Jones (6–12) potential with the BB–BP hybrid function $\Phi_x = \Phi_{BB} + x[\Phi_{BP} - \Phi_{BB}]$. $x = 0.25$, $\sigma = 3.405$ Å and $\epsilon/k = 119.8$ °K.

although marginally better than a straight LJ 6–12 inter-action. A linear interpolation between the BB potential and another potential due to Barker and Pompe[(75)] (BP) was formed, according to

$$\Phi_x(r) = \Phi_{BB}(r) + x[\Phi_{BP}(r) - \Phi_{BB}(r)].$$

It was found that such a linear combination of potentials with $x = 0.25$ gave excellent agreement with experimental pressures at high densities. This purely pair function is shown in fig. 5.8.6.

Barker, Fisher and Watts use this potential in a Monte Carlo calculation of the thermodynamic properties of liquid argon. There are no problems of principle in performing Monte Carlo or molecular dynamics computations in the presence of three-body potentials, but in practice the computer requirements would be very large since for two-body interactions the time varies as N, whilst for triplet potentials it varies as N^2. For that reason three-body effects are included by using a perturbative technique involving an expansion in powers of the three-body coefficient ν ($= 73.2 \times 10^{-84}$ erg cm^9)[(72)] with neglect of all terms beyond that of degree 1. Quantum effects are included by means of the usual expansion of the partition function in terms of $\hbar^2$; however, effects of statistics are negligible for liquid argon at all temperatures. The thermodynamic internal energy U and pressure P is then given as[61]:

$$U = \langle\Phi_2\rangle_0 + \langle\Phi_3\rangle_0 - \beta\{\langle\Phi_2\Phi_3\rangle_0 - \langle\Phi_2\rangle_0\langle\Phi_3\rangle_0\} \\ + 2\langle\Phi_{qu}\rangle_0 - \beta\{\Phi_{qu}\Phi_2\rangle_0 - \langle\Phi_{qu}\rangle_0\langle\Phi_2\rangle_0\}; \quad (5.8.4)$$

$$PV = NkT + \langle P_2\rangle_0 + \langle P_3\rangle_0 - \beta\{\langle\Phi_3 P_2\rangle_0 - \langle\Phi_3\rangle_0\langle P_2\rangle_0\} \\ + \langle P_{qu}\rangle_0 - \beta\{\langle\Phi_{qu}P_2\rangle_0 - \langle\Phi_{qu}\rangle_0\langle P_2\rangle_0\}. \quad (5.8.5)$$

$\langle\ \rangle_0$ implies canonical averaging for the classical system with only two-body interactions. Φ_2 and Φ_3 represent the two-body and three-body total potentials:

$$\Phi_N = \sum_{i<j=1}^{N} \Phi(ij) + \sum_{i<j<k}^{N} \Phi(ijk) \\ \equiv \Phi_2 + \Phi_3 \quad (5.8.6)$$

and $\Phi(ijk)$ has the Axilrod–Teller form[62]. Further,

$$P_3 = \tfrac{1}{3} \sum_{i<j<k} (r_{ij}\,\partial/\partial r_{ij} + r_{ik}\,\partial/\partial r_{ik} + r_{jk}\,\partial/\partial r_{jk})\,\Phi(ijk) \\ = \Phi_2 + \Phi_3, \quad (5.8.7)$$

this latter form having been obtained by Graben[73]. P_2 is given by the usual virial expression

$$P_2 = -\Sigma\tfrac{1}{3}(r_{ij}\,\partial/\partial r_{ij})\,\Phi(ij) \quad (5.8.8)$$

and the quantum corrections

$$\Phi_{qu} = \frac{\beta\hbar^2}{12m} \sum_{i<j} \nabla_i^2 \Phi(ij), \quad (5.8.9)$$

$$P_{qu} = -\frac{\beta\hbar^2}{12m} \sum_{i<j} \tfrac{1}{3}(r_{ij}\,\partial/\partial r_{ij})\,\nabla_i^2\Phi(ij). \quad (5.8.10)$$

The three-body terms were averaged over a subset of the Monte Carlo chain; Barker *et al.* averaged the triplet term over every 1000th configuration. Some of the correction terms were calculated directly from the molecular dynamics calculations, when the three-body term was calculated after every 25 time-steps.

Thus, on the basis of the BB–0.25 BP interpolated pair potential shown in fig. 5.8.6 in conjunction with the Axilrod–Teller

TABLE 5.8.5. *Contributions to the internal energy of liquid argon on the basis of the BB–0.25 BP interpolated pair potential*

T (°K)	V (cm³ mole⁻¹)	U two-body (cal mole⁻¹)	U three-body (cal mole⁻¹)	U quantum (cal mole⁻¹)	U total (cal mole⁻¹)	U exptl (cal mole⁻¹)
100.00	27.04	−1525.2	87.1	15.6	−1423	−1432
100.00	29.66	−1393.6	67.9	12.5	−1313	−1324
140.00	30.65	−1284.7	62.8	9.3	−1213	−1209
140.00	41.79	−951.3	39.5	6.4	−1061	−1069
150.87	70.73	−603.8	26.6	4.6	−573	−591

TABLE 5.8.6. *Contributions to the system pressure of liquid argon on the basis of the BB–0.25 BP interpolated pair potential*

T (°K)	V (cm³ mole⁻¹)	P two-body (atm.)	P three-body (atm.)	P quantum (atm.)	P total (atm.)	P exptl (atm.)
100.00	27.04	239.9	364.2	42.2	646	652
100.00	29.66	−21.3	238.8	25.3	116	105
140.00	30.65	348.9	214.3	16.7	580	583
140.00	41.79	−33.7	49.0	2.7	18	37
150.87	70.73	34.5	13.2	1.2	49	49

triple-dipole term, excellent agreement is obtained between the machine calculations and the experimentally observed thermodynamic parameters. The contributions to the internal energy are shown in table 5.8.5. In table 5.8.6 we show the corresponding values for the pressure. The discrepancy between the experimental and calculated values is within the experimental error.

It is clear from these tables that the three-body interactions are quite large and must be included in a realistic calculation. The possibility that these contributions are to an appreciable extent cancelled by other many body interactions now seems remote. The quantum corrections are relatively small, but not negligible at this level of accuracy.

Barker *et al.*[61] observe that they have thus been led to a pair potential (BB–0.25 BP) which is *in itself* quite accurate. By contrast the *initial* assumption that *all* many-body interactions may be neglected led Dymond and Alder[63] to an effective pair potential which is somewhat at variance with experiment.

5.9 Transport phenomena

In the last section a qualitative association between the form of the pair potential and the nature of the velocity autocorrelation and its time integral, the diffusion coefficient, was established. In this section we shall consider a little more closely the nature of the various transport phenomena *per se*.

The molecular dynamics approach allows us to follow in microscopic detail the temporal evolution of a classical system to an extent denied the experimental techniques. Non-equilibrium phenomena can be followed provided the relaxation time for the process is substantially less than the length of the entire chain. Macroscopic consequences of hydrodynamic flow are, for example, therefore excluded from these studies. Steady-state time-dependent phenomena such as self-diffusion and velocity autocorrelation are, however, particularly well suited to the molecular dynamic technique. We have seen (§5.2) that a non-equilibrium system rapidly approaches a Maxwellian velocity distribution, although it is found that a hard sphere assembly equilibrates within two to four collisions per particle depending upon density, whilst a square-well fluid appears to develop a Maxwell–Boltzmann distribution quite quickly; but the *mean velocity* of the distribution approaches its equilibrium relatively slowly, taking about 60 collisions per particle. This is understood in terms of configurational impedance of the total energy distribution in the case of the square-well fluid: the system has to exchange potential and kinetic energy in coming to equilibrium and this is governed by configurational adjustment – a problem which does not arise in the hard sphere system.

For a Markovian evolution it may be shown that the growth of the mean square atomic displacement is a linear function of time.

However, for a non-Markovian system with a finite memory this may be shown to be only asymptotically correct:

$$\operatorname*{Lt}_{t\to\infty} \langle r^2 \rangle = 6Dt + c, \tag{5.9.1}$$

where c is a constant and D is the diffusion coefficient. Again, for a Markovian system c is zero, whilst for the non-Markovian evolution c is generally non-zero and furthermore contains information on the initial history of the diffusing centres. Rahman(76) has studied the behaviour of $\langle r^2 \rangle$ as a function of time for a liquid argon system ($\rho = 1.374\,\text{g cm}^{-3}$, $T = 94.4\,°\text{K}$) by the method of molecular dynamics. His result is shown in fig. 5.9.1 (*a*). We see that the mean square displacement is essentially a linear function of time after about 1×10^{-12} s, having by then assumed its asymptotic form. The calculated value of D is $2.43 \times 10^{-5}\,\text{cm}^2\,\text{s}^{-1}$, which is ~ 15 per cent lower than the experimental value at the same temperature and density. Agreement is apparently improved when an exp-6 potential is used. The constant c is found by Rahman to be $0.2 \times 10^{-16}\,\text{cm}^2$. This constant tells us whether the diffusion is ahead or behind the corresponding Markovian process. Rearrangement of (5.9.1) gives

$$\frac{\langle r^2 \rangle - c}{6t} = D;$$

c being positive we conclude that diffusion is impeded relative to a Markovian evolution. The diffusing particles are delayed about their initial positions in their early history: this is easily understood when we come to consider the velocity autocorrelation function. Then it is found that the particles are effectively 'contained' by a relaxing cage of nearest neighbours. It may be seen from fig. 5.9.1 that the root mean square displacement is only 1.9 Å after 2.5×10^{-12} s. This is approximately one half the nearest neighbour distance in the liquid. Thus, even after this time we would expect that the identity of the first shell of neighbours is not completely lost. If, with van Hove(77), we resolve the time-dependent pair distribution function $G(r, t)$ into its 'self' and 'distinct' components

$$G(r, t) = G_s(r, t) + G_d(r, t) \tag{5.9.2}$$

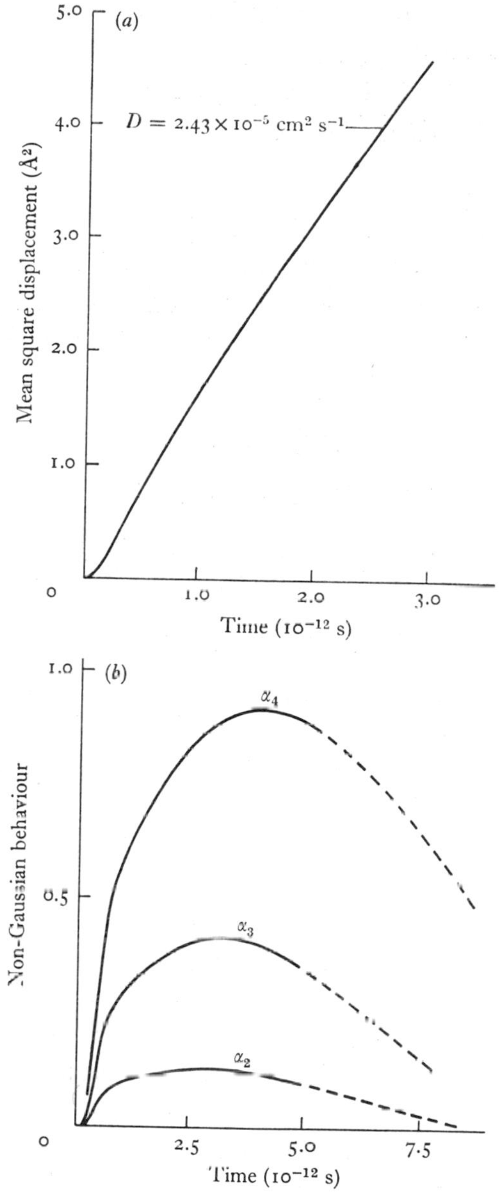

Fig. 5.9.1. (*a*) The mean square atomic displacement determined by Rahman for liquid argon $\rho = 1.374$ g cm^{-3}, $T = 94.4$ °K. The asymptotic form of the curve is given by (5.9.1), with $c = 0.2$ Å^2. (*b*) The departure of the van Hove self-function $G_s(r, t)$ from a Gaussian form, as determined by Rahman. Initially, $G_s(r, t)$ shows a Gaussian behaviour lasting for $\sim 0.15 \times 10^{-12}$ s and, on extrapolation, a Gaussian form is recovered at about 10^{-11} s. Maximum departure of $\langle r^4 \rangle$ from its Gaussian value is only $\sim$ 13 per cent. (Redrawn by permission from Rahman, *Phys. Rev.* **136**, A405 (1964).)

in an obvious notation, then we may consider the relaxation of the 'cage' of nearest neighbours $G_d(r,t)$ and the self-diffusion of the origin molecule $G_s(r,t)$. We may remark here that $G_d(r,t)$ gives the temporal decay of the radial distribution function: furthermore $G_d(r,0) \equiv g_{(2)}(r)$.

If the trajectory of the 'self' (origin) particle evolved subject to a stochastically fluctuating force field then the decay of the self peak $G_s(r,t)$ would have a Gaussian form $[4\pi\rho(t)]^{-\frac{3}{2}} \exp[r^2/4\rho(t)]$ familiar from the theory of random errors. Furthermore we should have the following relationships:

$$\langle r^2 \rangle = 6\rho(t), \tag{5.9.3}$$

$$\langle r^{2n} \rangle = c_n \langle r^2 \rangle^n, \tag{5.9.4}$$

$$c_n = 1 \times 3 \times 5 \times 7 \times \ldots (2n+1)/3^n \tag{5.9.5}$$

for $n = 1, 2, 3, \ldots$. Thus, a departure of $G_s(r,t)$ from a Gaussian form can be expressed in terms of the functions $\alpha_n(t)$, defined by

$$\alpha_n(t) = (\langle r^{2n} \rangle / C_n \langle r^2 \rangle^n) - 1 \tag{5.9.6}$$

which vanishes for a Gaussian, but not otherwise, as a simple insertion of relations (5.9.3), (5.9.4), (5.9.5) will show. In fig. 5.9.1 (*b*) we show α_2, α_3 and α_4 as determined by Rahman[(76)]. Since the values are all positive we conclude that $G_s(r,t)$ goes to zero with increasing r more slowly than a Gaussian: this was apparent from the form of the mean square displacement, fig. 5.9.1 (*a*). The flatness of the curves near the origin reflects the Maxwellian distribution of velocities, because at short times, during the traversal of the cage, $\langle r^{2n} \rangle$ tends to $\langle v^{2n} \rangle t^{2n}$. It is clear from fig. 5.9.1 that the Markovian form is eventually regained, and non-Gaussian behaviour should presumably disappear as $t \to \infty$. Fig. 5.9.1 (*b*) shows that α_2, α_3 and α_4 begin to decrease after 3.0×10^{-12} s. Rahman extrapolates the curves to the right, and concludes that G_s becomes Gaussian again after approximately 10^{-11} s. Rahman goes on to express the self function in terms of a Gaussian in product with a linear expansion of the even Hermite polynomials, $H_{2n}(x)$, thus

$$G_s(r,t) = [4\pi\rho(t)]^{-\frac{3}{2}} \exp(-r^2/4\rho(t)) \times \{1 + b_6(t) H_6(\alpha r) + b_8 H_8(\alpha r) + \ldots\}. \tag{5.9.7}$$

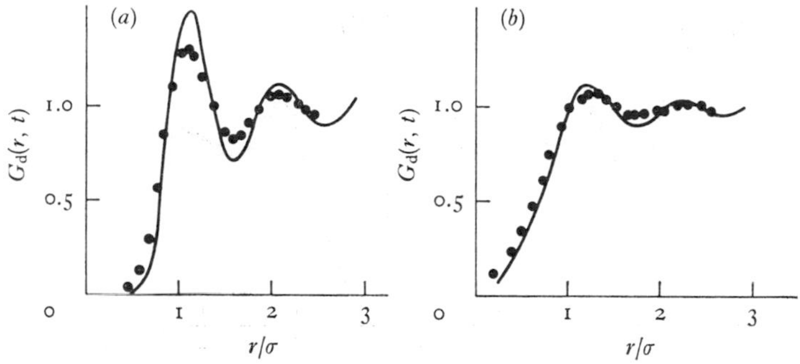

Fig. 5.9.2. Comparison of the relaxation of Rahman's distinct function $G_d(r, t)$ at (a) $t = 1 \times 10^{-12}$ s, and (b) $t = 2.5 \times 10^{-12}$ s and the Vineyard convolution approximation. Vineyard's approximation is seen to give a too rapid decay of G_d. ● indicates convolution. (Redrawn by permission from Rahman, *Phys. Rev.* **136**, A405 (1964).)

The coefficients b_6, b_8, etc., may be very simply related to α_2, α_3, α_4, etc., and in fact $b_6 = 0.0027$, $b_8 = -0.0003$ and $b_{10} = 0.00003$, showing that the first few terms in the expansion give a very good description of the non-Gaussian behaviour of $G_s(r, t)$.

To describe the time-dependence of $G_d(r, t)$ Vineyard[78] has suggested an approximation which makes G_d a convolution between $g_{(2)}(r)$ and $G_s(r, t)$. First, Vineyard writes the formal equality

$$G_d(r, t) = \int g_{(2)}(\boldsymbol{r}') \mathscr{H}(\boldsymbol{r}-\boldsymbol{r}', t)\, d\boldsymbol{r}', \tag{5.9.8}$$

where $\mathscr{H}(\boldsymbol{r}-\boldsymbol{r}')$ denotes the transition probability of a particle from $\boldsymbol{r} \rightarrow \boldsymbol{r}'$ in time t, given that another particle was situated at the origin at $t = 0$. The Vineyard approximation consists in setting $\mathscr{H} \equiv G_s$ in the above equation. However, the motions of particles in the first shell are strongly correlated with the occupation of the origin at $t = 0$, and this model overlooks this fact. Thus the $\boldsymbol{r} \rightarrow \boldsymbol{r}'$ transition probability is too great and $g_{(2)}(r)$ decays too quickly in consequence. This is shown in fig. 5.9.2 where the actual $G_d(r, t)$ as calculated by Rahman is compared with the Vineyard approximation at $t = 1 \times 10^{-12}$ s, and 2.5×10^{-12} s. Rahman[76] has suggested a delayed convolution approximation defined by

$$G_d(r, t) = \int g_{(2)}(\boldsymbol{r}')\, G_s(\boldsymbol{r}-\boldsymbol{r}', t')\, d\boldsymbol{r}', \tag{5.9.9}$$

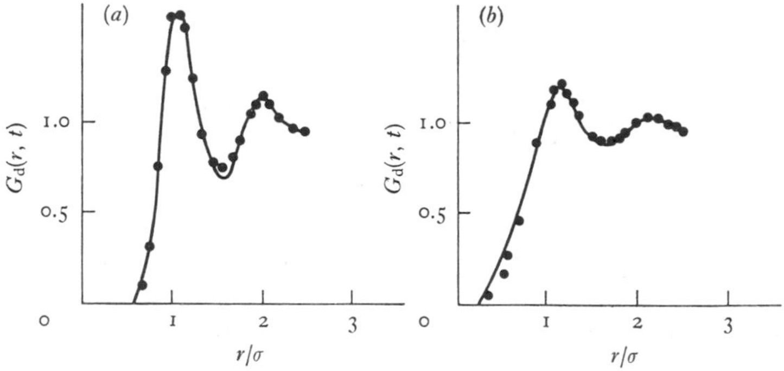

Fig. 5.9.3. Comparison of the molecular dynamic form of the distinct function $G_d(r, t)$ at (a) $t = 0.8 \times 10^{-12}$ s, and (b) 2.3×10^{-12} s; with the delayed convolution of $g_{(2)}(r)$ and $G_s(r, t)$ with $t' < t$ ($t' = 0.5 \times 10^{-12}$ and 1.5×10^{-12} s, respectively). The delayed convolution approximation is seen to be an improvement on the Vineyard convolution (fig. 5.9.2). (Redrawn by permission from Rahman, *Phys. Rev.* **136**, A405 (1964).)

where $t' \geqslant t$. Thus at a given t an earlier form of the self-distribution $G_s(t)$ is chosen. Thus, for a given convolution there will be a pair of times (t, t') which yield a best fit. Rahman finds the following pairs (t, t') in units of 10^{-12} s: (0.2, 0.4), (0.5, 0.8), (1.0, 1.6), (1.5, 2.3), (2.0, 2.9), (2.5, 3.5). This can be described as a functional relation between t' and t, and Rahman suggests the following simple one-parameter function:

$$t' = t - \tau[1 - \exp(-t/\tau) - (t^2/\tau^2)\exp(-t^2/\tau^2)]. \quad (5.9.10)$$

Taking $\tau = 1.0 \times 10^{-12}$ s, we obtain the following pairs:

(0.21, 0.4), (0.59, 0.8), (1.0, 1.6),

(1.4, 2.3), (2.0, 2.9), (2.5, 3.5).

The results of the delayed convolution approximation of Rahman are shown in fig. 5.9.3. The height of the first peak in $G_d(r, t)$ at $t = 0$, 1 and 2.5×10^{-12} s is, respectively, 2.8, 1.5 and 1.1. Remnants of the first shell of neighbours therefore persist for at least 2.5×10^{-12} s.

This persistence in local order is reflected in the form of the velocity autocorrelation function shown for liquid argon from Rahman's molecular dynamics computations in fig. 5.9.4. The

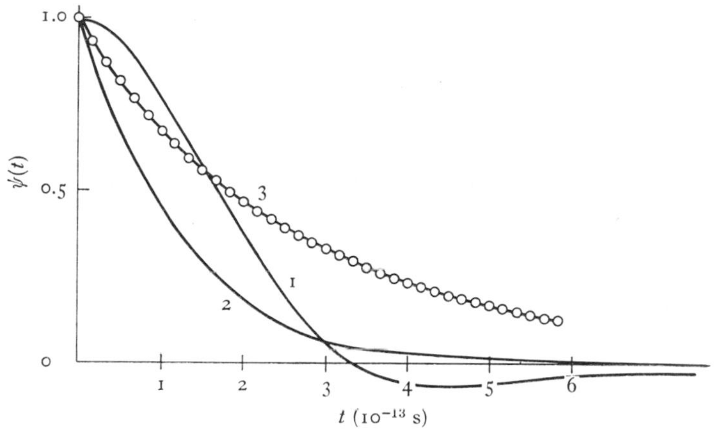

Fig. 5.9.4. The normalized velocity autocorrelation function for liquid argon (ρ = 1.374 g cm^{-3}, T = 94.4 °K) as determined by Rahman (curve 1). Also shown is the Langevin function exp $(-kT/mD)\, t$ (curve 2), and Nelson's square-well autocorrelation (curve 3), neither of which show a negative region. Nelson's Monte Carlo simulation was at ρ = 1.30 g cm^{-3}, T = 120.0 °K.

Lennard-Jones (6–12) (Rahman): T = 94.4 °K, V = 29.33 cm^3 mole^{-1}, 864 particles. Langevin function: T = 94.4 °K, D = 2.43 × 10^{-5} cm^2 s^{-1}. Square-well model (Nelson): T = 120 °K, V = 30 cm^3 mole^{-1}, 256 particles.

negative region is clearly to be attributed to the 'backscattering' of the diffusing particle within its cage of nearest neighbours. Such an event can only occur subject to persistence in the local atomic environment. A Langevin type of velocity autocorrelation, also shown in fig. 5.9.4, arising from a completely uncorrelated or Markovian velocity evolution remains positive for all t. In this respect they are radically different. However, for small t, during which the particle is executing its first traversal of the atomic cage the motion is gas-like, and to this extent the Langevin function gives quite a good representation of the velocity autocorrelation. Nelson[(110)] has reported a velocity autocorrelation for a square-well fluid whose parameters are quite closely related to those of argon. The function is shown in fig. 5.9.4: it shows no oscillations and appears to be of Langevin form. The liquid autocorrelation is quite clearly intermediate between that of a dilute gas (Langevin) and the damped oscillatory form of a solid. However, Croxton and Ferrier have reported anisotropic forms of the velocity autocorrelation at the liquid argon surface, ranging from liquid-like to damped

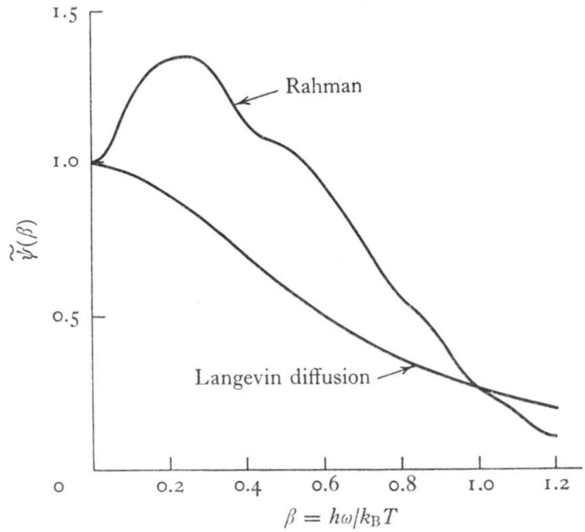

Fig. 5.9.5. Rahman's spectral density of liquid argon determined from the Fourier inversion of the velocity autocorrelation shown in fig. 5.9.4. The Lorentzian spectrum of a Langevin-type autocorrelation is also shown, and is seen to be quite inadequate to represent the high frequency vibratory modes. (Redrawn by permission from Rahman, *Phys. Rev.* **136**, A405 (1964).)

oscillatory forms more characteristic of the solid (see § 5.11). Rahman takes the Fourier transform of the velocity autocorrelation in liquid argon and obtains the spectral density shown in fig. 5.9.5. From the non-zero form of $\psi(\beta)$ as $\beta \to 0$ the existence of low-frequency diffusive modes is apparent, whilst the maximum at $\beta = 0.25$ may be identified with the vibratory component absent in the phonon spectrum of a gas.

The diffusion constant is very simply related to the normalized velocity autocorrelation function as

$$D = \int_0^\infty \psi(t)\,\mathrm{d}t. \tag{5.9.11}$$

Again it is clear that the negative region of $\psi(t)$, absent from a Markovian system, is responsible for a lowering in the diffusion coefficient. This was apparent from (5.9.1). On the basis of these molecular dynamics studies we are able to form at least a qualitative picture of the transport phenomena occurring in a simple dense liquid. The existence of both diffusive and oscillatory modes is

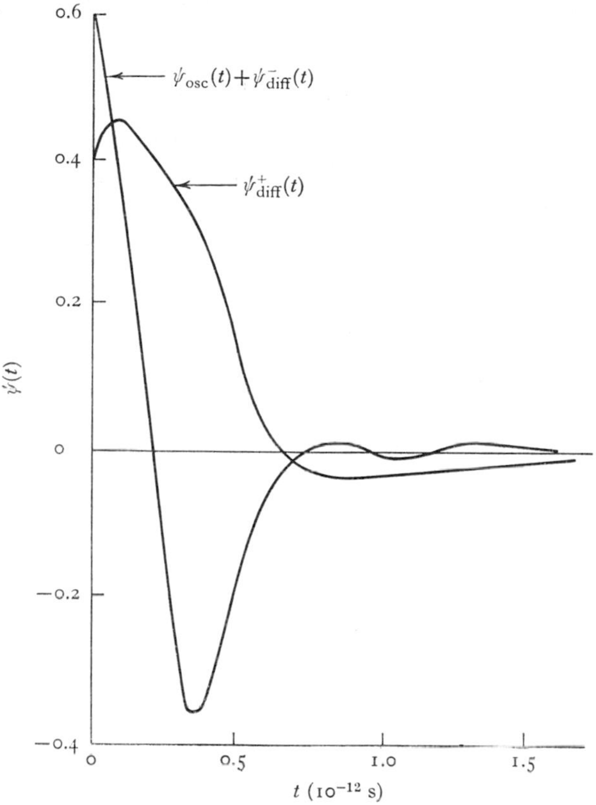

Fig. 5.9.6. Components of the velocity autocorrelation function for liquid argon corresponding to the diffusive and vibratory motion of the atoms ($\rho = 1.407$ g cm^{-3}, $T = 85.5$ °K). This molecular dynamic simulation of Rahman is based upon a modified Buckingham exp-6 interaction.

evidenced by the form of the velocity autocorrelation function and its Fourier transform. Rahman[79] resolves the velocity autocorrelation into its diffusive and oscillatory components, thus

$$\psi(t) = \psi_{diff}(t) + \psi_{osc}(t), \tag{5.9.12}$$

where $t \leqslant \tau$, the decay time. $\psi_{diff}(t)$ is further resolved into two parts, $\psi^{+}_{diff}(t)$ and $\psi^{-}_{diff}(t)$, which represent the diffusive contributions of particles whose motions at $t = 0$ over the interval τ are in a positive or negative sense. Both $\psi^{-}_{diff}(t)$ and $\psi^{-}_{osc}(t)$ represent an oscillatory type of motion. Fig. 5.9.6 shows the resolution for liquid argon at 85.5 °K using an exp-6 potential. The sum

$$[\psi_{\text{osc}}(t) + \psi^{-}_{\text{diff}}(t)]$$

decays to zero rapidly after executing an oscillatory motion: this component contributes little to the diffusion, the net area beneath the curve being virtually zero (see (5.9.11)). The Fourier transform of this component will, quite clearly, be a broad maximum displaced from the origin, whilst the transform of the approximately Gaussian diffusive component $\psi^{+}_{\text{diff}}(t)$ is another Gaussian, and we regain the phonon spectrum shown in fig. 5.9.5. This latter component evidently contributes principally to the coefficient of diffusion. This approach requires, however, a somewhat arbitrary division to be made between high frequency oscillatory and low frequency diffusive modes. In fact the two are strongly coupled and not independent as (5.9.12) implies.

5.10 The melting transition

The computer experiments described thus far have provided only quantitative information for hard spheres in the transition region, and although in the systems studied both phases could be generated separately, the ensembles were evidently too small for the solid and fluid phases to coexist. Most commonly a transition would proceed only in one direction, and then only after many thousands of collisions per particle; any attempt to crystallize the fluid would result in a metastable 'glassy' extension of the fluid isotherm. Even in large systems of macroscopic size glasses and other metastable states are encountered, but sufficiently large fluctuations, generally excluded in the finite-periodic simulations, are always available to annihilate the metastable states. In systems of finite size, such as very small droplets or computer ensembles, both the region over which metastable phases can occur and the time for which they persist can be large. Precisely these difficulties have complicated the search for the hard sphere phase transition in small systems.

Since the solid–fluid branches of the isotherm on a pressure–volume plot generated by machine calculation were separated by about 10 per cent in density, a first-order phase transition between the two seems likely, and indeed has been extensively anticipated. The problem is to locate the tie-line between the two branches. Only a thermodynamic comparison of the chemical potential of the two phases could settle the point, the two-phase equilibrium co-existence being characterized by an equality of pressure and chemical potential in the solid and fluid phases. The determination of the chemical potential, however, is not an easy task in a computer experiment, involving first of all an absolute determination of the entropy.

In the fluid phase both the pressure and the entropy are known. Up to two-thirds of close packing the pressure and the entropy are given within 1 per cent and 0.01 Nk, respectively, by various approximate expressions based on the virial series – the Padé approximants. In the solid phase the pressure is known quite well[(56)], and by an integration the entropy in the solid phase is also known, *to within an additive constant*. It is this additive constant which is crucial, and Hoover and Ree[(57)] have established a very elegant means of determining it.

The method is extremely simple, and involves the seemingly paradoxical expedient of studying a system which cannot melt. By confining the centre of each particle in an N-particle system to its own cell of volume V/N at *all* densities, the solid phase can be artificially extended to cover the entire density range. Particles in the artificial single-occupancy solid can collide both with the walls which confine the particles and with the other nearby particles. At high density, particles are usually confined by their neighbours alone, rather than by cell walls. Each particle stays near the centre of its cell, and the single-occupancy cell system faithfully represents the properties of a perfect solid. At low densities, collisions with cell walls become appreciable – these collisions prevent the artificial solid from melting, that is prevent the diffusive motions which characterize the normal fluid. Instead of melting, the solid phase thermodynamic properties are artificially extended to low density.

At low enough density all of the thermodynamic properties of either the constrained cell system or the unconstrained real system can be calculated exactly. Hoover and Ree show how this low

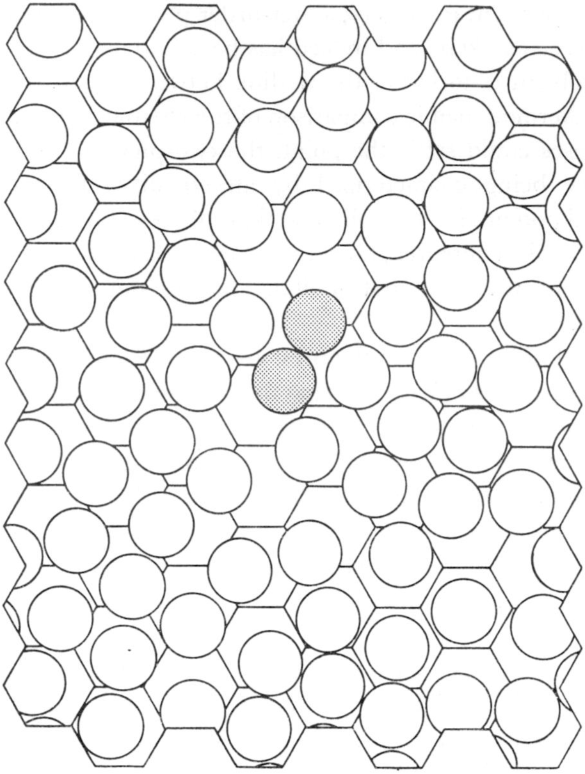

Fig. 5.10.1. The 'Wigner–Seitz' single occupancy cells for hard discs. (Redrawn by permission from Hoover and Ree, *J. Chem. Phys.* **47**, 4873 (1967).)

density limit, coupled with computer-generated thermodynamic properties for the *artificial* cell system spanning the whole density range, can be used to calculate the entropy in the *real* solid phase. It should be realized why these authors are adopting this single-occupancy scheme, for without it they could not avoid the development of metastable contributions as the density is decreased. At high densities, however, in the vicinity of the phase transition, the cells make no contribution since the communal entropy defect incurred in partitioning the free volume no longer arises. The view that the entire communal entropy makes its appearance at melting thereby 'explaining' the entropy of fusion is by now obsolete[22, 58].

Thus, Hoover and Ree determine the entropy constant by

extending the solid phase through its region of metastability and on to low density. In three dimensions each particle was confined to a dodecahedral 'Wigner–Seitz' cell of a cubic lattice. For hard discs the geometry is simpler, and the hexagonal cells are shown in fig. 5.10.1. Notice that adjacent particles may nevertheless interact. Apart from the confinement to cells, a Monte Carlo hard sphere calculation of the pressure proceeded in the usual way.

$$(\partial S/\partial V)_{N,T} = P/T$$

was measured up to three-quarters of the close-packed density, which was found to be safely within the stable solid phase, and then integrated from the known low density limit to obtain the value of the previously unknown entropy constant in the solid, with an expected error of 0.015 Nk. That is, the *real* entropy at high density is given by

$$S(\rho) = S_0(\rho_0) + N\int_{\rho}^{\rho_0} \frac{P}{T\rho^2}\,d\rho, \qquad (5.10.1)$$

where $S_0(\rho_0)$ is the *artificial* solid entropy at some low density ρ_0, and this is taken to be known.

In fig. 5.10.2(*a*) we show the hard sphere equations of state taken from Hoover and Ree's paper. The fluid and solid isotherms are joined by a tie-line connecting states of equal chemical potential. In the thermodynamic limit the equilibrium pressure for solid–fluid coexistence is found to be $(8.27 \pm 0.13)\,NkT/V_0$, where V_0 is the volume at close packing. The corresponding relative densities are 0.736 ± 0.003 (solid) and 0.667 ± 0.003 (fluid). The fluid density seems to agree well with Ross and Alder's empirical rule[59] that the tie-line should be associated with the highest density at which the solid melts in computer experiments. A similar situation arises for hard discs.

Fig. 5.10.2 shows that the sphere (and, in fact, the disc) single-occupancy isotherm lies very close to the fluid curve. The sphere results suggest a cusp in the single-occupancy isotherm at a reduced density of about 0.6: at this density the cell walls begin to become important in preserving the stability of the crystal. At this cusp the solid is in fact mechanically unstable and without the walls would rapidly disintegrate. A theory of melting based on the properties of

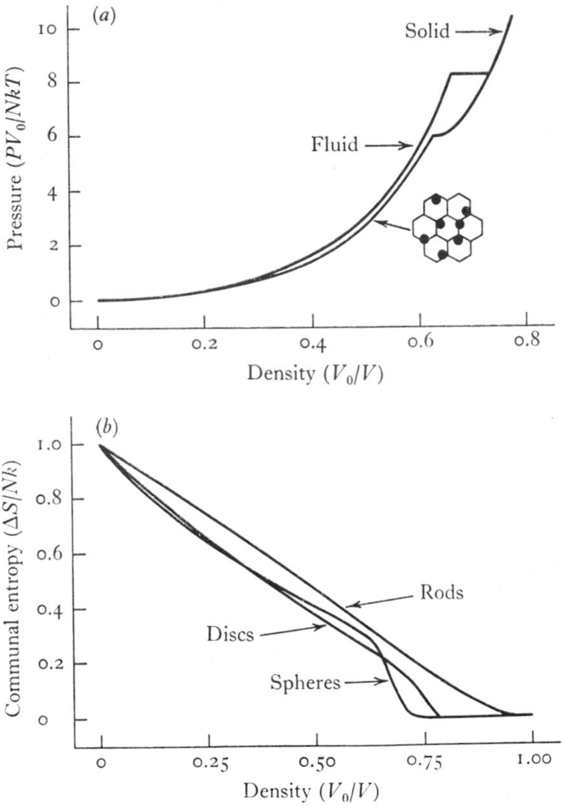

Fig. 5.10.2. (*a*) The hard sphere equation of state together with the single-occupancy isotherm. (*b*) The communal entropy defect incurred for one-, two- and three-dimensional systems as a function of density for a single-occupancy geometry. (Redrawn by permission from Hoover and Ree *J. Chem. Phys.* **47**, 4873 (1967).)

the solid phase would, presumably, predict fusion at the cusp density instead of at the higher density determined by *thermodynamic* rather than *mechanical* instability. Alder and Wainwright[22] have already identified the van der Waals loop in the equation of state of a hard disc fluid (see fig. 1.4.1): Ree and Hoover have since shown that the two states joined by the loop do indeed have the same chemical potential and are therefore in thermodynamic equilibrium.

The difference in entropy between an unconstrained system of

TABLE 5.10.1. *The 'real' (Padé) and single-occupancy* (SO) *entropies for a system of hard spheres*

Density	S (Padé)$/Nk$	S_{SO}/Nk	$\Delta S/Nk$
0.00	0.000	−1.000	1.000
0.05	−0.155	−1.061	0.906
0.10	−0.326	−1.155	0.829
0.15	−0.516	−1.278	0.762
0.20	−0.726	−1.428	0.702
0.25	−0.961	−1.608	0.647
0.30	−1.225	−1.816	0.591
0.35	−1.524	−2.061	0.537
0.40	−1.864	−2.352	0.488
0.45	−2.253	−2.695	0.442
0.50	−2.702	−3.101	0.399
0.55	−3.226	−3.583	0.357
0.60	−3.843	−4.155	0.312
0.65	−4.577	−4.806	0.229
0.66	−4.741	−4.933	0.192
0.67	−4.911	−5.058	0.148
0.68	−5.088	−5.184	0.107
0.69	−5.272	−5.311	0.072
0.70	−5.463	−5.440	0.044
0.71	−5.662	5.570	0.023
0.72	−5.870	−5.703	0.009
0.73	−6.087	−5.838	0.001
0.74	−6.313	−5.978	0.000

particles and a constrained one with one particle per cell is the so-called 'communal entropy', the determination of which has been a fundamental problem in the theory of liquids. The situation has been somewhat confused by a vagueness in the definition of the communal entropy. Here we shall assume, with Hoover and Ree, the Kirkwood definition[60] given above. Designating this quantity ΔS, the communal entropy is immediately given as

$$\Delta S/Nk = (S - S_{SO})/Nk,$$

and is tabulated for hard spheres in table 5.10.1. The entropies given are relative to an ideal gas at the same temperature and density. The 'real' entropies were determined via a Padé approximant to the equation of state for the fluid isotherm: the agreement between the two is very close. The communal entropy is the difference between the first two columns except in the two-phase region. The communal entropy for one-, two- and three-dimensional hard spheres is shown in (fig. 5.10.2(*b*)). One feature is particularly apparent: the communal entropy defect seems to be a linear function of density, at least up to the phase transition, and up to a relative density ~0.6. ΔS is largely independent of dimensionality. This suggests that in cell-model calculations a linear, rather than a constant correction term should be added to simulate the effect of many-body correlations.

5.11 Surface phenomena

In § 4.3 the liquid–vapour density transition profile was established from a solution of the one-dimensional Born–Green–Yvon equation by Croxton and Ferrier[(94, 95)]. These authors go on to perform a molecular dynamics simulation of the transition zone for a Lennard-Jones system of argon molecules. The greater complexity of surface studies necessitates computation either for fewer atomic centres, resulting in poor statistics, or for a two-dimensional instead of a three-dimensional assembly. Croxton and Ferrier[(96, 97)] decided upon the latter programme as representing the better compromise between expediency and significance of the computed correlation. It must be borne in mind, however, that a two-dimensional system will tend to accentuate a given correlation in magnitude, although not in concept.

The two-dimensional molecular dynamic array of 200 atoms is topologically identical to a cylinder of height and circumference 100 Å (fig. 5.11.1(*a*)). By making the replication distance large (100 Å) communal entropy defects will be small, and since one of the objectives is to identify the low entropy surface states described in § 4.6, this provision is essential although it does in practice restrict the computations to two dimensions. Elastic reflection of the vapour atoms from the top boundary is perfectly adequate. However, the lower boundary with its intimate bearing upon the dynamical and configurational coupling of the bulk and surface 'phases' must be considered in more detail. Ideally, one would wish to allow a simulated bulk to interact with a simulated surface, whereupon coupling

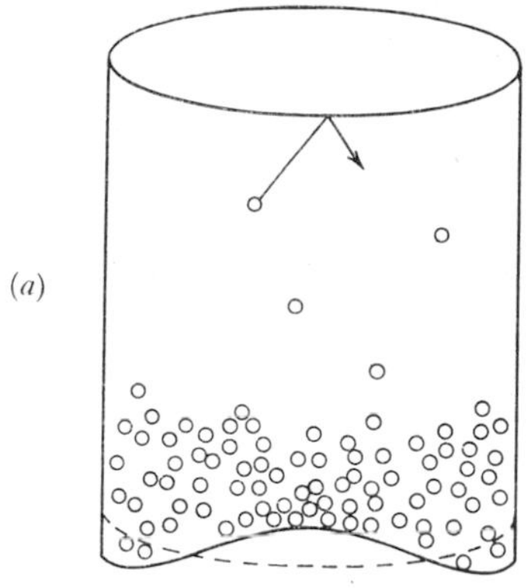

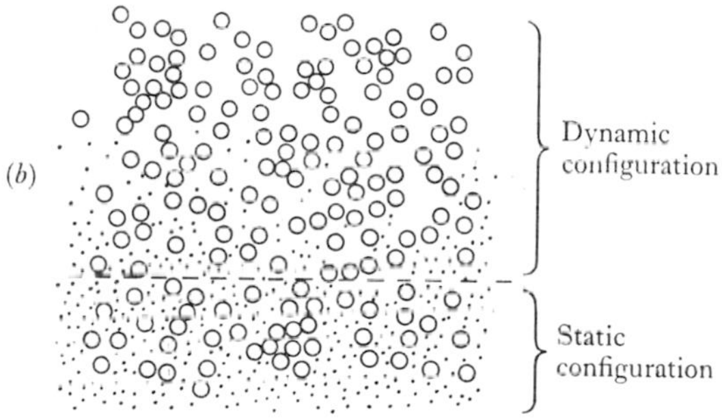

Fig. 5.11.1. (a) The molecular dynamic array in the two-dimensional simulation of Croxton and Ferrier is topologically identical to a cylinder of height and circumference 100 Å. Elastic reflection of the atoms occurs at the upper boundary whilst a thermal disruption matrix at the lower boundary simulates the coupling of the surface and bulk states. (b) The lowest 10 Å of the dynamic array is subjected to the field of a further static array whose force contours are subjected to a space–time modulation, thereby simulating the presence of a strongly coupled bulk phase.

would take care of itself. An expedient embodying the characteristics of the coupling can be used instead, by setting up a matrix of disruptive elements whose magnitudes depend on time and space in a way characteristic of the bulk.

The thermal disruption matrix (TDM) is set up as follows: a static system of a further 200 argon 'atoms' are placed at random positions below the dynamic array, and the planar force contours which permeate the latter system are calculated at 0.5 Å intervals normal to the lower boundary from $z = 0$ to $z = 10$ Å, beyond which the direct attractive influence of the 'bulk' static array is assumed negligible. Thus, the lowest 10 Å of the dynamic array is immersed in the static field of the 'bulk'. These force contours are then subjected to a sinusoidal space–time modulation having characteristic eigenvalues λ, τ, where λ is the characteristic correlation length, or range of spatial coherence of a cluster of particles in the liquid (~ 10 Å), while τ is the characteristic lifetime of such a cluster ($\sim 10^{-11}$ s). Such a modulation ensures that on average the dynamic system neither gains nor loses energy, and certainly there is no change in the number of particles. The trajectories of the atoms are determined in the usual way, although now subject to the additional term due to the TDM.

The dynamic system eventually develops a transition profile of the form shown in fig. 4.3.3. The layered configuration is stable, although only statistically defined: the trajectory plots of individual atoms show continuous excursions from one layer to another. The density oscillations, of period ~ 3.8 Å, are seen to be about the mean density, until the last few ångströms when the singlet function tails off into the vapour.

The pronounced anisotropy at the surface suggests that there should be some associated dynamical anisotropy. The two normalized velocity autocorrelations

$$\psi(t)_{\parallel} = \frac{\langle \dot{x}(0)\, \dot{x}(t) \rangle}{\langle \dot{x}(0)^2 \rangle}, \tag{5.11.1}$$

$$\psi(t)_{\perp} = \frac{\langle \dot{z}(0)\, \dot{z}(t) \rangle}{\langle \dot{z}(0)^2 \rangle}, \tag{5.11.2}$$

were determined, and we show $\psi(t)_{\perp}$ in the vicinity of the liquid surface in fig. 5.11.2(*a*). The pronounced oscillatory nature of this function is immediately evident, and implies the existence of phonon lifetimes up to 4.5×10^{-12} s. These characteristics have been observed to a much lesser extent in the bulk studies of Rahman[(76)], and in the neutron data of Cocking[(98)] and Randolph[(99)]. Certainly the relaxation of the solvent cage of nearest neighbours at

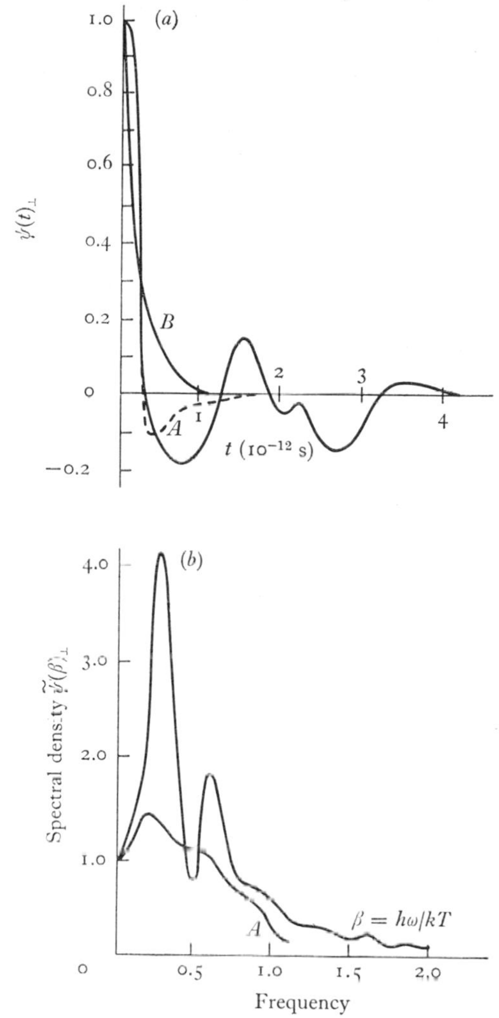

Fig. 5.11.2. (*a*) The transformed velocity autocorrelation for motions normal to the liquid surface. The autocorrelations show a very pronounced oscillation which is interpreted in terms of a quasi-crystalline or structured interphase. Curve A is Rahman's bulk autocorrelation and curve B is the Langevin function $\exp(-kT/mD_\perp)\,t$. (*b*) The transformed phonon spectrum for motions normal to the liquid surface. The strong peaks are indicative of a structured interphase of the type shown in fig. 4.3.3. Curve A is Rahman's spectral distribution for bulk argon.

the liquid surface is anisotropic. The $\psi(t)_{\parallel}$ autocorrelation is essentially 'liquid-like'. The very short time behaviour seems to be adequately described by the Langevin model,

$$\psi(t)_{\perp} = \exp(-kTt/mD_{\perp}),$$

where $T = 94.4\,°K$, $D_{\perp}$ (normal diffusion coefficient) $= 1.05 \times 10^{-4}$ $cm^2\,s^{-1}$, although any one of a variety of models will describe the 'gas-like' behaviour within the cage until the anti-correlating effect of the cage impresses oscillations on $\psi(t)$. Croxton and Ferrier conclude[96, 97] that the intraplanar dynamic correlations are effectively liquid-like, whilst the interplanar characteristics are those of a more permanent quasi-crystalline system.

These results can be even more convincingly displayed in terms of the spectral density of the autocorrelation function. The Fourier transform of (5.11.2) gives the frequency distribution $f(\beta)_{\perp}$ ($\beta = \hbar\omega/kT$) of the atomic motions normal to the surface. This function is shown in fig. 5.11.2(*b*). The series of sharp peaks are seen to be located on the broad inversion and inflection of the bulk spectrum of Rahman[76], with which the present spectral density is compared. TDM oscillations cannot be detected on this frequency spectrum: such a peak would be located at $\beta < 0.01$. The harmonic nature of the spectrum, having maxima at $\beta \simeq 0.25n$, $n = 1, 2, 3, \ldots$, suggests a very simple lattice structure. Of importance is the non-zero value of $f(0)$ indicating the existence of low frequency diffusive modes.

The mean square displacement normal and parallel to the surface is shown in fig. 5.11.3, and is compared to Rahman's bulk curve[76]. The gradient of the asymptotically linear region of the curve is proportional to the diffusion coefficient. The asymptotic behaviour does not set in until 3×10^{-12} s due to protracted correlative effects in the initial evolution. The time-independent (asymptotic) value of the two diffusion coefficients is $D_{\perp} = 1.05 \times 10^{-4}\,cm^2\,s^{-1}$ and $D_{\parallel} = 5.22 \times 10^{-5}\,cm^2\,s^{-1}$ at 94.4 °K obtained by Rahman[76].

That the normal diffusion coefficient $D_{\perp}$ is larger than the lamellar coefficient $D_{\parallel}$ is at first sight surprising. However, on the basis of an atomic hopping or jump diffusion model we should

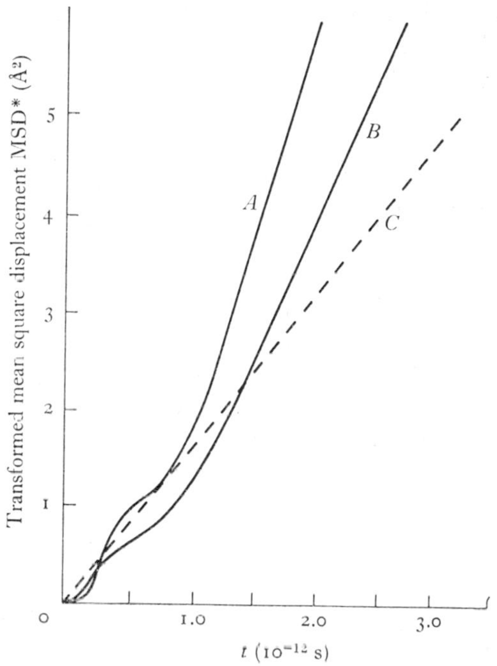

Fig. 5.11.3. The transformed mean square atomic displacement for motions (A) normal and (B) parallel to the liquid surface. The small t behaviour indicates correlated motions whilst the asymptotic form is the familiar linear function $6Dt+c$. The mean square displacement as obtained by Rahman is also shown (C).

expect the diffusion coefficient normal to the liquid surface to be given by

$$D_{\perp} = \frac{\bar{l}^2}{\tau_0},$$

where $\bar{l}^2$ is the mean square diffusive step length, and τ_0 is the phonon lifetime during which it defines the lattice site. If we assume interplanar hopping $\bar{l} \simeq 3.8$ Å, and from the diffusion and autocorrelation functions we assume a mean phonon lifetime of approximately 2.5×10^{-12} s, the above relation yields a diffusion constant of about $10^{-4}\,\mathrm{cm^2\,s^{-1}}$, substantially in agreement with the computed value of $D_{\perp} = 1.05 \times 10^{-4}\,\mathrm{cm^2\,s^{-1}}$. The choice of $\bar{l}$ for a crystalline solid has to be close to the interatomic spacing, approximately 3.8 Å, but in this case the diffusion coefficient is of the order of $10^{-9}\,\mathrm{cm^2\,s^{-1}}$ yielding a phonon lifetime of about 10^{-7} s. In this case the lattice is precisely defined. For lamellar diffusion both $\bar{l}$ and τ_0 would be smaller, but the jump diffusion relation above is seen to depend more sensitively upon the diffusive step length than upon τ_0, and undoubtedly the mean lamellar diffusive step length will be considerably smaller than 3.8 Å. In consequence $D_\perp > D_\parallel$ at the liquid surface. Freeman and McDonald(114) have very recently evaluated the KB relations (4.2.11), (4.2.12) by a Monte Carlo method.

5.12 Quantum liquids

The interacting Bose gas(80–84) has been the subject of intensive theoretical investigation as a microscopic model for the behaviour of liquid ^{4}He. The existence of superfluid phenomena, phonons and quantized vortices in the interacting Bose gas has been demonstrated(80, 81), but it does not permit a quantitative calculation for liquid ^{4}He.

The machine methods are not easily applied to quantum-mechanical systems: indeterminism of the particle distributions and conditions on exchange symmetry of the particle wave functions are not easily handled in the conventional molecular dynamic and Monte Carlo approaches.

A variational method in which the trial wave function is expressed as a product of pair functions of the Bijl–Dingle–Jastrow type has been used extensively(85–88) to describe the ground state of the interacting Bose system. This method can be applied directly to the intermediate density hard sphere gas or to a realistic Hamiltonian for liquid ^{4}He. McMillan(89) reports the quantitative calculation of the properties of the ground state of liquid ^{4}He using this variational method, the integrals being evaluated by a Monte Carlo technique.

The trial wave function was taken to be a product of pair functions, the product being taken over all pairs:

$$\psi = \prod_{i<j=1}^{N} f(r_{ij}), \quad r_{ij} = |\boldsymbol{r}_i - \boldsymbol{r}_j|. \tag{5.12.1}$$

If the atomic diameter is taken as $\sigma = 2.556$ Å(90) the pair function can be made small for $r < \sigma$, and approach a constant for large r, thereby incorporating correlation. Since we are dealing with bosons, the wave function should be totally symmetric under exchange of any two-particle coordinates; the trial function has the proper symmetry. McMillan chooses

$$f(r) = \exp\left[-(a_1/r)^{a_2}\right] \tag{5.12.2}$$

as the trial pair function with variable parameters a_1 and a_2. This function has been used by Wu and Feenberg. The expectation value of the Hamiltonian for spinless bosons interacting through a two-body potential may be easily shown to be

$$\frac{\int \psi \mathcal{H} \psi \, d\tau}{\int \psi^2 \, d\tau} = \frac{\int \sum_{i<j} \left\{ -\frac{\hbar^2}{2m} \nabla_i^2 \ln f(r_{ij}) + \Phi(r_{ij}) \right\} \psi^2 \, d\tau}{\int \psi^2 \, d\tau}. \tag{5.12.3}$$

This may be written in the form,

$$\langle \mathcal{H}(a_1, a_2) \rangle = \frac{\int \mathcal{H}(\mathbf{r}_1 \dots \mathbf{r}_N) P(\mathbf{r}_1 \dots \mathbf{r}_N) \, d\tau}{\int P(\mathbf{r}_1 \dots \mathbf{r}_N) \, d\tau}, \tag{5.12.4}$$

where $$\mathcal{H}(\mathbf{r}_1 \dots \mathbf{r}_N) = \sum_{i<j} \left\{ -\frac{\hbar^2}{2m} \nabla_i^2 \ln f(r_{ij}) + \Phi(r_{ij}) \right\} \tag{5.12.5}$$

$$P(\mathbf{r}_1 \dots \mathbf{r}_N) = \psi^2(\mathbf{r}_1 \dots \mathbf{r}_N) \tag{5.12.6}$$

which is seen to be in the classical form (5.3.7) suitable for Monte Carlo evaluation. McMillan takes the pair potential $\Phi(r)$ to be of LJ 6–12 form with the parameters $\sigma = 2.556$ Å, $\epsilon = 10.22$ °K [90]. Thus, for fixed values of the variational parameters a_1 and a_2, the expectation value of the Hamiltonian, (5.12.4), may be determined. Preliminary calculations were performed with 32 particles in a cubic box with periodic boundary conditions for 41 sets of the parameters (a_1, a_2). The ground-state energy as a function of a_1 and a_2 for the equilibrium density of ^{4}He ($\rho = 2.20 \times 10^{-22}$ atoms cm^{-1}) possesses a minimum for $a_1 \sim 2.6$ Å and $a_2 \sim 5$. The energy at the minimum is $-0.77 \pm 0.09 \times 10^{-15}$ erg atom^{-1} compared with the experimental ground-state energy of -0.988×10^{-15} erg atom^{-1}, a discrepancy of about 20 per cent. At other densities, ranging to beyond the solid–liquid transition at a pressure of 25 atm., a_2 was

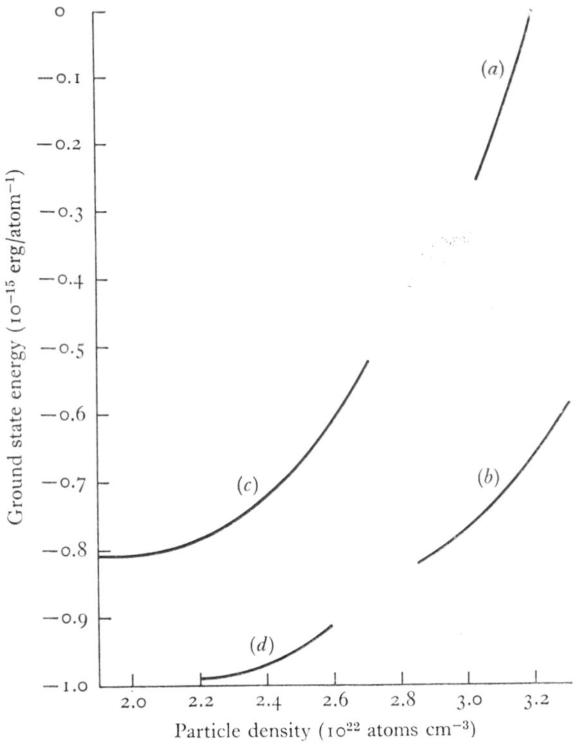

Fig. 5.12.1. The experimental and theoretical ground state energy of solid and liquid ^{4}He as a function of density. (*a*) and (*b*) represent the theoretical and experimental solid curves respectively, and (*c*) and (*d*) represent the theoretical and experimental liquid branches. (Redrawn by permission from McMillan, *Phys. Rev.* **138**, A442 (1965).)

TABLE 5.12.1. *Comparison of experimental and theoretical estimates of the properties of liquid ^{4}He (after McMillan)*

	Particle density (10^{22} atoms cm^{-1})	Theory	Expt
Minimum energy (10^{-15} erg atom^{-1})		-0.810 ± 0.015	-0.988
Equilibrium density (10^{22} atoms cm^{-1})		1.95 ± 0.02	2.20
Energy (10^{-15} erg atom^{-1})	2.20	-0.781 ± 0.015	-0.988
Energy (10^{-15} erg atom^{-1})	1.59	-0.616 ± 0.025	-0.911
Pressure (atm.)	2.20	13 ± 2	0
Pressure (atm.)	2.59	40 ± 6	25
Velocity of sound (m s^{-1})	2.20	267 ± 40	238
Velocity of sound (m s^{-1})	2.59	316 ± 60	365
$\langle \Phi \rangle$ (10^{-15} erg atom^{-1})	2.20	-2.736	
$\langle KE \rangle$ (10^{-15} erg atom^{-1})	2.20	1.955	

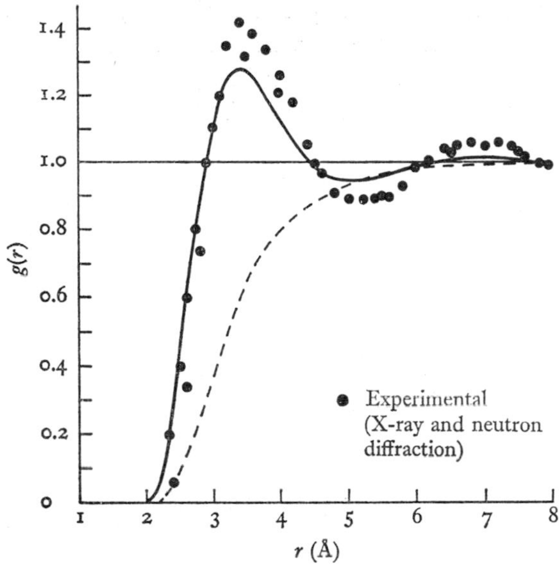

Fig. 5.12.2. Comparison of the Monte Carlo and experimental radial distribution functions of liquid ^{4}He. The pair function squared $f^2(r)$ for $a_1 = 2.6$ Å, $a_2 = 5$ is shown by the broken curve. (Redrawn by permission from McMillan, *Phys. Rev.* **138**, A442 (1965).)

held fixed at 5, and a_1 adjusted to minimize $\langle \mathscr{H} \rangle$ at each density. A break in the resulting a_1 versus density curve was identified as the liquid–solid transition. The corresponding energy–density curve determined by McMillan is shown in fig. 5.12.1 together with the experimental data. The pressure can then be computed from $P = -\partial E/\partial V$, and the velocity of sound from $c^2 = \partial P/m\,\partial \rho$. The theoretical and experimental quantities are compared in table 5.12.1. The agreement between theory and experiment is seen to be reasonably good. The calculation shows that the liquid is the stable phase at zero temperature and pressure. However, the location of the liquid–solid phase transition is only tentative. The theoretical correlation function, fig. 5.12.2 agrees well with the experimental one in the position and sharpness of the cut-off near 2.6 Å; the two variational parameters in the pair function adjust the position and sharpness of the cut-off in the pair function. The experimental curves have a higher peak at 3.5 Å and larger oscillations at large r.

This seems to indicate that the pair function $f(r)$ should peak up somewhat in the region of greatest attraction, but there is no freedom to do this in the two-parameter function used by McMillan. The one-particle exchange density matrix defined by

$$\rho_{(1)}(\boldsymbol{r}_1-\boldsymbol{r}_1') \equiv N\int \psi(\boldsymbol{r}_1,\boldsymbol{r}_2,\ldots,\boldsymbol{r}_N)\,\psi(\boldsymbol{r}_1',\boldsymbol{r}_2,\ldots,\boldsymbol{r}_N)\,\mathrm{d}\boldsymbol{r}_2\,\mathrm{d}\boldsymbol{r}_3\ldots\mathrm{d}\boldsymbol{r}_N\bigg/\int\psi^2\,\mathrm{d}\tau \tag{5.12.7}$$

is also calculated. For small r, $\rho_{(1)}(r)$ approaches the density of particles; Penrose and Onsager(91) have shown that for large r, $\rho_{(1)}(r)$ approaches the density of particles in the zero-momentum state:

$$\rho_{(1)}(r)\rightarrow\rho_0 \quad (r\rightarrow\infty). \tag{5.12.8}$$

According to Penrose and Onsager it is characteristic of the superfluid phase of the boson fluid that ρ_0 is some finite fraction of ρ, and these authors estimate that $\rho_0 \sim 0.08\rho$, that is, 80 per cent of the particles are condensed into the zero momentum state. The momentum distribution is obtained by Fourier transformation:

$$n_k = \int [\rho_{(1)}(\boldsymbol{r})-\rho_0]\,\mathrm{e}^{i\boldsymbol{k}\cdot\boldsymbol{r}}\,\mathrm{d}\boldsymbol{r} \quad (\boldsymbol{k}\neq 0), \tag{5.12.9}$$

where n_k is the average number of particles in momentum state $\boldsymbol{k}$. The one-particle density matrix as determined by McMillan is shown in fig. 5.12.3. The ratio approaches an asymptotic value $\rho_{(1)}(r)/\rho = 0.11\pm0.01$ at about 4 Å. This, according to the Penrose–Onsager criterion(91), is evidence that the system is in the condensed or superfluid phase. The corresponding momentum distribution, n_k, is shown in fig. 5.12.4. n_k approaches a constant ~ 0.75 atoms per momentum state for small k. Also plotted is k^2n_k which is proportional to the number of atoms with momentum k. This function exhibits two peaks, with a minimum at 2 Å^{-1}. 15 per cent of the atoms are in the peak at 2.5 Å^{-1}, and these carry 45 per cent of the kinetic energy.

The variational procedure used by McMillan above, clearly becomes difficult at temperatures above 0 °K, when excited states would have to be included. A formal description of a more general

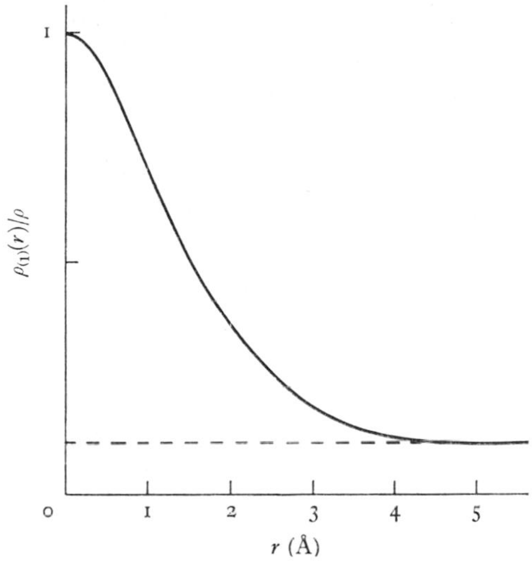

Fig. 5.12.3. The single-particle density matrix as a function of separation. The dashed curve represents the asymptotic limit for large r, $\rho_{(1)}(r) \sim \rho_0 = 0.11\rho$. (Redrawn by permission from McMillan, *Phys. Rev.* **138**, A442 (1965).)

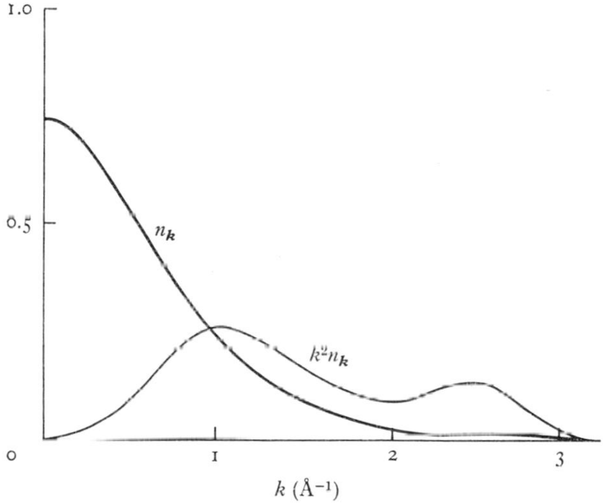

Fig. 5.12.4. The momentum distribution function n_k as determined by McMillan, and n_k multiplied by k^2 to emphasize the behaviour at large k. (Redrawn by permission from McMillan, *Phys. Rev.* **138**, A442 (1965).)

Monte Carlo scheme for the estimation of quantum statistical partition functions based upon the Wiener integral formulation has been given by Fosdick[92]. The procedure is difficult to apply in practice and is fraught with computational difficulties. Fosdick and Jordan[93] have, however, determined the partition function for two particles interacting through a LJ 6–12 potential. The low-temperature second virial coefficient is also calculated, and is in satisfactory agreement with the usual phase-shift analysis of the problem.

5.13 Liquid water

We now make a brief digression from our study of simple liquids to consider the application of simulation techniques to solid and liquid water, for apart from its intrinsic importance, the treatment is of considerable methodological interest in the study of asymmetric molecular systems. The problem cannot, of course, be treated in terms of the simple spherically-symmetric central potentials, effective or otherwise, which form the analytical basis of the theory of fluids reported in this book. Under these circumstances the phase integral is to be taken not only over all the positions and translational momenta, but also over the conjugate coordinates that describe the rotational and vibrational states. The phase integral cannot now be reduced to a product of molecular partition functions and the configurational integral unless the total energy is independent of the orientations and rotational energy, and of the internal vibrational states of the molecule. This is a reasonably good approximation for a simple quasi-spherical molecule such as methane, but is quite inappropriate for a system such as the water molecule. Therefore, we shall first consider the structure of the isolated water molecule, and identify the electropolar components of its potential function.

The available atomic orbitals[100] for the formation of a stable ground-state water molecule are the two 1s hydrogen orbitals and the $2p_x$, $2p_y$ orbitals of the oxygen atom. The $2p_z$ orbital is not available for bonding on symmetry grounds. This directed valence arises by virtue of the orthogonality of the localized $2p_x$, $2p_y$ orbitals, and we might conclude that the HOH bond angle should therefore be 90°. In fact it is experimentally determined to be 104.5°, and this cannot be entirely accounted for in terms of the HH repulsion deriving from Coulombic and exchange repulsive effects. It follows that the bonding orbitals are not pure p-modes but instead are sp-hybrids arising from hybridization of the p orbitals with the oxygen 2s orbital. The condition for bonding is now that the *hybrid* orbitals be orthogonal, and this accounts well for the bond angle of ~105°. Moreover, hybridization of the bonding orbital results in a shift of the centroid of electronic charge away from the oxygen nucleus, resulting in the very strong dipole moment of water, 1.84 debye. We adopt the Cartesian frame of Glaeser and Coulson[101], the oxygen atom defining the origin of coordinates and the z-axis bisecting the HOH bond angle.

It is quite apparent that such a charge distribution will have dipole, quadrupole and octopole moments, and Glaeser and Coulson have determined these on the basis of a number of hybrid orbital wave functions, taking the bond angle to be 105°. Perhaps the most striking result of their calculations is the extensive agreement in the estimates of the electropole moments amongst the various wave functions. Taking the calculated dipole and quadrupole tensors,

$$\left.\begin{aligned} \begin{bmatrix} \mu_x \\ \mu_y \\ \mu_z \end{bmatrix} &= \begin{bmatrix} 0 \\ 0 \\ 1.84 \end{bmatrix} \quad \text{(debye)} \\ \begin{bmatrix} Q_{xx} & Q_{xy} & Q_{xz} \\ Q_{yx} & Q_{yy} & Q_{yz} \\ Q_{zx} & Q_{zy} & Q_{zz} \end{bmatrix} &= \begin{bmatrix} -6.56 & 0 & 0 \\ 0 & -5.18 & 0 \\ 0 & 0 & -5.51 \end{bmatrix} \end{aligned}\right\} \qquad (5.13.1)$$

and an experimental water polarizability of $\alpha = 1.59 \times 10^{-24}\,\text{cm}^3$, Coulson and Eisenberg[102] have calculated the fields and energies arising from a static three-dimensional ice-like array of 85 nearest neighbours, these being represented by the multipolar components in (5.13.1). They find that the mean field[107] has the direction of the dipole moment of the central molecule and is sufficiently strong to increase its total dipole moment to ~2.60 debye. The energy of interaction of the molecules was calculated by taking the electropolar and polarization energies into account. The dipole–dipole,

dipole–quadrupole and quadrupole–quadrupole terms account for 72, 36 and 7.5 per cent of their total electrostatic energy respectively, and clearly in any investigation of water the dipole–quadrupole terms cannot be ignored. Indeed, Fletcher[103] suggests that these terms might be responsible for the spatially-ordered surface structures in ice and water observed under certain conditions. It does not appear worthwhile at the present time, however, to extend the multipole series any further, so as to include dipole–octopole terms, for example.

A direct quantum-mechanical attack on extended arrays of water molecules is clearly out of the question if a rigorous estimate of the total energy is required, although del Bene and Pople[104] have obtained some interesting results for simple linear and chain configurations of water molecules. These calculations have, to date, extended only to six-particle arrays, and do not appear to afford a promising approach to the interaction energy of larger general arrays.

An iterative simulation technique recently developed by Barnes[105] for extended three-dimensional static arrays of asymmetric molecules consists in determining the local field $\boldsymbol{F}$ at the ith molecule due initially to a simple permanent dipole–quadrupole array. The dipolar increments $\Delta\mu_{x_i}$, $\Delta\mu_{y_i}$, $\Delta\mu_{z_i}$ are then calculated from a knowledge of the second rank polarizability tensor, as follows:

$$\begin{bmatrix}\Delta\mu_x\\ \Delta\mu_y\\ \Delta\mu_z\end{bmatrix}_i = \begin{bmatrix}\alpha_{xx} & \alpha_{xy} & \alpha_{xz}\\ \alpha_{yx} & \alpha_{yy} & \alpha_{yz}\\ \alpha_{zx} & \alpha_{zy} & \alpha_{zz}\end{bmatrix} \begin{bmatrix}F_x\\ F_y\\ F_z\end{bmatrix}_i. \tag{5.13.2}$$

This will in turn of course, modify the field within the system, and the dipolar increments of the remaining molecules throughout the array arising from the modified field are now calculated. This self-consistent iterative procedure continues until the electropolar energy attains its final asymptotic value. The energies of polarization together with a spherically symmetric component may be finally added, giving the total configurational energy of the molecular array. Presumably, for recognized ice structures the dipolar enhancements should develop in accordance with the Bernal–Fowler rules[106] if, indeed, the asymptotic total free energy is a minimum. In the case of liquid water, which is currently being investigated, these rules are, in all probability, only statistically satisfied.

In the absence of any knowledge of a quadrupole polarizability tensor $[\alpha_{ijk}]$, calculation of quadrupolar increments is not feasible at present.

Rahman and Stillinger[115] using a Ben-Naim–Stillinger effective (modified Bjerrum) pair potential have performed a series of molecular dynamic simulations of an assembly of 216 water molecules over a range of temperatures. Reasonable agreement with experimental data is obtained in the cases where direct comparison may be made, although the potential appears to be tetrahedrally too directional to represent real water. A scaling of the potential by a factor of 1.06 is also found to improve the agreement. Rahman and Stillinger observe that the liquid structure consists of a highly strained random hydrogen-bond network which appears to show little structural resemblance to known aqueous crystals. Diffusion appears to consists of a continuous and co-operative movement of molecules rather than a hopping process, and increasing temperature results in a breakdown of hydrogen-bond order and an increase in molecular freedom.

Barker and Watts[116] criticize the molecular dynamics computations on the grounds that a set of charges will develop on the surface of truncation of the molecular interaction (cf. fig. 5.2.1 (*a*)) for a system of particles possessing a permanent dipole moment. One must therefore incorporate the effects of the Onsager reaction field which develops, and whilst this contribution diminishes with increasing truncation radius, it must necessarily be incorporated at the Rahman Stillinger cut-off (3.25σ, $\sigma = 2.82$ Å). The inclusion of the reaction field appears to make little difference to the structural and thermodynamic properties, but Barker and Watts show that for an assembly of 64 particles the reaction field is essential if a correct qualitative description of the dielectric properties is to be given.

CHAPTER 6

TRANSPORT PROCESSES

6.1 Introduction

Thus far our interest has been primarily centred on equilibrium time-independent phenomena in simple dense fluids. In general we have dealt with isotropic systems in which there have been no thermal or density gradients with their associated fluxes of energy, momentum and matter. Now, however, we must consider how a non-equilibrium distribution in phase evolves as function of time, and furthermore why it is always observed to evolve towards a stable equilibrium, subject to applied constraints. Clearly we have arrived at the very root of the reversibility paradox. We might anticipate on the basis of the equilibrium theory, as developed in this book, that any attempt to describe the complete phase function $f_{(N)}(\boldsymbol{p}_N, \boldsymbol{q}_N, t)$ would be a hopelessly impossible task. Instead, we shall be primarily interested in the lower order distributions $f_{(1)}, f_{(2)}$; these, it will turn out, are most directly related to the observable transport coefficients; the coefficients of self-diffusion, viscosity, thermal conductivity, etc. Furthermore, it is intuitively clear that for small deviations from equilibrium the lower order distributions evolve rapidly in comparison to the higher order functions, and in fact, an approximation which we shall later invoke assumes that $f_{(N-1)}(\boldsymbol{p}_{N-1}, \boldsymbol{q}_{N-1}, t)$ remains essentially constant during the evolution $f_{(1)}(\boldsymbol{p}, \boldsymbol{q}, t)$. The essential continuity in phase leads directly to the Liouville equation, which is basically no more than a statement of the zero divergence of the phase flux in a conservative system:

$$\frac{\partial f_{(N)}}{\partial t} + \sum_{i=1}^{N} \left\{ \boldsymbol{F}_i \frac{\partial f_{(N)}}{\partial \boldsymbol{p}_i} + \boldsymbol{X}_i \frac{\partial f_{(N)}}{\partial \boldsymbol{p}_i} + \frac{\boldsymbol{p}_i}{m} \frac{\partial f_{(N)}}{\partial \boldsymbol{q}_i} \right\} = 0, \tag{6.1.1}$$

and any phase evolution must proceed subject to this equation. $\boldsymbol{F}_i$ is the net interparticle force, and $\boldsymbol{X}_i$ an external force acting on particle i. Should we require the hth order *reduced* distribution in terms we merely integrate over the phase coordinates of the remaining $(N-h)$ particles. Remembering that $f_{(h)}$ vanishes whenever the phase variables move outside the phase available to the system, the equation for $f_{(h)}$ is found to have the form

$$\frac{\partial f_{(h)}}{\partial t}+\sum_{i=1}^{h}\frac{\boldsymbol{p}_i}{m}\frac{\partial f_{(h)}}{\partial \boldsymbol{q}_i}+\sum_{i=1}^{h}\boldsymbol{X}_i\frac{\partial f_{(h)}}{\partial \boldsymbol{p}_i}$$
$$=-\sum_{i=1}^{h}\frac{\partial}{\partial \boldsymbol{p}_i}\int\cdots\int \boldsymbol{F}(i,h+1)f_{(h+1)}\,\Pi\,\mathrm{d}\boldsymbol{p}_{(h+1)}\,\mathrm{d}\boldsymbol{q}_{(h+1)}, \quad (6.1.2)$$

i.e.

$$\frac{\partial f_{(h)}}{\partial t}+\sum_{i=1}^{h}\frac{\boldsymbol{p}_i}{m}\frac{\partial f_{(h)}}{\partial \boldsymbol{q}_i}+\sum_{i=1}^{h}\boldsymbol{X}_i\frac{\partial f_{(h)}}{\partial \boldsymbol{p}_i}$$
$$=+\sum_{i=1}^{h}\frac{\partial}{\partial \boldsymbol{p}_i}\int\cdots\int \frac{\partial\Phi(i,h+1)}{\partial \boldsymbol{q}_i}f_{(h+1)}\,\Pi\,\mathrm{d}\boldsymbol{p}_{(h+1)}\,\mathrm{d}\boldsymbol{q}_{(h+1)}, \quad (6.1.3)$$

in the pair approximation $\Phi(ij)$ is the interatomic pair potential. This non-linear integral equation is seen to express the hth order distribution function in terms of the distribution $f_{(h+1)}$, and this is precisely the same hierachical problem encountered in the Born–Green–Yvon development (§2.5). Indeed, the configurational projection of (6.1.3) is the hth order BGY integro-differential equation, prior to closure. For our purposes we shall find the lowest order distribution functions of greatest importance. The singlet evolution is therefore described as

$$\frac{\partial f_{(1)}}{\partial t}+\left\{\frac{\boldsymbol{p}_1}{m}\frac{\partial f_{(1)}}{\partial \boldsymbol{q}_1}+\boldsymbol{X}_i\frac{\partial f_{(1)}}{\partial \boldsymbol{p}_1}\right\}=-\frac{\partial}{\partial \boldsymbol{p}_1}\int \boldsymbol{F}_{12}\,f_{(2)}\,\mathrm{d}\boldsymbol{p}_2\,\mathrm{d}\boldsymbol{q}_2. \quad (6.1.4)$$

This then is the single-particle Liouville equation whose components may be identified as

$$\frac{\partial f_{(1)}}{\partial t}+\left(\frac{\partial f_{(1)}}{\partial t}\right)_{\text{drift}}-\left(\frac{\partial f_{(1)}}{\partial t}\right)_{\text{collision}}. \quad (6.1.5)$$

The term on the right hand side, generally designated J, the collision integral, describes the scattering to and from the phase element at $(\boldsymbol{p}, \boldsymbol{q})$. This equation as it stands tells us nothing about the scattering population or depopulation of the state $(\boldsymbol{p}, \boldsymbol{q})$ – all we do know is that scattering proceeds so as to preserve a zero divergence in phase. Clearly, the details of this process are contained within the collision integral, and the determination of this term and subsequent solution of the Liouville equation would, in principle, enable us to determine the phase evolution.

The simplest phase evolution is described by the collisionless Boltzmann equation which consists in setting the collision integral to zero:

$$\frac{\partial f_{(1)}}{\partial t}+\frac{\boldsymbol{p}_1}{m}\frac{\partial f_{(1)}}{\partial \boldsymbol{q}_1}+\boldsymbol{X}_1\frac{\partial f_{(1)}}{\partial \boldsymbol{p}_1}=0. \tag{6.1.6}$$

A more realistic description however, will inevitably require an explicit determination of the form of the collision integral.

One approximation which has been proposed is the so-called 'relaxation approximation' which sets

$$\left(\frac{\partial f_{(1)}}{\partial t}\right)_{\text{collision}}=-\frac{f_{(1)}-f^0_{(1)}}{\tau}, \tag{6.1.7}$$

where τ is some characteristic relaxation time of the system, and $f^0_{(1)}$ is the equilibrium distribution. The Boltzmann transport equation then becomes

$$\frac{\partial f_{(1)}}{\partial t}+\frac{\boldsymbol{p}_{(1)}}{m}\frac{\partial f_{(1)}}{\partial \boldsymbol{q}_1}+\boldsymbol{X}_1\frac{\partial f_{(1)}}{\partial \boldsymbol{p}_1}=-\frac{f_{(1)}-f^0_{(1)}}{\tau} \tag{6.1.8}$$

which is simply a linear partial differential equation for $f_{(1)}$. This equation has been expressed in the 'path-integral formulation'

$$f_{(1)}(\boldsymbol{p},\boldsymbol{q},t)=\int_0^\infty f^0_{(1)}(\boldsymbol{p}_0,\boldsymbol{q}_0,t-t')\exp\left(-\int_{t-t'}^{t}\frac{ds}{\tau(s)}\right)\frac{dt'}{\tau(t-t')} \tag{6.1.9}$$

which may be quite easily shown to be identical to (6.1.8). In this form it allows one to compute $f_{(1)}$ from a knowledge of the collision time $\tau(s)$ and from an assumed function f^0 describing the local equilibrium distribution immediately after a collision. Furthermore, it assumes that the same distribution $f^0_{(1)}$ is always restored after collisions, and hence disregards persistence of velocity effects, that is the fact that the post-collision velocity of the molecule is not independent of its pre-collision velocity. Whilst this approach has been used with great success in many problems, it is quite clearly inapplicable at liquid densities, and will be pursued no further here.

6.2 The Boltzmann transport equation: the collision integral

It is quite clear from the preceding discussion that any realistic application of the Boltzmann equation entails a determination of the collision integral. The number of limiting cases in which analytic expressions for the collision integral may be determined is small. Nevertheless, one important case which we shall investigate here concerns the strongly repulsive interaction between molecules involved in large momentum binary exchanges. Of course, the concept of a binary collision is meaningful only when the macroscopic fluid density is sufficiently low (dilute gas). In this limit, in which we shall also necessarily invoke the hypothesis of molecular chaos, we may consider the transport equation for a high temperature dilute gas. Our objective, as always, will be to determine the collision integral.

We shall make a number of assumptions:

(i) In general the gas will be assumed sufficiently dilute that binary encounters need only be considered. However, for the high density hard sphere fluid the collision process may be resolved as a rapid succession of binary encounters provided the time-graining is fine enough. In this case, then, the density of the fluid does not impede the determination of the collision integral. As we shall see, however (assumption (ii)), we are nevertheless restricted to low density determinations.

(ii) The mean free path is considered to be much greater than the range of the molecular forces. This eliminates velocity correlations and persistence of velocity effects; the concept of 'molecular chaos' is thus invoked. It is this assumption in particular which ensures irreversibility of the transport equation. The evolution is essentially Markovian in this case, and clearly the collision integral derived on the basis of this assumption is applicable only at low densities.

(iii) Any possible effects of the external force $\boldsymbol{X}$ on the magnitude of the collision cross section can be ignored.

(iv) The distribution function $f_{(1)}(\boldsymbol{p}, \boldsymbol{q}, t)$ does not vary appreciably during a time interval of the order of a molecular collision, nor does it vary appreciably over a spatial distance of the order of the range of intermolecular forces.

If we focus attention on the phase element $\Delta\boldsymbol{p}\Delta\boldsymbol{q}$ located at $(\boldsymbol{p}_1, \boldsymbol{q}_1)$, then in the period between t and $t+\Delta t$ a number of particles will be scattered out of the element by virtue of collisions with other molecules. We denote this *decrease* by $J^{(-)}\,\mathrm{d}\boldsymbol{p}_1\,\mathrm{d}\boldsymbol{q}_1\,\mathrm{d}t$. Similarly, molecules initially outside the phase element will be scattered into the state $(\boldsymbol{p},\boldsymbol{q})$ by virtue of collisions with other particles: we denote the resulting *increase* in time $\mathrm{d}t$ of the number of molecules scattered into this region of phase by $J^{(+)}\,\mathrm{d}\boldsymbol{p}_1\,\mathrm{d}\boldsymbol{q}_1\,\mathrm{d}t$. Thus, scattering to and from the phase element $\mathrm{d}\boldsymbol{p}_1\,\mathrm{d}\boldsymbol{q}_1$ in time $\mathrm{d}t$ is given as

$$J\,\mathrm{d}\boldsymbol{p}_1\,\mathrm{d}\boldsymbol{q}_1\,\mathrm{d}t = [J^{(+)}-J^{(-)}]\,\mathrm{d}\boldsymbol{p}_1\,\mathrm{d}\boldsymbol{q}_1\,\mathrm{d}t. \tag{6.2.1}$$

Depopulation of the state $(\boldsymbol{p}_1,\boldsymbol{q}_1)$ is most easily envisaged as follows. The number, or flux, of scattering particles of unit mass in the range $\boldsymbol{p}_2 \to \boldsymbol{p}_2+\mathrm{d}\boldsymbol{p}_2$ incident on particles in the state $(\boldsymbol{p}_1,\boldsymbol{q}_1)$ in time $\mathrm{d}t$ is

$$(|\boldsymbol{p}_2-\boldsymbol{p}_1|)f_{(1)}(\boldsymbol{p}_2,\boldsymbol{q}_2,t)\,\mathrm{d}\boldsymbol{p}_2\,\mathrm{d}t.$$

The number of particles *initially* in the phase element is, of course, $f_{(1)}(\boldsymbol{p}_1,\boldsymbol{q}_1,t)\,\mathrm{d}\boldsymbol{p}_1\,\mathrm{d}\boldsymbol{q}_1$. Hence, the total number of particles scattered *out* of the element $\mathrm{d}\boldsymbol{p}_1\,\mathrm{d}\boldsymbol{q}_1$ in time $\mathrm{d}t$ is

$$\begin{aligned} J^{(-)}\,\mathrm{d}\boldsymbol{p}_1\,\mathrm{d}\boldsymbol{q}_1\,\mathrm{d}t = \iiint_{p_2\,p_2'\,p_1'} & [(|\boldsymbol{p}_2-\boldsymbol{p}_1|)f_{(1)}(\boldsymbol{p}_2,\boldsymbol{q}_2,t)\,\mathrm{d}\boldsymbol{p}_2\,\mathrm{d}t] \\ & \times [f_{(1)}(\boldsymbol{p}_1,\boldsymbol{q}_1,t)\,\mathrm{d}\boldsymbol{p}_1\,\mathrm{d}\boldsymbol{q}_1] \\ & \times [\sigma(\boldsymbol{p}_1,\boldsymbol{p}_2\to\boldsymbol{p}_1',\boldsymbol{p}_2')\,\mathrm{d}\boldsymbol{p}_1'\,\mathrm{d}\boldsymbol{p}_2'], \end{aligned} \tag{6.2.2}$$

where $\sigma(\boldsymbol{p}_1,\boldsymbol{p}_2\to\boldsymbol{p}_1',\boldsymbol{p}_2')\,\mathrm{d}\boldsymbol{p}_1'\,\mathrm{d}\boldsymbol{p}_2'$ is the scattering cross-section for the process $\boldsymbol{p}_1,\boldsymbol{p}_2\to\boldsymbol{p}_1',\boldsymbol{p}_2'$, and the integration is over all possible values of $\boldsymbol{p}_2,\boldsymbol{p}_2',\boldsymbol{p}_1'$. Likewise, population of the phase element proceeds as

$$\begin{aligned} J^{(+)}\,\mathrm{d}\boldsymbol{p}_1\,\mathrm{d}\boldsymbol{q}_1\,\mathrm{d}t = \iiint_{p_2'\,p_2\,p_1'} & [(|\boldsymbol{p}_2'-\boldsymbol{p}_1'|)f_{(1)}(\boldsymbol{p}_2',\boldsymbol{q}_2',t)\,\mathrm{d}\boldsymbol{p}_2'\,\mathrm{d}t] \\ & \times [f_{(1)}(\boldsymbol{p}_1',\boldsymbol{q}_1',t)\,\mathrm{d}\boldsymbol{p}_1'\,\mathrm{d}\boldsymbol{q}_1'] \\ & \times [\sigma'(\boldsymbol{p}_1',\boldsymbol{p}_2'-\boldsymbol{p}_1,\boldsymbol{p}_2)\,\mathrm{d}\boldsymbol{p}_1\,\mathrm{d}\boldsymbol{p}_2], \end{aligned} \tag{6.2.3}$$

where $\sigma'(\boldsymbol{p}_1',\boldsymbol{p}_2'\to\boldsymbol{p}_1,\boldsymbol{p}_2)$ is the cross-section for the analogous process. Before combining $J^{(+)}$ and $J^{(-)}$ to form the collision integral, we make one or two simplifying observations. First, the probability σ has various useful symmetry properties with regard to the collision process. The equations of motion must remain invariant under reversal of the sign of time from t to $-t$. Under such a time reversal, which implies of course a corresponding reversal of all the velocities, one obtains the 'reverse' collision in which the particles simply retrace their paths in time. Further, the equations of motion must be invariant under the transformation which reverses the sign of all spatial coordinates so that $\boldsymbol{q} \to -\boldsymbol{q}$. Under such a 'space inversion', the sign of all velocities change but the time order does not. The consequence of these two operations is that

$$\sigma(\boldsymbol{p}_1, \boldsymbol{p}_2 \to \boldsymbol{p}_1', \boldsymbol{p}_2') \equiv \sigma'(\boldsymbol{p}_1', \boldsymbol{p}_2' \to \boldsymbol{p}_1, \boldsymbol{p}_2).$$

Secondly, conservation of energy for a two-body elastic collision implies that

$$|\boldsymbol{p}_2 - \boldsymbol{p}_1| = |\boldsymbol{p}_2' - \boldsymbol{p}_1'|$$

whereupon, in an obvious notation, the collision integral may be written

$$J = \iiint_{p_2 p_2' p_1'} |\boldsymbol{p}_1 - \boldsymbol{p}_2| \, \sigma\{f_{(1)}'(2) f_{(1)}'(1) - f_{(1)}(1) f_{(1)}(2)\} \, \mathrm{d}\boldsymbol{p}_2 \, \mathrm{d}\boldsymbol{p}_2' \, \mathrm{d}\boldsymbol{p}_1'. \tag{6.2.4}$$

This development of the collision integral follows the original derivation of Boltzmann, but we could equally well have taken the right hand side of (6.1.4)

$$-\frac{\partial}{\partial \boldsymbol{p}_1} \iint \boldsymbol{F}_{12} f_{(2)} \, \mathrm{d}\boldsymbol{p}_2 \, \mathrm{d}\boldsymbol{q}_2$$

and given a statistical argument. The force term may be written in terms of the corresponding momentum increment $\Delta \boldsymbol{p}_1$, in a time interval s, so that the integrand above becomes of the form

$$\Delta \boldsymbol{p}_1 (\partial / \partial \boldsymbol{p}_1) \{f_{(2)}(\boldsymbol{p}_1, \boldsymbol{p}_2, \boldsymbol{q}_1, \boldsymbol{q}_2, t+s)\}.$$

The position increment $\Delta \boldsymbol{q}$ during the interval s will be negligibly small, and may be disregarded in its effect on the distribution.

For the time interval τ, which is macroscopically small but microscopically large, and as it turns out, does not need to be specified precisely, we may write the transport equation as

$$\frac{\partial \bar{f}_{(1)}}{\partial t} + \frac{\boldsymbol{p}_1}{m} \frac{\partial \bar{f}_{(1)}}{\partial \boldsymbol{q}_1} = \frac{N}{\tau} \iint \{f_{(2)}(t-\tau) - f_{(2)}(t)\} \, \mathrm{d}\boldsymbol{p}_2 \, \mathrm{d}\boldsymbol{q}_2 \tag{6.2.5}$$

where the system is subject to no external forces. $\boldsymbol{X}$. This 'time-smoothed' equation for the evolution is denoted by the bar: $\bar{f}_{(1)}$.

The concept of a time-smoothed evolution introduced by Kirkwood, does not affect the applicability of the Liouville or Boltzmann equations – indeed, the smoothed distribution satisfies these as well as the unsmoothed function. In the limit $\tau \to 0$ the original unsmoothed evolution is regained. We now introduce the hypothesis of molecular chaos in the form

$$f_{(2)}(\boldsymbol{p}_1, \boldsymbol{q}_1, \boldsymbol{p}_2, \boldsymbol{q}_2, t) = f_{(1)}(\boldsymbol{p}_1, \boldsymbol{q}_1, t) f_{(1)}(\boldsymbol{p}_2, \boldsymbol{q}_2, t) \qquad (6.2.6)$$

which asserts the statistical independence of each of the two particles of the pair. This is, in fact, contrary to our initial assumption that the particles correlate through binary collisions. However, the collision period $\tau_c \ll \tau_r$, the relaxation period, and consequently over the period τ in (6.2.5) the assumption of molecular chaos is a reasonable approximation. Substitution of (6.2.6) in (6.2.5) yields

$$\frac{\partial \bar{f}_{(1)}}{\partial t} + \frac{\boldsymbol{p}_1}{m} \frac{\partial \bar{f}_{(1)}}{\partial \boldsymbol{q}_1} = \frac{N}{\tau} \iint \{\bar{f}_{(1)}(1)(t-\tau) \bar{f}_{(1)}(2)(t-\tau) - \bar{f}_{(1)}(1)(t) \bar{f}_{(1)}(2)(t)\} \, d\boldsymbol{p}_2 \, d\boldsymbol{q}_2, \qquad (6.2.7)$$

where the time-smoothed distribution has been inserted in the right hand side. As written, the assertion has been made that τ is so small that there is essentially no difference between $\bar{f}_{(1)}$ and $f_{(1)}$. This is demonstrably the case for the impulsive interaction considered here, but clearly this simplification is inapplicable at higher densities under protracted interaction. The phase integral in (6.2.7) over the entire phase of particle 2 represents the net population or depopulation of the phase element and this enables us to express (6.2.7) in terms of the scattering cross-section:

$$\frac{\partial \bar{f}_{(1)}}{\partial t} + \frac{\boldsymbol{p}_1}{m} \frac{\partial \bar{f}_{(1)}}{\partial \boldsymbol{q}_1} = \frac{N}{\tau} \iiint |\boldsymbol{p}_1 - \boldsymbol{p}_2| \, \sigma \{f'_{(1)}(1) f'_{(1)}(2) - f_{(1)}(1) f_{(1)}(2)\} \, d\boldsymbol{p}_2 \, d\boldsymbol{p}'_2 \, d\boldsymbol{p}'_1 \qquad (6.2.8)$$

which is seen to be identical to (6.2.4), although in time-smoothed form. It might be mentioned that a knowledge of the time-smoothed function can be used to find information about the corresponding unsmoothed distribution, but clearly short-time information is irretrievably lost. The exact trajectory patterns cannot be identified, and the phase motion must be treated statistically.

6.3 The equilibrium state of a dilute gas

We may define the equilibrium distribution as a solution of the Boltzmann equation which is independent of time, that is, one whose collision integral is zero. We shall also see that it is also the distribution assumed by the system in the limit $t \to \infty$. If we recall the definition of the Boltzmann functional

$$H(t) = \int f_{(N)}(\boldsymbol{p}, t) \ln [f_{(N)}(\boldsymbol{p}, t)] \, \mathrm{d}\boldsymbol{p}, \tag{6.3.1}$$

where independence of the velocity distribution upon position has been assumed, then Boltzmann's H-theorem tells us that if f satisfies the Boltzmann equation

$$\frac{\mathrm{d}H(t)}{\mathrm{d}t} \leqslant 0 \tag{6.3.2}$$

and since H is, by definition, the negative of the entropy, (6.3.2) is a statement of the second law. The time differential of (6.3.1) is

$$\frac{\mathrm{d}H(t)}{\mathrm{d}t} = \int \frac{\partial f_{(N)}(\boldsymbol{p}, t)}{\partial t} [1 + \ln f_{(N)}(\boldsymbol{p}, t)] \, \mathrm{d}\boldsymbol{p}, \tag{6.3.3}$$

where $(\partial f_{(N)}/\partial t)$ is the collisional derivative. To simplify the discussion we quite reasonably assume that the distribution is independent of position, and that no external forces act. Moreover, if we assume $f_{(N)}(\boldsymbol{p}, t) = N\bar{f}_{(1)}(\boldsymbol{p}_1, t)$ we have, from (6.2.8),

$$\frac{\mathrm{d}H(t)}{\mathrm{d}t} = \iiiint \sigma |\boldsymbol{p}_1 - \boldsymbol{p}_2| \{f'(1)f'(2) - f(1)f(2)\} \times [1 + \ln f_{(1)}(\boldsymbol{p}_1, t)] \, \mathrm{d}\boldsymbol{p}_1 \, \mathrm{d}\boldsymbol{p}_2 \, \mathrm{d}\boldsymbol{p}_1' \, \mathrm{d}\boldsymbol{p}_2', \tag{6.3.4}$$

where we have dropped the subscript (1) without ambiguity. Following Huang, we observe that interchanging $\boldsymbol{p}_1$ and $\boldsymbol{p}_2$ in this integrand leaves the integral invariant since σ is invariant under such an interchange. Making this change of variables of integration, and taking one half of the resulting expression and (6.3.4), we obtain

$$\frac{\mathrm{d}H(t)}{\mathrm{d}t} = \frac{1}{2} \iiiint \sigma |\boldsymbol{p}_2 - \boldsymbol{p}_1| \{f'(1)f'(2) - f(1)f(2)\} \times [2 + \ln f(1)f(2)] \, \mathrm{d}\boldsymbol{p}_1 \, \mathrm{d}\boldsymbol{p}_2 \, \mathrm{d}\boldsymbol{p}_1' \, \mathrm{d}\boldsymbol{p}_2'. \tag{6.3.5}$$

Furthermore, this integral is invariant to interchange of

$$(\boldsymbol{p}_1, \boldsymbol{p}_2 \rightarrow \boldsymbol{p}'_1, \boldsymbol{p}'_2)$$

since for every collision there is an inverse collision having the same cross-section. We may therefore write, after (6.3.5),

$$\frac{dH(t)}{dt} = \frac{1}{2}\iiiint \sigma' |\boldsymbol{p}'_2 - \boldsymbol{p}_1| \{f(1)f(2) - f'(1)f'(2)\} \times [2 + \ln f'(1)f'(2)]\, d\boldsymbol{p}'_1\, d\boldsymbol{p}'_2\, d\boldsymbol{p}_1\, d\boldsymbol{p}_2. \quad (6.3.6)$$

Since $d\boldsymbol{p}_1\, d\boldsymbol{p}_2 = d\boldsymbol{p}'_1\, d\boldsymbol{p}'_2$, $|\boldsymbol{p}_1 - \boldsymbol{p}_2| = |\boldsymbol{p}_2 - \boldsymbol{p}_1|$ and $\sigma = \sigma'$ (c.f. §6.2), then we may take half the sum of (6.3.5) and (6.3.6), giving finally

$$\frac{dH(t)}{dt} = \frac{1}{4}\iiiint \sigma |\boldsymbol{p}_2 - \boldsymbol{p}_1| \{f'(1)f'(2) - f(1)f(2)\} \times [\ln f(1)f(2) - \ln f'(1)f'(2)]\, d\boldsymbol{p}_1\, d\boldsymbol{p}_2\, d\boldsymbol{p}'_1\, d\boldsymbol{p}'_2. \quad (6.3.7)$$

From this we see that the integrand is never positive ensuring the monotonic decrease of H with time. Furthermore, $(dH/dt) = 0$ only when the integrand in (6.3.7) vanishes, i.e. when the collision integral inserted in (6.3.3) is zero:

$$f'_0(1)f'_0(2) = f_0(1)f_0(2), \quad (6.3.8)$$

where the subscript 0 denotes that this is the equilibrium distribution. Incidentally, this also shows that $f(\boldsymbol{p}, \boldsymbol{q}, t) \rightarrow f_0(\boldsymbol{p}, \boldsymbol{q}, \infty)$ from any initial distribution.

It is not very difficult to go on from here to demonstrate that the equilibrium velocity distribution is indeed Maxwell–Boltzmann, but this would take us too far afield, and instead the reader is directed to the several excellent statistical mechanics texts on the subject.

Our primary concern, of course, is with dense fluid systems; however the case of the dilute gas occupies a special position in that it represents possibly the best known and most successful branch of non-equilibrium statistical theory. We shall, therefore, consider one or two aspects of the dilute system for which there is no corresponding satisfactory description at liquid densities. Firstly we observe that the monotonic decrease of H $(= -S/k)$ with time has been shown only for the case where molecular chaos prevails: any departure from this condition by virtue of propagation of correlation either in time on space prevents us from drawing any general conclusions concerning the temporal evolution of the H-function. It may however be very easily shown that H is at a local peak when the system is in a state of molecular chaos. Furthermore H is at its minimum value when the distribution function is strictly Maxwell–Boltzmann and this is independent of the assumption of molecular chaos. Thus, evolution of the distribution is adequately described by the Maxwell–Boltzmann equation only when conditions of molecular chaos apply for most of the time. In this way the evolution (6.3.7) is to be accepted as a true representation of the non-equilibrium condition of a dilute gas only in statistical terms.

A means of solving the transport equation in the limit of low densities has been proposed by Enskog. Suppose we express the collision integral in the following notation:

$$J(f_i, f_j) \equiv \iiint (f_i' f_j' - f_i f_j)\, |\boldsymbol{p}_i - \boldsymbol{p}_j|\, \sigma\, \mathrm{d}\boldsymbol{p}_i\, \mathrm{d}\boldsymbol{p}_j\, \mathrm{d}\boldsymbol{p}_i'\, \mathrm{d}\boldsymbol{p}_j',$$

whereupon the Boltzmann transport equation is written

$$\frac{\partial f_i}{\partial t} + \frac{\boldsymbol{p}_i}{m}\frac{\partial f_i}{\partial \boldsymbol{q}_i} + \boldsymbol{X}_i \frac{\partial f_i}{\partial \boldsymbol{p}_i} = J(f_i, f_j). \qquad (6.3.9)$$

The Chapman–Enskog (CE) method then solves (6.3.9) by successive approximation. This method will not yield the most general solution of the Boltzmann equation, but only solutions that depend on time implicitly through the local density, velocity and temperature. It is appropriate to point out a special characteristic of the Chapman–Enskog solution; a solution not initially of this type becomes of this type in a time of the order of a collision time. Since the collision time for ordinary gases is of the order of 10^{-11} seconds solutions of the type mentioned above are the relevant ones in most physical applications.

According to the method proposed by Enskog, the non-equilibrium distribution $f(\boldsymbol{p}_i, \boldsymbol{q}_i, t)$ is expanded in terms of a parameter ϵ, thus

$$f(t) = f_0 + \epsilon f_1(t) + \epsilon^2 f_2(t) + \epsilon^3 f_3(t) + \ldots = \sum_{k=0}^{\infty} \epsilon^k f_k(t), \qquad (6.3.10)$$

where $(1/\epsilon)$ is a collision-frequency parameter. Thus, when collisions are frequent and $1/\epsilon$ is large the zeroth-order equilibrium distribution f_0 is rapidly attained and the higher order terms rapidly disappear. Conversely, for a low collision frequency equilibrium is only slowly achieved. Equation (6.3.10) presupposes that the deviation from local equilibrium is small. This series expansion may then be inserted in (6.3.9) with the modification that the collision term be replaced by $(1/\epsilon)J$, so that the transport equation now reads

$$\frac{\partial f_i}{\partial t}+\frac{\boldsymbol{p}_i}{m}\frac{\partial f_i}{\partial \boldsymbol{q}_i}+\boldsymbol{X}_i\frac{\partial f_i}{\partial \boldsymbol{p}_i}=\frac{1}{\epsilon}J(f_i,f_j) \tag{6.3.11}$$

in Enskog form. This, of course, becomes identical with the original transport equation for the condition $\epsilon = 1$. The functions f_k in (6.3.10) are to be obtained by inserting (6.3.10) into (6.3.11), and then equating the coefficients of ϵ successively to zero in the resulting equation. As a result (6.3.11) is replaced by the scheme of component equations:

$$0 = J(f_{i0},f_{j0}), \tag{6.3.12}$$

$$\frac{\partial f_{i0}}{\partial t}+\left(\frac{\boldsymbol{p}_i}{m}\frac{\partial f_{i0}}{\partial \boldsymbol{q}_i}\right)+\left(\boldsymbol{X}_i\frac{\partial f_{i0}}{\partial \boldsymbol{p}_i}\right)=J(f_{i0},f_{j1})+J(f_{i1},f_{j0}) \tag{6.3.13}$$

$$\frac{\partial f_{i1}}{\partial t}+\left(\frac{\boldsymbol{p}_i}{m}\frac{\partial f_{i1}}{\partial \boldsymbol{q}_i}\right)+\left(\boldsymbol{X}_i\frac{\partial f_{i1}}{\partial \boldsymbol{p}_i}\right)=J(f_{i0},f_{j2})+J(f_{i1},f_{j1})+J(f_{i2},f_{j0}) \tag{6.3.14}$$

and so on. The first equation, (6.3.12), involves the unknown function f_0, the second, (6.3.13), involves f_0 and f_1. The third, f_0, f_1, f_2, etc. In principle a knowledge of the solutions preceding a given equation enables that equation to be expressed in terms of a single unknown function. Thus, in (6.3.12) a knowledge of f_0 enables us to determine f_1. In (6.3.13) a knowledge of f_0 and f_1 enables us to determine f_2, and so on. We shall not consider the convergence of the Enskog series here.

The limitations of the Maxwell–Boltzmann transport equation are contained within the restriction to molecular chaos with the form of collision integral adopted here. Provided the scattering to and from an elemental phase volume may be essentially described in terms of impulsive binary interactions, then a collision integral of the form discussed above will be appropriate. And provided the system is below some critical density, the exact value of which depends upon the details of the interparticle force, then the evolution of the system in phase tends to an equilibrium distribution under the action of the streaming terms and the collision integral.

6.4 The Fokker–Planck equation

Our principal interest, of course, is with high density low temperature systems of particles in continuous interaction, where the concept of a binary interaction is no longer meaningful. (Nevertheless, as we have observed in §6.3, we may, in the case of a high density hard sphere system, still resolve the evolution of the system as a rapid series of impulsive binary encounters as in the development of the Boltzmann collision integral. The assumption of molecular chaos, however, must be abandoned under these circumstances.) Furthermore, at these densities the momentum exchanges for a system in continuous interaction are generally small, such that $\boldsymbol{p} \gg \Delta \boldsymbol{p}$. The large momentum binary exchanges discussed above no longer govern the evolution. Moreover, whilst in the case of binary impulsive encounters the colliding centres execute dynamically complementary conservative trajectories, a fundamental assymmetry arises in the case of the high density extended interaction system, for whilst the dynamical interaction between a particle and its neighbours is collectively both reciprocal and conservative, the net effect on the neighbourhood is insignificant. The motion of the particle in question may therefore be taken as being essentially stochastic with an associated correlation time τ, after which its motion is no longer correlated with its initial phase. In other words, the force autocorrelation taken over the remaining $N-1$ particles vanishes for some time $s > \tau$ as follows:

$$\Xi(s) = \langle \boldsymbol{F}_1(t) . \boldsymbol{F}_1(t+s) \rangle = 0 \quad \text{for} \quad s > \tau \qquad (6.4.1)$$

does not have to be defined precisely, other than that it should be 'sufficiently long', that is, greater than the relaxation time, but less than the Poincaré period when the periodicity of the phase evolution becomes apparent. This essentially dissipative condition replaces the molecular chaos hypothesis introduced in the case of the dilute gas, and, moreover, establishes the irreversible approach to equilibrium in the dense fluid. The 'loss of memory' implied by $\Xi(s) = 0$ for $s > \tau$ prevents the otherwise mechanically reversible Liouville equation retracing the phase trajectory.

Given the initial distribution in phase $f(\boldsymbol{p}, t)$, continuity in momentum implies that

$$f_{(1)}(\boldsymbol{p}+\Delta\boldsymbol{p}, t+\Delta t) = -\frac{1}{\tau}\int_0^\tau \iint \frac{1}{m}\sum_{i=1}^{N}\left(\boldsymbol{F}_i\frac{\partial \bar{f}_{(N)}}{\partial \boldsymbol{p}_i}\right)\prod_{j=2}^{N} \mathrm{d}\boldsymbol{p}_j\,\mathrm{d}\boldsymbol{q}_j\,\mathrm{d}s$$

$$= \int \overset{p+\Delta p}{\underset{p}{W}} f_{(1)}(\boldsymbol{p}, t)\,\mathrm{d}(\Delta\boldsymbol{p}) \tag{6.4.2}$$

(cf. (6.1.4)) where $\overset{p+\Delta p}{\underset{p}{W}}$ is the time-independent transition probability in time Δt of the transition of particle 1 from state $(\boldsymbol{p}, t)$ to state $(\boldsymbol{p}+\Delta\boldsymbol{p}, t+\Delta t)$. The final distribution $f_{(1)}(\boldsymbol{p}+\Delta\boldsymbol{p}, t+\Delta t)$ is then determined as the integral over all momentum exchanges $\Delta\boldsymbol{p}$. Provided these latter are small, we may Taylor expand the above integrand and obtain directly for the total evolution:

$$\frac{\partial f_{(1)}}{\partial t}+\frac{\boldsymbol{p}_1}{m}\frac{\partial f_{(1)}}{\partial \boldsymbol{q}_1} = \frac{\partial}{\partial \boldsymbol{p}_1}\left\{\frac{\langle\Delta\boldsymbol{p}_1\rangle}{\Delta t}\bar{f}_{(1)}+\frac{1}{2}\frac{\partial}{\partial \boldsymbol{p}_1}\frac{\langle\Delta\boldsymbol{p}_1\Delta\boldsymbol{p}_1\rangle}{\Delta t}\bar{f}_{(1)}\right\}+\boldsymbol{X}_1\frac{\partial f_{(1)}}{\partial \boldsymbol{p}_1}, \tag{6.4.3}$$

where

$$\langle\Delta\boldsymbol{p}_1\rangle = \int \overset{p+\Delta p}{\underset{p}{W}} \Delta\boldsymbol{p}_1\,\mathrm{d}(\Delta\boldsymbol{p}_1); \quad \langle\Delta\boldsymbol{p}_1\Delta\boldsymbol{p}_1\rangle = \int \overset{p+\Delta p}{\underset{p}{W}} \Delta\boldsymbol{p}_1\Delta\boldsymbol{p}_1\,\mathrm{d}(\Delta\boldsymbol{p}_1). \tag{6.4.4}$$

Including the effects of an external force, $\boldsymbol{X}_1$, (6.4.3) is known as the Fokker–Planck equation and the expression in the curly brackets represents the small momentum exchange analogue to the collision integral in the Boltzmann transport equation. The equation may be written in time-smoothed form provided the time dependence of $f_{(1)}$ is not too strong

$$\frac{\partial \bar{f}_{(1)}}{\partial t}+\frac{\boldsymbol{p}_1}{m}\frac{\partial \bar{f}_{(1)}}{\partial \boldsymbol{q}_1}-\frac{1}{\tau}\frac{\partial}{\partial \boldsymbol{p}_1}\left\{\langle\Delta\boldsymbol{p}_1\rangle\bar{f}_{(1)}+\frac{1}{2}\frac{\partial}{\partial \boldsymbol{p}_1}\langle\Delta\boldsymbol{p}_1\Delta\boldsymbol{p}_1\rangle\bar{f}_{(1)}\right\}-\boldsymbol{X}_1\frac{\partial \bar{f}_{(1)}}{\partial \boldsymbol{p}_1} = 0, \tag{6.4.5}$$

whereupon the remaining problem becomes one of identifying the quantities $\langle\Delta\boldsymbol{p}_1\rangle$, $\langle\Delta\boldsymbol{p}_1\Delta\boldsymbol{p}_1\rangle$. From (6.4.4) we see that this implies the determination of the transition probability $\overset{p+\Delta p}{\underset{p}{W}}$. Alternatively, but equivalently, we may require that the conditional probability $f_{(N-1:1)}$ be specified in detail. This latter formally isolates particle 1 from the remaining $N-1$ particles as follows

$$f_{(N)}(\boldsymbol{p}_N, \boldsymbol{q}_N, t) = f_{(1)}(\boldsymbol{p}_1, \boldsymbol{q}_1, t) f_{(N-1:1)}(\boldsymbol{p}_{N-1}, \boldsymbol{q}_{N-1} : \boldsymbol{p}_1, \boldsymbol{q}_1) \quad (6.4.6)$$

and it is clear that $\overset{p_1+\Delta p_1}{\underset{p_1}{W}}$ is closely related to the conditional probability $f_{(N-1:1)}$. In this approximation that the distribution $f_{(N-1:1)}$ is essentially independent of time, the evolution of $f_{(N)}$ is seen to arise from the time dependence of $f_{(1)}$. This is a simplifying assumption of the theory and enables us to identify $W(\Delta \boldsymbol{p}_1)$ as a time-independent transition probability for the process $\boldsymbol{p}_1 \to \boldsymbol{p}_1 + \Delta \boldsymbol{p}_1$ in time $\boldsymbol{r}$. If we are prepared to accept this identification we may rewrite expressions (6.4.4) for the mean and mean square momentum increment as

$$\langle \Delta \boldsymbol{p}_1 \rangle = \iint \Delta \boldsymbol{p}_1 f_{(N-1:1)}(\boldsymbol{p}_{N-1}, \boldsymbol{q}_{N-1} : \boldsymbol{p}_1, \boldsymbol{q}_1) \prod_{i=2}^{N} \mathrm{d}\boldsymbol{p}_i \, \mathrm{d}\boldsymbol{q}_i, \quad (6.4.7)$$

$$\langle \Delta \boldsymbol{p}_1 \Delta \boldsymbol{p}_1 \rangle = \iint \Delta \boldsymbol{p}_1 \Delta \boldsymbol{p}_1 f_{(N-1:1)}(\boldsymbol{p}_{N-1}, \boldsymbol{q}_{N-1} : \boldsymbol{p}_1, \boldsymbol{q}_1) \prod_{i=2}^{N} \mathrm{d}\boldsymbol{p}_i \, \mathrm{d}\boldsymbol{q}_i. \quad (6.4.8)$$

We suppose that the conditional probability $f_{(N-1:1)}$ is independent of time and is given by a local equilibrium value applying to an initial time $t = 0$:

$$f_{(1)}(\boldsymbol{p}_1, \boldsymbol{q}_1, t) f_{(N-1:1)}(\boldsymbol{p}_{N-1}, \boldsymbol{q}_{N-1}, 0) = f_{(N)}(\boldsymbol{p}_N, \boldsymbol{q}_N, t). \quad (6.4.9)$$

The momenta occur in the total distribution $f_{(N)}$ through the factor

$$\exp\left\{-\sum_{j=1}^{N} \boldsymbol{p}_j^2 / 2mkT\right\},$$

and the momenta at time t are given by

$$\boldsymbol{p}_j = \boldsymbol{p}_j^0 + \Delta \boldsymbol{p}_j,$$

where $\boldsymbol{p}_j^0$ are the momenta at the initial time $t = 0$.

Writing,

$$f_{(N)}(\boldsymbol{p}_N, \boldsymbol{q}_N, t) = f^0_{(N)} \exp\left\{-\sum_{j=1}^{N} (\boldsymbol{p}^0_j + \Delta\boldsymbol{p}_j)^2/2mkT\right\}, \quad (6.4.10)$$

where $f^0_{(N)}$ is the local total equilibrium distribution at $t = 0$, (6.4.9) can immediately be written,

$$f_{(N-1:1)}(\boldsymbol{p}_{N-1}, \boldsymbol{q}_{N-1}, 0) = f^0_{(N)}\left\{1 - \frac{1}{mkT}\sum_{j=1}^{N} (\boldsymbol{p}^0_j . \Delta\boldsymbol{p}_j) + \dots\right\} \quad (6.4.11)$$

to first order in $\Delta\boldsymbol{p}_j$. If $\boldsymbol{F}_j$ is the total force on particle j due to the remaining $N-1$ particles, we have

$$\Delta\boldsymbol{p}_j = \int_0^\tau \boldsymbol{F}_j(s)\,\mathrm{d}s, \quad (6.4.12)$$

where $\Delta\boldsymbol{p}_j$ is understood to be the time average of the force over the interval τ. Equation (6.4.7) for the mean momentum increment may now be written, from (6.4.11), (6.4.12), as

$$\langle\Delta\boldsymbol{p}_1\rangle = \iiint_0^\tau \boldsymbol{F}_1(s)\,\mathrm{d}s\, f^0_{(N)}\left\{1 - \frac{1}{mkT}\sum_{j=1}^{N} \boldsymbol{p}^0_j \boldsymbol{F}_j(0)\,\tau + \dots\right\} \prod_{i=2}^{N} \mathrm{d}\boldsymbol{p}_i\,\mathrm{d}\boldsymbol{q}_i, \quad (6.4.13)$$

where $\Delta\boldsymbol{p}_j$ appearing in the summation has been given in terms of its value at $t = 0$. The integral over the equilibrium distribution will vanish, whereas the integral over the second term will not, but leads to collective effect arising from the mean force due to the neighbours. The second term does not entirely vanish, one term only remaining,

$$\langle\Delta\boldsymbol{p}_1\rangle = \iiint_0^\tau \{\boldsymbol{F}_1(s) . \boldsymbol{F}_1(0)\}\,\mathrm{d}s\, f^0_{(N)} \prod_{i=2}^{N} \mathrm{d}\boldsymbol{p}_i\,\mathrm{d}\boldsymbol{q}_i\,\tau\boldsymbol{p}^0_1. \quad (6.4.14)$$

If we now define

$$\beta_{(1)} = \frac{1}{3mkT}\int_0^\tau \langle\boldsymbol{F}_1(t) . \boldsymbol{F}_1(t+s)\rangle_{N-1}\,\mathrm{d}s = \frac{\Xi(s)}{3mkT} \quad (6.4.15)$$

(where

$$\langle\boldsymbol{F}_1(t) . \boldsymbol{F}_1(t+s)\rangle_{N-1} = \iint \{\boldsymbol{F}_1(t) . \boldsymbol{F}_1(t+s)\} f^0_{(N-1:1)} \Pi\,\mathrm{d}\boldsymbol{p}_i\,\mathrm{d}\boldsymbol{q}_i), \quad (6.4.16)$$

then

$$\langle\Delta\boldsymbol{p}_1\rangle = -\int_0^\tau \frac{\tau}{3mkT}\Xi(s)\,\mathrm{d}s\,\boldsymbol{p}^0_1, \quad (6.4.17)$$

where $\Xi(s)$ is the phase correlation function of the force over the conditional probability distribution. If τ is larger than s then we may suppose from (6.4.1) that $\beta_{(1)}$ is a constant (provided τ is less than the Poincaré period); furthermore it never vanishes, even for an equilibrium distribution, since there is always a net force acting on particle 1. Equation (6.4.17) becomes, finally

$$\langle \Delta \boldsymbol{p}_1 \rangle = -\beta_{(1)} \boldsymbol{p}_1^0 \tau. \tag{6.4.18}$$

By similar arguments we find for the mean square increment of momentum

$$\langle \Delta \boldsymbol{p}_1 \Delta \boldsymbol{p}_1 \rangle = 6mkT\beta_{(1)}\tau. \tag{6.4.19}$$

Accepting the expressions (6.4.18), (6.4.19), and inserting them into (6.4.5) we have, in time-smoothed form

$$\frac{\partial \bar{f}_{(1)}}{\partial t} + \frac{\boldsymbol{p}_1}{m}\frac{\partial \bar{f}_{(1)}}{\partial \boldsymbol{q}_1} = \beta_{(1)} \frac{\partial}{\partial \boldsymbol{p}_1}\left\{\left(\frac{\boldsymbol{p}_1}{m} - \boldsymbol{u}\right)\bar{f}_{(1)} + kT\frac{\partial \bar{f}_{(1)}}{\partial \boldsymbol{p}_1}\right\} \equiv M, \tag{6.4.20}$$

where $\boldsymbol{u}$ is the mean local fluid velocity. This is the Fokker–Planck equation describing the smoothed single-particle evolution for a system in continuous interaction. It is evident that an irreversible approach to equilibrium is nevertheless consistent with the mechanically reversible Liouville equation. The constant, irretrievable loss of information expressed through the dissipative 'friction constant' $\beta_{(1)}$ ensures an irreversible evolution in phase. As we have observed above, the right hand side of (6.4.20), M, represents the analogue of the collision integral, J, appropriate to impulsive binary encounters. The Fokker–Planck equation applies to systems in extended interaction in which large momentum impulsive exchanges are negligible – Coulomb forces in plasma physics and gravitational interaction in astrophysics represent other examples of the application of the Fokker–Planck equation.

Chandrasekhar generalized the Fokker–Planck equation to include the spatial derivative $(\partial f(\boldsymbol{p}_1, \boldsymbol{q}_1, t)/\partial \boldsymbol{q}_1)$: if for the purposes of the present discussion we assume that $f(\boldsymbol{p}_1, \boldsymbol{q}_1, t)$ is independent of the spatial coordinate, and that no external forces apply, then (6.4.20) simplifies to give

$$\frac{\partial \bar{f}_{(1)}}{\partial t} = \beta_{(1)} \frac{\partial}{\partial \boldsymbol{p}_1}\left(\frac{\boldsymbol{p}_1}{m}\bar{f}_{(1)}\right) + kT\beta_{(1)}\frac{\partial^2 \bar{f}_{(1)}}{\partial \boldsymbol{p}_1^2} \tag{6.4.21}$$

in the absence of any local mean fluid velocity. We may attempt a solution of this equation subject to the initial condition that

$$f_{(1)}(\boldsymbol{p}_1, t) \to \delta(\boldsymbol{p}_1 - \boldsymbol{p}_1^0) \quad \text{for} \quad t \to 0,$$

where $\boldsymbol{p}_1^0$ is the initial momentum. $\beta_{(1)}$ is assumed to be a constant defined by (6.4.15). As such, the value of the friction constant is intimately related to the microscopic small-time evolution of the system: this will be considered in greater detail in §6.7. For present purposes it is enough to assume that $\beta_{(1)}$ is a scalar, independent of $\boldsymbol{p}_1$ and t.

We assume a solution of the form

$$f(\boldsymbol{p}_1, t) = e^{\beta_{(1)}t/m}\, Q(\tilde{\boldsymbol{p}}_1, t), \tag{6.4.22}$$

where $\tilde{\boldsymbol{p}}_1 = \boldsymbol{p}_1 e^{\beta_{(1)}t/m}$. Then we have

$$\left.\begin{aligned} \frac{\partial f_{(1)}}{\partial \boldsymbol{p}_1} &= e^{\beta_{(1)}t/m} \frac{\partial Q}{\partial \tilde{\boldsymbol{p}}_1} \frac{\partial \tilde{\boldsymbol{p}}_1}{\partial \boldsymbol{p}_1} = e^{2\beta_{(1)}t/m} \frac{\partial Q}{\partial \tilde{\boldsymbol{p}}_1}, \\ \frac{\partial^2 f_{(1)}}{\partial \boldsymbol{p}_1^2} &= e^{2\beta_{(1)}t/m} \frac{\partial^2 Q}{\partial \tilde{\boldsymbol{p}}_1^2} \frac{\partial \tilde{\boldsymbol{p}}_1}{\partial \boldsymbol{p}_1} = e^{3\beta_{(1)}t/m} \frac{\partial^2 Q}{\partial \tilde{\boldsymbol{p}}_1^2}. \end{aligned}\right\} \tag{6.4.23}$$

For the time derivative we have

$$\begin{aligned} \frac{\partial f_{(1)}}{\partial t} &= \frac{\beta_{(1)}}{m} e^{\beta_{(1)}t/m} Q + e^{\beta_{(1)}t/m} \left\{ \frac{\partial Q}{\partial t} + \frac{\partial Q}{\partial \tilde{\boldsymbol{p}}_1} \frac{\partial \tilde{\boldsymbol{p}}_1}{\partial t} \right\} \\ &= \frac{\beta_{(1)}}{m} e^{\beta_{(1)}t/m} Q + e^{\beta_{(1)}t/m} \left\{ \frac{\partial Q}{\partial t} + \boldsymbol{p}_1 \beta_{(1)} e^{\beta_{(1)}t/m} \frac{\partial Q}{\partial \tilde{\boldsymbol{p}}_{(1)}} \right\}. \end{aligned} \tag{6.4.24}$$

Substitution of these derivatives into the Fokker–Planck equation (6.4.21) yields

$$\frac{\partial Q}{\partial t} = kT\beta_{(1)} e^{2\beta_{(1)}t/m} \frac{\partial^2 Q}{\partial \tilde{\boldsymbol{p}}_1^2} \tag{6.4.25}$$

and if we set $dt = e^{-2\beta_{(1)}t/m}\, d\theta$,

i.e.
$$\theta = \frac{m}{2\beta_{(1)}} (e^{2\beta_{(1)}t/m} - 1), \tag{6.4.26}$$

then (6.4.25) simplifies to the standard diffusion equation

$$\frac{\partial Q}{\partial \theta} = C \frac{\partial^2 Q}{\partial \tilde{\boldsymbol{p}}_1^2}, \tag{6.4.27}$$

where $$C = kT\beta_{(1)}.$$

Therefore $$Q = (4\pi C\theta)^{\frac{1}{2}} \exp\{-(\tilde{p}_1 - \tilde{p}_1^0)^2/4C\theta\} \tag{6.4.28}$$

and this satisfies the condition that $Q \to \delta(\tilde{p} - \tilde{p}^0)$ as $\theta \to 0$. In terms of the original variables we obtain

$$f(p_1, t) = \left\{\frac{1}{2\pi kT(1 - e^{-2\beta_{(1)}t/m})}\right\}^{\frac{1}{2}} \exp\left\{\frac{(p_1 - p_1^0 e^{-\beta_{(1)}t/m})^2}{2mkT(1 - e^{-2\beta_{(1)}t/m})}\right\}. \tag{6.4.29}$$

We see that as $t \to \infty$ a Maxwellian distribution is approached, as it should, and that for intermediate times (6.4.29) is a Gaussian with a mean value
$$\overline{p_1(t)} = p_1^0 e^{-\beta_{(1)}t/m}$$

where $\beta_{(1)}/m$ has the significance of a time constant.

The theory of Brownian motion, which describes transport processes in dilute solution, is based upon the Langevin equation describing the motion of a molecule in an environment in statistical equilibrium. In an attempt to show how the Langevin equation may be obtained from statistical mechanics and applied to microscopic evolution, we may write the equation of motion of the ith particle as
$$\frac{dp_i}{dt} = X_i + F_i, \tag{6.4.30}$$

where X_i is an external force and F_i the intermolecular force acting on molecule i. The interaction force F_i must be affected by the motion of the particle in such a way that F_i itself also contains a slowly varying part $\bar{F}_i$ tending to restore the particle to equilibrium. Hence we may write

$$F_i = \bar{F}_i + \langle\tilde{F}_i\rangle, \tag{6.4.31}$$

where $\tilde{F}_i$ is some rapidly fluctuating force whose average over a period τ vanishes for a system in thermodynamic equilibrium. The slowly varying component $\bar{F}_i$ must be dependent on p_i such that

$$\bar{F}_i(\bar{p}_i) = 0$$

at equilibrium. Provided $\bar{p}_i$ is not too large, it should be possible to expand $\bar{F}_i$ in a power series whose first non-vanishing term is linear in momentum, thus

$$\bar{F}_i = -\beta_{(1)}\frac{\bar{p}_i}{m}, \tag{6.4.32}$$

where $\beta_{(1)}$, the friction constant, acts so as to dissipate the 'fluctuation' $\bar{\boldsymbol{F}}_i$. $\beta_{(1)}$ is, therefore, an as yet unspecified positive constant independent of $\boldsymbol{p}_i$. The equation of motion of the ith particle is, then,

$$\frac{\mathrm{d}\boldsymbol{p}_i}{\mathrm{d}t} = -\frac{\beta_{(1)}}{m_i}\boldsymbol{p}_i + \boldsymbol{X}_i + \langle\tilde{\boldsymbol{F}}_i\rangle. \tag{6.4.33}$$

This latter equation becomes identical with the Langevin equation provided $\beta_{(1)}$ is independent of $\boldsymbol{p}_i$: when $\beta_{(1)}$ depends upon $\boldsymbol{p}_i$, a generalized Langevin equation containing a friction constant dependent upon the velocity of the molecule is obtained. Having established the Langevin equation, the details of the conventional theory of Brownian motion may be developed, and in the absence of external forces the evolution (6.4.29) may be regained. Thus we come to understand the generalized statistical mechanical theory of transport processes as a generalized theory of Brownian motion in which the friction constant is explicitly related to the intermolecular forces acting in the system. The friction constant has to be calculated subject to the nature of the environmental force field of the ith particle – however, in the case of an essentially *macro*scopic Brownian particle the hydrodynamic drag on a spherical particle of radius a in a fluid of viscosity η immediately yields a friction constant $6\pi\eta a$, whereupon (6.4.33) reduces to Stokes's equation.

Clearly the details of the evolution are contingent upon the specification of the friction constant and this we shall consider in some detail in §6.7.

6.5 Realistic interactions – The Rice–Allnatt equations

Thus far we have determined the single-particle phase evolution in the two extreme limits of large-momentum binary exchanges appropriate to impulsive encounters between rigid repulsive particles, and in the small momentum collective exchanges for a system of particles in extended collective interaction. In both cases the mechanically reversible Liouville equation was found to develop characteristics of irreversibility in establishing the collision integral J, in the Boltzmann transport equation, and the Fokker–Planck analogue, M. For high density low temperature fluids neither J nor M in itself adequately represents the dynamical evolution of the system. Rice and Allnatt, however, have suggested that the streaming terms in the Liouville equation might be equated to a combination of the processes J and M, provided these two processes proceed without destructive coupling:

$$\frac{\partial \bar{f}_{(1)}}{\partial t} + \frac{\boldsymbol{p}_1}{m} \frac{\partial \bar{f}_{(1)}}{\partial \boldsymbol{q}_1} + \boldsymbol{X}_1 \frac{\partial \bar{f}_{(1)}}{\partial \boldsymbol{p}_1} = J + M, \tag{6.5.1}$$

where $\boldsymbol{X}_1$ is any external field acting on the particles of the system. The impulsive collision process J must be presumed not to interfere with the essentially stochastic process described by M. We therefore imagine infrequent impulsive binary encounters followed by a Brownian motion of the random type in the fluctuating force field of the environment. It does not seem unreasonable to suppose that the Brownian motion will destroy collisional correlations rapidly so that the J and M components are essentially uncoupled. The direct encounter has therefore been resolved into a short time scale process associated with a large momentum exchange impulsive encounter, and a longer time process during which there is a succession of small momentum increments. It must be emphasized that the assumption is that the approach to equilibrium is due to two independent processes to each of which may be ascribed a corresponding friction constant. For J encounters the 'hard' friction constant describes the collisional approach to equilibrium, whilst for the more extended interactions due to the weak, long range attractive, and possibily the softer part of the repulsive interaction an independent 'soft' friction constant will apply. In a real fluid, of course, these two constants will be related, but it has been shown that provided the stochastic repulsive impulsive events are the same as those between hard spheres then the J and M processes, with their associated friction constants, remain entirely separate. In the first instance, therefore, the analysis was applied by Rice and Allnatt to a model monatomic dense fluid in which the intermolecular potential has the form of a rigid core repulsion, superimposed on an arbitrary soft potential. Subsequent analysis has shown that the extension of the model to include more realistic potentials presents no formal difficulty.

6.6 Extension to the doublet distribution $f_{(2)}(\boldsymbol{p}_1, \boldsymbol{p}_2, \boldsymbol{q}_1, \boldsymbol{q}_2, t)$

The Chandrasekhar generalization of the single-particle Fokker–Planck equation (6.4.20) enables us in principle to determine the phase evolution of a dilute gas together with its associated macroscopic transport coefficients. Whilst the time-smoothed singlet evolution is adequate at low densities it is, however, necessary to appeal to the doublet distribution for the calculation of the macroscopic properties of a dense fluid. Just as we cannot profitably consider a liquid as a dense gas, so some recognition of both the spatial and dynamical correlation between particles in the fluid at high densities must be recognized.

The most immediate approach is to consider the singlet distribution $f_{(1)}$ as being a *relative* distribution, one particle being taken as origin. Under these circumstances the friction constant $\beta_{(1)}$ must be understood to involve the correlation of the relative force between the pair of particles. This treatment has been of some use in the determination of the transport properties of dense fluids. We may, however, make an explicit derivation of the Fokker–Planck equation in terms of the doublet reduced distribution,

$$f_{(2)}(\boldsymbol{p}_1, \boldsymbol{p}_2, \boldsymbol{q}_1, \boldsymbol{q}_2, t).$$

The development from the Liouville equation proceeds exactly as before: in the absence of external forces the streaming terms are written

$$\frac{\partial \bar{f}_{(2)}}{\partial t} + \left(\frac{\boldsymbol{p}_1}{m} \cdot \frac{\partial \bar{f}_{(2)}}{\partial \boldsymbol{q}_1}\right) + \left(\frac{\boldsymbol{p}_2}{m} \cdot \frac{\partial \bar{f}_{(2)}}{\partial \boldsymbol{q}_2}\right) \tag{6.6.1}$$

whilst the Fokker–Planck operator in the weak-coupling limit is now given as

$$-\frac{1}{\tau}\int_0^\tau \iint \frac{1}{m}\sum_{i=1}^{N}\left(\boldsymbol{F}_i \frac{\partial \bar{f}_{(N)}}{\partial \boldsymbol{p}_i}\right)\prod_{j=3}^{N} \mathrm{d}\boldsymbol{p}_j\, \mathrm{d}\boldsymbol{q}_j\, \mathrm{d}s \tag{6.6.2}$$

in time-smoothed form (cf. (6.4.2)). Some immediate simplification of (6.6.2) is possible with respect to integration, for $3 \leqslant j \leqslant N$. The two remaining integrals involving $\boldsymbol{F}_1$ and $\boldsymbol{F}_2$, however, are far from trivial since we have to deal with the force on the correlated particles 1 and 2 arising from their interaction with the remaining $(N-2)$ particles in the system. Furthermore, it is clear that we shall have to reconsider the nature of the friction constant. We shall instead have to deal with a quantity $\beta_{(2)}$ given in terms of the interparticle forces as

$$\beta_{(2)} = \frac{1}{3mkT}\int_0^T \langle[\boldsymbol{F}_{(2)}(t).\boldsymbol{F}_{(2)}(t+s)]\rangle_{N-2}\,\mathrm{d}s \qquad (6.6.3)$$

whilst we already have for its singlet counterpart

$$\beta_{(1)} = \frac{1}{3mkT}\int_0^T \langle[\boldsymbol{F}_{(1)}(t).\boldsymbol{F}_{(1)}(t+s)]\rangle_{N-1}\,\mathrm{d}s, \qquad (6.6.4)$$

where the former phase average of the force on the representative pair is taken over the remaining $(N-2)$ particles of the system, whilst the latter represents the phase average of the force on a single particle due to the remaining $(N-1)$ centres. In the case of the interacting pair we note that the force of one particle on the other over the interval s will be proportional to s, and is consequently to be neglected since $\beta_{(2)}$ is to be a constant for dissipation of the fluctuation. Incidentally, the friction constants are generally tensor quantities; nevertheless, the bulk isotropy of the fluid in the absence of external forces enables us to write

$$\boldsymbol{\beta}_{(1)} = \hat{\beta}_{(1)}\mathbf{1}, \quad \boldsymbol{\beta}_{(2)} = \hat{\beta}_{(2)}\mathbf{1},$$

where $\mathbf{1}$ is the unit tensor and ^ denotes a purely scalar function.

The integral (6.6.2) may be developed in an analogous manner to that of the single-particle Fokker–Planck operator: the resulting equation is however, too lengthy and complicated to be reproduced here. It is evident, however, that the magnitude of the mathematical task involved in its application is very considerable. Fortunately we shall not need to refer to it explicitly in what follows.

Rice and Allnatt have extended the doublet Fokker–Planck equation to include both strong and weak coupling (J and M) terms.

6.7 The friction constant

We have seen that the evolution in phase of a small subsystem of molecules may be approached from a statistical mechanical point of view via the Liouville and Fokker–Planck equations, or alternatively, from a mechanical standpoint in terms of the Langevin equation. In both cases a generalized theory of Brownian motion arises, determined to within the n-body friction tensor $\beta_{(n)}$. Kirkwood's outstanding contribution to the field was to provide a relation between the friction constant and the intermolecular forces acting in the system. A number of attempts have been made subsequently to establish a satisfactory relationship between $\beta_{(n)}$ and the molecular parameters, but all the methods thus far proposed involve various approximations, and undoubtedly this will be a point for future development.

In definitions such as (6.6.3), (6.6.4), we intentionally avoid identifying $\beta_{(n)}(\tau)$, for sufficiently long τ, with its asymptotic value $\beta_{(n)}(\infty)$, since this latter value may be shown to vanish for systems confined to a finite region of configuration space. The apparent paradox encountered here is precisely the paradox between dynamical reversibility and thermodynamic irreversibility. The dissipative processes appearing to operate in thermodynamic systems provide a valid description of the macroscopic behaviour only if the period over which the properties of the system are observed is long with respect to the relaxation time of a microscopic dynamical event, and short with respect to the Poincaré period, within which secular changes in state may be spontaneously observed. For these reasons Kirkwood proposed that a 'plateau value' for $\beta_{(n)}(\tau)$ should be adopted for the purposes of transport phenomena, rather than its asymptotic value $\beta_{(n)}(\infty)$. Whether such a plateau value exists for the force autocorrelation of either pairs or single particles depends upon the dynamical properties of the system, in particular, its Hamiltonian. Kirkwood observes, for example, that the friction constant appears to have no plateau value for a crystal with *harmonic* lattice vibrations; this is apparently due to the fact that the motion is multiply periodic. This is not to say that there is no dissipative exchange between a particle moving within the lattice, and the lattice vibrational modes, but rather that such an exchange occurs through the anharmonic terms.

On the other hand it is found that $\beta_{(1)}$ does have a plateau value for values of τ long relative to the representative duration of one collision, but still, of course, short relative to the Poincaré period.

Allowing $\tau \to \infty$ in the case of a dilute gas, $\beta_{(1)}(\infty)$ is vanishing or indeterminate since after a time of the order of the Poincaré cycle, contributions to the friction constant would be cancelled by a collision in which the phase trajectory was traversed in the reverse sense. Kirkwood has demonstrated that for an arbitrary central force, the only invariants being energy, total linear momentum and total angular momentum, $\beta_{(1)}(\infty)$ always vanishes.

The evaluation of the force correlations (6.6.3) and (6.6.4) involve averages of $(N-2)$ and $(N-1)$ particles respectively, and as such represents a formidable mathematical task. Early optimism that reasonably accurate estimates of the friction constants would be relatively forthcoming has now disappeared, and it is clear that considerable ingenuity and approximation is going to be necessary before a statisfactory determination of $\beta_{(1)}$ and $\beta_{(2)}$ is to be made. However, it turns out that if apparently reasonable simplifications are made, such as that the particle moves in the average force field of its neighbours, the result is obtained that either the friction constant is zero, whereupon the system does not proceed to its equilibrium state showing no irreversibility, otherwise it is indefinitely large, in which case the system moves instantaneously to equilibrium.

The two-body friction constant $\beta_{(2)}$ is even more difficult to evaluate, and present calculations of the friction constant $\beta_{(2)}$ are based upon the single-particle friction constant – the doublet function being obtained by simple addition. This is most unlikely to be satisfactory since $\beta_{(2)}$ has no reason to show the same temperature dependence as $\beta_{(1)}$ and is even less likely to have the same numerical magnitude. $\beta_{(1)}$, for example, would possibly be expected to show an Arrhenius temperature dependence since the correlation time τ over which the force correlation is integrated to yield the single-particle friction constant will be temperature dependent. On the basis of the cell model, for example, the correlation time τ that an atom remains essentially caged in the cell field of its neighbours will vary as $\exp(U/kT)$, where U is the cell potential. On the other hand, the doublet friction constant $\beta_{(2)}$ will be largely independent of temperature since the correlation is not necessarily broken by the pair leaving the cell, only indirect forces being involved.

Assuming a Gaussian decorrelation of the singlet force autocorrelation $\langle \boldsymbol{F}_1(t).\boldsymbol{F}_1(t+s)\rangle_{N-1}$, Kirkwood[1] provided the first estimate of the single-particle friction constant:

$$\beta_{(1)} \sim \frac{(2\pi)^{\frac{1}{2}}\langle \boldsymbol{F}_1^2\rangle}{6kT\omega}. \qquad (6.7.1)$$

$\langle \boldsymbol{F}_1^2\rangle$ is the mean square of the force on the single particle and ω is a characteristic frequency describing the time change of the force. This approximation establishes a single-particle friction constant which is independent of momentum. Evidently momentum dependence is negligible provided the momentum $\boldsymbol{p}_1$ is in the vicinity of the equilibrium root mean square momentum – that is, provided the system is only slightly displaced from its equilibrium distribution.

An alternative, but closely related, expression has been given by Kirkwood *et al.*[4]. If we write the single-particle force autocorrelation in the form

$$\langle \boldsymbol{F}_1(t).\boldsymbol{F}_1(t+s)\rangle_{N-1} = \langle \boldsymbol{F}_1(t).\boldsymbol{F}_1(t)\rangle\, \psi(s), \qquad (6.7.2)$$

where $\psi(s)$ is a monotonic decreasing function of time defined to be such that $\psi(0) = 1$, then we may write

$$\beta_{(1)} = \frac{1}{3mkT}\langle \boldsymbol{F}_1^2\rangle \int_0^T \psi(s)\, ds. \qquad (6.7.3)$$

Appealing now to dimensional arguments, we might propose that $\psi(s)$ will decay as the average momentum of particle 1, whereupon we are able to replace the time integral of (6.7.3) by $m/\beta_{(1)}$. If the environment is in local equilibrium, then the mean square force on particle 1 will be, for purely radial interactions

$$\frac{\langle \boldsymbol{F}_1\rangle}{kT} = 4\pi\rho \int_0^\infty \nabla^2\Phi(r)\, g_{(2)}(r)\, r^2\, dr, \qquad (6.7.4)$$

since the mean square force may be shown to be related to the average of the Laplacian of the total potential energy by

$$\langle [\nabla_1 \Phi_N]^2\rangle = kT\langle [\nabla_1^2 \Phi_N]\rangle.$$

$\langle [\nabla_1^2 \Phi_N]\rangle$ may be measured directly by studying isotope separation[19]. $g_{(2)}(r)$ is the equilibrium radial distribution function. The final expression for the friction constant then becomes

$$\beta_{(1)}^2 = \frac{4\pi m}{3} \int_0^\infty g_{(2)}(r)\, \nabla^2\Phi(r)\, r^2\, dr. \qquad (6.7.5)$$

Rice[3] has shown that (6.7.5), or (6.7.3), arises as a consequence of assuming that the momentum autocorrelation has the form of a Gaussian in time: under these circumstances the expression (6.7.1) for the friction constant is regained, but for a factor of $(2/\pi)^{\frac{1}{2}}$. It may easily be shown that a relationship

$$\frac{\partial^2}{\partial s^2}\langle \boldsymbol{p}_1(t)\,.\,\boldsymbol{p}_1(t+s)\rangle = -\langle \boldsymbol{F}_1(t)\,.\,\boldsymbol{F}_1(t+s)\rangle \tag{6.7.6}$$

exists in a stationary state, this being defined as

$$\frac{\partial}{\partial s}\langle \boldsymbol{p}(t)\,.\,\boldsymbol{p}(t+s)\rangle = 0. \tag{6.7.7}$$

The assumption of a Gaussian form for $\langle \boldsymbol{p}_1(t)\,.\,\boldsymbol{p}_1(t+s)\rangle$ leads inevitably to a Gaussian form for the force autocorrelation, this being the primitive Kirkwood[1] assumption in his original determination of the single-particle friction constant. However (6.7.3), or equivalently (6.7.5), has also been shown to be the consequence of identifying the time-smoothing period τ with the relaxation time of the force correlation[4] for which $\tau \sim 1/\omega$ (cf. (6.7.1)). It also follows if the non-equilibrium singlet density has the same relaxation time as the velocity autocorrelation[5]. Collins and Raffel[6] derived (6.7.5) by assuming that the momentum of a molecule as a function of time can be represented as having equal but opposite curvatures in the steady state, and in the initial decay of the fluctuation. Ross[7], exploiting the relatively weak nature of the total van der Waals force exerted on its neighbours by a molecule, established a relation between the friction constant and the force autocorrelation which is essentially a version of the fluctuation–dissipation theorem; it relates the dissipative (friction) coefficient to the integral of force autocorrelation exerted by the environment on the molecule:

$$\beta_{(1)} = \frac{1}{3mkT}\int_0^\infty \langle \boldsymbol{F}_1(t)\,.\,\boldsymbol{F}(t+s)\rangle\,\mathrm{d}s \tag{6.7.8}$$

and differs from Kirkwood's expression in the upper limit. Helfand[8] subsequently evaluated the correlation term in the linear trajectory approximation, according to which all molecular trajectories are restricted *by definition* to the linear motions corresponding to the initial phase. The neglect of the effect of soft forces on the molecular trajectories means that there is essentially no feedback between the motion and the forces, with the result that the integral does *not* vanish as the upper limit tends to infinity. Alternatively we may say that the negative region of the schematic force autocorrelation shown in fig. 6.13.1 (*a*) is not taken into account. Helfand finally obtains for the single-particle friction constant

$$\beta_{(1)} = \frac{\rho}{3}\left(\frac{\pi m}{kT}\right)^{\frac{1}{2}}(2\pi)^{-2}\int_0^\infty \tilde{\Phi}_S(k)\,\tilde{g}_{(2)}(k)\,k^3\,\mathrm{d}k, \qquad (6.7.9)$$

where

$$\tilde{\Phi}_S(k) = \int_0^\infty \Phi_S(r)\,\mathrm{e}^{\mathrm{i}\boldsymbol{k}.\boldsymbol{r}}\,\mathrm{d}\boldsymbol{r},$$

$$\tilde{g}(k) = \int_0^\infty g_{(2)}(r)\,\mathrm{e}^{\mathrm{i}\boldsymbol{k}.\boldsymbol{r}}\,\mathrm{d}\boldsymbol{r}$$

and the subscript S on the potential function restricts the interaction to the soft component. Helfand has inverted (6.7.9) for a modified 'soft' Lennard–Jones interaction for which

$$\Phi_S(r) = \begin{cases} 0 & r < \sigma, \\ 4\epsilon\left\{\left(\dfrac{\sigma}{r}\right)^{12} - \left(\dfrac{\sigma}{r}\right)^{6}\right\} & r \geqslant \sigma. \end{cases} \qquad (6.7.10)$$

However, Helfand's linear trajectory approximation is still sufficiently unrepresentative of the actual molecular dynamics as to yield a friction constant some 50 per cent too low. Nonetheless, this approach is particularly convenient for application to the Rice–Allnatt model for which (6.7.9) gives the soft friction coefficient β_S. The total friction constant is then the sum of the soft and hard components $\beta = \beta_H + \beta_S$, and using (6.7.5) with the hard core potential, Rice and Allnatt[9] obtain

$$\beta_H = \tfrac{3}{8}\rho\sigma^3 g_{(2)}(\sigma)\,(\pi mkT)^{\frac{1}{2}}, \qquad (6.7.11)$$

where σ is the atomic diameter. In view of the discrepancy in the Helfand estimate of β_S, it is considered that a better test of the Rice–Allnatt theory would be to deduce empirical values of the soft friction coefficient from experimental measurements of the diffusion coefficient, D, giving

$$\beta_S = \left(\frac{kT}{mD}\right) - \beta_H, \qquad (6.7.12)$$

TABLE 6.7.1.* β_H *and* β_S *for liquid argon*†

T (°K)	ρ (g cm^{-1})	P (atm.)	$(kT/D)\times 10^{10}$ (g s^{-1})	$\beta_H \times 10^{10}$ (g s^{-1})	$\beta_S \times 10^{10}$ (g s^{-1})
90	1.38	1.3	5.11	0.64	4.47
128	1.12	50	2.94	0.94	2.00
133.5	1.12	100	3.13	1.00	2.13
185.5	1.12	500	3.20	1.52	1.68

* Taken from Rice and Gray[10].

† Self-diffusion coefficient, D, estimated from the measurements of Naghizadeh and Rice[11]. β_S deduced from (6.7.12).

where β_H is given in (6.7.11). Estimates of β_S and β_H thuswise estimated are shown in table 6.7.1. We recall that in the Rice–Allnatt model the basic assumption is that the Brownian motion induces a sufficiently rapid relaxation to equilibrium of the momentum of a molecule that successive hard core collisions may be regarded as almost independent. Under these circumstances the collision operator is a direct sum of the Chapman–Enskog (J) and Fokker–Planck (M) terms, (6.5.1). From table 6.7.1 we see that the soft friction constant, β_S, is about twenty times more effective in bringing about equilibrium, thus vindicating the initial assumption[12]. Inspection of table 6.7.1 shows that the importance of the soft component of the pair interaction diminishes with decreasing ϵ/kT such that at sufficiently high temperatures $T \gg \epsilon/k$,

$$\beta_H \gg \beta_S.$$

In this limit the Fokker–Planck operators M on the right hand sides of the kinetic equations become small and impulsive strong coupling effects predominate through the operation of the hard core friction coefficient, β_H.

If, however, we consider the separation of forces already achieved, the total intermolecular force autocorrelation may be split up in the form

$$\begin{aligned}\langle \boldsymbol{F}_1(t).\boldsymbol{F}_1(t+s)\rangle = &\langle \boldsymbol{F}_1^{(H)}(t).\boldsymbol{F}_1^{(H)}(t+s)\rangle + \langle \boldsymbol{F}_1^{(H)}(t).\boldsymbol{F}_1^{(S)}(t+s)\rangle \\ &+\langle \boldsymbol{F}_1^{(S)}(t).\boldsymbol{F}_1^{(H)}(t+s)\rangle \\ &+\langle \boldsymbol{F}_1^{(S)}(t).\boldsymbol{F}_1^{(S)}(t+s)\rangle\end{aligned}$$

leading to a friction constant of the form

$$\beta = \beta_{\mathrm{H}} + \beta_{\mathrm{S}} + \beta_{\mathrm{HS}},$$

where the cross-correlation term β_{HS} is generally considered to be negligibly small, the hard and soft interactions being assumed essentially uncoupled and hence uncorrelated. Davis and Palyvos[20] have tried incorporating the cross-coefficient β_{HS} in the Helfand linear trajectory approximation. The improvement, however, is only marginal, as we shall see later.

Davis, Rice and Sengers[21] have derived a formula for β_{S} for the special case of a square-well fluid by comparing the exact kinetic equation for the square-well fluid with the corresponding Fokker–Planck version. Correlation effects are properly included, but the final equation must be integrated numerically owing to its complicated form: the details of the calculation are tedious but straightforward, and we shall not reproduce them here. Longuet-Higgins and Valleau[22] have determined the self-diffusion coefficient for a square-well fluid:

$$\left.\begin{aligned} D &= \frac{3\rho}{32}\left(\frac{kT}{m}\right)^{\frac{1}{2}}\{\sigma_1^2 g_{(2)}(\sigma_1) + \sigma_2^2 g_{(2)}(\sigma_2)\,F(\epsilon/kT)\}^{-1}, \\ F(\epsilon/kT) &= \exp(\epsilon/kT) - \frac{\epsilon}{2kT} + \frac{\epsilon}{2kT}\exp(\epsilon/kT)\,J_1(\epsilon/2kT), \end{aligned}\right\} \quad (6.7.13)$$

where $J_1(\epsilon/2kT)$ is the first-order Bessel function. σ_1, σ_2 and ϵ are parameters of the square well. Equation (6.7.13) is of much simpler form than that of Davis, Rice and Sengers; this is the case since Longuet-Higgins and Valleau *assume* an exponential decay of the momentum autocorrelation.

Arguments have been given by Rice[13] for supposing that several of the models proposed above provide a lower bound for the value of the friction constant in that they underestimate the extent of correlations between forces, momenta, or coordinates. In the estimates of Kirkwood *et al.*[4], (6.7.3), and Rice and Kirkwood[2], (6.7.5), a relaxation period τ is assumed after which force autocorrelation and momentum autocorrelation have essentially vanished: in other words, within a time τ an increment of force or momentum is rendered independent of the initial state. Atomic forward and backscattering within the cage of nearest neighbours makes the above assumption only approximately correct at liquid densities. Rice[13] observes that in the Rice–Kirkwood model[2], (6.7.5), the underestimate of correlation effects enters through the spatial decoupling effected when the pair diffusion tensor is set equal, at all distances, to the direct sum of singlet diffusion tensors. It is clear that this assumption is valid as $r_{12} \to \infty$, but, of course, at small distances this cannot be correct. Unfortunately, the derivations also require consideration of the pair diffusion tensor for small pair displacements. Rice[13] points out that the Collins and Raffel[23] derivation which requires that the time derivative of the force acting on a molecule be equal and opposite in the limits $t \to 0$, $t \to \infty$ places a severe restriction upon the rate at which velocity can decay, and in fact the coupling of the transient fluctuation with the medium is not as complete as in the steady state. Finally, Gray's[5] treatment requires the velocity and coordinate relaxation times to be the same. This is at variance with what is known of relaxation phenomena in dilute gases and dense fluids. In the latter case, for a dense hard sphere fluid it has been shown that the relaxation time for the momentum equilibration corresponds to no more than one or two collisions per particle[14] (see § 5.2), whilst the configurational relaxation must inevitably be a much slower process because of the cage of nearest neighbours tending to restrict the motion of a particle.

Rice[15] continues to consider a model in which an upper bound may be placed upon the single-particle friction constant. It is the existence of molecular fluctuations which limit the range, both spatial and temporal, of intermolecular correlation. If the molecularly coarse environment is replaced by a continuum, then it is easy to see that in this case the correlations will be overestimated. It is clear that in this model a strong correlation is always maintained between the origin of the distributed force and fluid displacements, independent of the separation of the displacement from the origin. This correlation arises from the coherence of the acoustic wave which is taken to be non-dissipatively propagated. The natural damping of correlations is thereby destroyed, and the molecular friction overestimated. Rice attempts to incorporate into the continuum the essential features of a real molecular system. Thus, the density of matter as a function of distance from the molecule is taken to agree with the pair correlation function, and the molecule and the accompanying density variations about it are taken to constitute a distributed force field affecting the amplitude and propagation of acoustic waves. This is in effect to study the propagation of phonons with speed c in an inhomogeneous continuum[(16)]. c here is the velocity of propagation of high frequency density fluctuations, and may not equal the velocity of propagation of low frequency sound[(17)]. It is found in fact that the dependence of c on liquid density parallels that of the velocity of sound on density. The result of these calculations is the expression:

$$\beta_S = \frac{\rho_m}{36\pi m^2 c^3}\left\{\int \nabla^2\Phi_S(r)\, g_{(2)}(r)\, \mathrm{d}\mathbf{r}\right\}^2 = \frac{\rho_m}{36\pi m^2 c^3}\langle\nabla^2\Phi_S(r)\rangle^2, \quad (6.7.14)$$

where ρ_m is the *mass* density of the fluid, and, on the continuum model, $\Phi_S(r)/m$ is interpreted as the potential of the force acting between a molecule and a unit mass of the surrounding medium. Further, the density variation surrounding a molecule is described by the equilibrium distribution $g_{(2)}(r)$, now interpreted in terms of the mass density of the continuum. Friedman *et al.*[(18)] and Boato *et al.*[(19)] have shown that the quantity $\langle\nabla^2\Phi(r)\rangle$ appearing in (6.7.14) can be determined from isotope separation data, the first-order quantum correction to the classical description of equilibrium being proportional to $\langle\nabla^2\Phi_N\rangle$. Direct calculation of β_S using (6.7.14) and the experimental data leads to a value of the diffusion constant $D = kT/(\beta_S + \beta_H)$ for liquid argon at 90 °K which is about 2.7 times too small. However, as we have observed above[(17)], care must be exercised in the choice of c in (6.7.14).

Before we go on to compare the theoretical estimates of the friction constant from the Einstein relation $\beta = (kT/mD)$ with experiment, we shall first make a brief digression to consider the qualitative features we might expect from the single-particle momentum autocorrelation function $\psi(s)$: the velocity autocorrelation will be treated in more detail in §6.13. In some of the preceding models the momentum or force correlation has been taken to be either of Gaussian or exponential form. It is intuitively clear, and indeed has been admirably demonstrated by Rahman, that at high liquid densities we might expect the representative molecule to be backscattered towards the centre of its cage of neighbouring molecules after a time $\sim \surd(md^2/kT)$ where d is the mean atomic separation. The momentum or force autocorrelation function therefore must become negative – a result confirmed by analysis of the incoherent scattering of neutrons from liquids. The presence of a negative contribution to the force or momentum autocorrelation is a consequence of the correlation in molecular motion imposed by the strong interaction in a dense fluid. Whilst these details are undoubtedly of importance – the Gaussian and exponential models, for example, exhibit no caging effect – it is of course the *time integral* over these autocorrelations which determines the value of the friction constant. In particular, the area under the peak is determined largely by the curvature at $t = 0$, i.e. by $\ddot{\psi}(0)$ or, from (6.7.6), the mean square force. We might anticipate, therefore, that the long time form of the autocorrelation has a negligible effect on the final value of the friction constant. It must be observed, however, that the momentum autocorrelation of a harmonic oscillator is long range oscillatory, yet its net time integral, and hence diffusion constant, is virtually zero.

Three models of the normalized velocity autocorrelation of liquid argon at 90 °K are shown in fig. 6.13.1, in addition to Rahman's[24] 'exact' molecular dynamics determination. The only curve showing a negative region is that based on the so-called 'relaxation approximation', discussed in §6.13, and forms the basis of one other estimate of the friction constant (see table 6.8.1).

6.8 The friction constant – comparison with experiment

A complete *a priori* determination of the friction constant requires the solution of the N body problem, β arising from the strong coupling and its $N-1$ neighbours. Whilst this as such is a hopelessly impossible task it appears that the association of this constant with other dissipative parameters of the liquid state is relatively easy. Experimental determination of, say, the diffusion constant

TABLE 6.8.1. *Comparison of calculated and experimental self-diffusion coefficients for liquid argon*

	Source	$D \times 10^5$ ($cm^2\ s^{-1}$) 84 °K	90 °K	100 °K
1	Experiment	1.84	2.35	3.45
2	Relaxation model (6.13.10)	0.84	1.16	1.69
3	Gaussian momentum autocorrelation (6.7.5)	—	3.05	3.21
4	Exponential autocorrelation: diffusion theory (6.7.5)	3.91	4.11	—
5	Isotope separation (6.7.5)	2.25	2.49	2.92
6	Longuet-Higgins and Valleau exponential correlation (square-well) (6.7.13)	1.43	1.8	2.25
7	Davis, Rice and Sengers (square-well)	1.81	2.31	3.39
8	Helfand linear trajectory (6.7.9)	2.80	3.25	3.85
9	Davis and Palyvos (linear trajectory + cross correlations)	2.50	2.75	3.40
10	Acoustic continuum (6.7.14)	—	0.84	—

provides us with a direct route to the friction constant with which we may compare the various theoretical estimates.

From table 6.8.1 and fig. 6.8.1 we see that the assumption of a Gaussian decay of the velocity autocorrelation is inadequate to describe a dense Lennard-Jones fluid. The predicted temperature derivative is one order of magnitude different from the experimental data of Naghizadeh and Rice[11], and the predicted magnitude of β is too small, in agreement with Rice's observations[13] in §6.7. It should be noted that the assumption of a Gaussian decorrelation of momentum implies an evolution composed of a series of many independent random changes of momentum subject only to the initial conditions – a result familiar from the theory of errors, and a consequence of the central limit theorem.

Undoubtedly, the best calculation in that it comes nearest to a solution of the N-body problem is that of Helfand[8] based upon an exact prescription for the friction constant in the linear trajectory approximation, (6.7.9). In this model the hard and soft friction coefficients are calculated separately, the diffusion constant being given by

$$D = kT/(\beta_S + \beta_H).$$

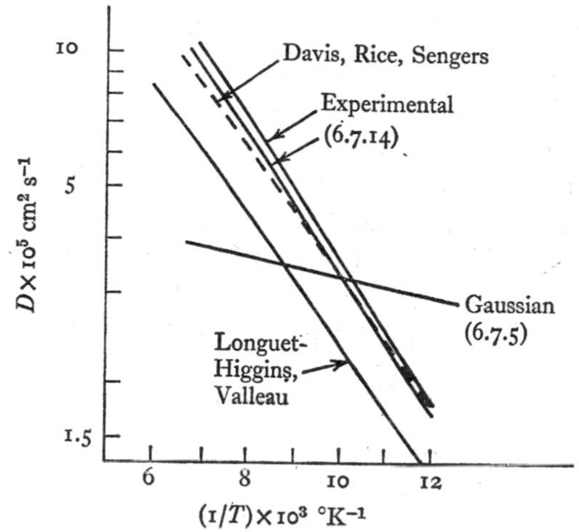

Fig. 6.8.1. Comparison of the temperature dependence of the self-diffusion coefficient D, on the basis of various models discussed in the text. See also table 6.8.1.

Using the theoretical pair correlation function computed by Kirkwood *et al.* on the basis of the superposition approximation, Davis and Palyvos[20] have extended the Helfand model in the linear trajectory approximation to include the cross term

$$D = kT/(\beta_S + \beta_H + \beta_{HS}).$$

As seen in table 6.8.1, the Davis–Palyvos theory gives a slight improvement upon the Helfand estimate, although these authors conclude from a consideration of the entire temperature range of liquid argon, krypton and xenon that the overall agreement between theory and experiment is not significantly improved by including the cross coefficient β_{HS}. They also concluded that the agreement between the Rice–Kirkwood and Douglass[25] theories (the latter differing from the Rice–Kirkwood estimate by a factor of $\pi/2$) and that of Helfand was equally good. It must be borne in mind, however, that the Davis–Palyvos estimate is based on an approximate value for the pair correlation function, and is therefore subject to change if a more reliable correlation function becomes available. Helfand's value is probably within 30 per cent of the value obtained using a radial distribution function yielding the correct pressure.

From table 6.8.1 it is seen that the square-well model gives the best agreement with experiment. The parameters of the pair interaction are fitted to gas data. In the Longuet-Higgins[22] model the temperature dependence is in good agreement with experiment (fig. 6.8.1), although the numerical value of the diffusion constant is about 30 per cent too low. Numerical discrepancy is largely removed, however, if the assumption of an exponential decay of the momentum autocorrelation is dropped. Thus, the Davis, Rice, Sengers[21] model yields a $D(T)$ curve in quite good agreement with the experimental data of Naghizadeh and Rice[11].

A calculation of the weak-coupling friction constant, β_S, on the basis of the acoustic continuum model yields a very much lower diffusion constant

$$D = kT/(\beta_S + \beta_H),$$

where $\beta_S/\beta_H > 20$. As we observed above (§6.7), however, this estimate was designed to provide an upper bound to β_S, resulting in a lower bound for the diffusion constant. Values are quoted in table 6.8.1 for the only temperatures for which the low frequency velocity of *sound* in argon is known. (The data from isotope separation for $\langle \nabla^2 \Phi_N \rangle$ are also used.) Rice observes that the velocity of propagation of the high frequency *density fluctuations* must be approximately $3\frac{1}{3}$ times greater than that of low frequency sound if this model is to produce a quantitatively correct description of a fluid.

Finally, we consider the flux of foreign molecules in a host liquid as a further significant test of the theory of molecular friction[27]. In particular, from an experimental point of view, we consider ion mobility and tracer diffusion. The situation envisaged is one wherein the foreign molecules are very few in number and do not interact, and, moreover, they are assumed not to disturb the equilibrium state of the host liquid. Davis, Rice and Meyer[26] have extended the Rice–Allnatt theory to the case of the mobility of positive ions, μ^+, in simple liquids, and obtain

$$\mu^+ = \frac{q}{\frac{8}{3}\rho_2 \sigma^2 g_{(2)}(\sigma)(2\pi\mu kT)^{\frac{1}{2}} + \beta_S}. \tag{6.8.1}$$

It is apparent that the critical problem is the specification of the soft friction coefficient. q is the ionic charge; $\sigma = \frac{1}{2}(\sigma_1 + \sigma_2)$, an effective atomic diameter, where the subscripts 1, 2 refer to the foreign and host molecules, respectively. In this case, of course, electrostriction modifies the radial deployment of neighbours about the ion, and Davis, Rice and Meyer separate the short and long ranged components to write

$$g_{(2)}(r) = g^0_{(2)}(r) + g^p_{(2)}(r), \tag{6.8.2}$$

where $g^0_{(2)}(r)$ is the unmodified distribution in the absence of polarization forces, $g^p_{(2)}(r)$ being the correction for polarization. This latter is estimated by considering the host fluid as a dielectric continuum in contact with a static ion. The density increase of the continuum due to electrostriction may then be written as a function of distance $\rho_1(r)$, whereupon

$$g^p_{(2)}(r) = \frac{\rho_1(r) - \rho_1(\infty)}{\rho_1(\infty)}. \tag{6.8.3}$$

Because the increase in density due to electrostriction is small, this continuum approximation is thought to be a reasonable estimate of the modification of the pair distribution. The electrostrictive perturbation leads to an increase in $g_{(2)}(r)$ for $r < 2\sigma$, whilst beyond this distance the distribution is virtually indistinguishable from that of the pure fluid. μ is the reduced mass $m_1 m_2/(m_1+m_2)$. The diffusion coefficient of the foreign molecules immersed in the host may be quite simply shown to be from the Einstein relation

$$D = \frac{kT}{\frac{8}{3}\rho_2 g_{(2)}(\sigma)\sigma^2(2\pi\mu kT)^{\frac{1}{2}} + \beta_S} \tag{6.8.4}$$

which reduces to the simple expression $kT/(\beta_S + \beta_H)$ for tracer diffusion of molecules of the same mass and atomic diameter as the host system.

Clearly the polarization interaction makes a significant contribution to the estimate of the friction coefficient. We see first that the polarization forces tend to increase the local density about the ion due to the Coulomb field, and secondly that there is presumably a mean square force proportional to $-\langle\nabla^2\Phi^p(r)\rangle$ acting. This latter may be shown to be negligible in comparison to the atomic interaction term $-\langle\nabla^2\Phi(r)\rangle$ arising in the determination of β_S. We

TABLE 6.8.2. *Positive ion (Ar_2^+) mobilities in liquid argon*(26)

	90 °K	120 °K
Expt. $\times 10^{-4}$ ($cm^2\,V^{-1}\,s^{-1}$)	6.10	17
Calc.	5.93	24

TABLE 6.8.3. *Comparison of neutral and ionic diffusion coefficients at 15 atm*(26)

	°K	$D \times 10^5$ ($cm^2\,s^{-1}$)	$D_i \times 10^5$ ($cm^2\,s^{-1}$)	D/D_i
Argon	90.1	2.35	0.474	5.0
Krypton	145.0	2.78	0.875	3.2
Xenon	184.3	2.48	0.442	5.6

therefore conclude that the effect of the polarization is to bring the atoms closer together in the vicinity of the ion when the usual short range dissipative interactions become effective.

We see from the entries in table 6.8.2 that at 90 °K the positive ion mobilities for Ar_2^+ in liquid argon are in quantitative agreement with the Rice–Allnatt theory of the soft friction coefficient. However, agreement is much less satisfactory if a different ionic species, Ar^+ for example, is postulated. It is found spectroscopically that the positive ionic species M_2^+ is stable (M = He, Ne, Ar, Xe), but has a hard core radius larger than the neutral atom M by a factor of about 1.5, whereupon the effective diameter $\sigma = \frac{1}{2}(\sigma_1 + \sigma_2)$ is larger than the σ_2–σ_2 diameter by a factor ~ 1.25. The ionic diffusion coefficients on account of the greater diameter of the diatom M_2^+ and the electrostrictive effects. The predicted ratio between the two coefficients, D/D_i, on the basis of the Rice–Allnatt theory of the soft friction constant is 4.8. A comparison of D and D_i for three liquids at 15 atm. has been made by Davis, Rice and Meyer(26), and the results are shown in table 6.8.3. The ratio D/D_i is seen to be $\sim$ 5. Furthermore, at constant temperature $g_{(2)}^{\text{p}}(r)$ in the vicinity of the ion would change negligibly with external pressure, whereupon D and D_i should show the same pressure dependence. This is found to be the case to within experimental error.

The study of negative ionic mobility is much more difficult because of impurity effects, and indeed, Davis *et al.* interpret their mobility data in terms of the properties of the O_2^- ion. If it may be assumed that the negative charge carriers in argon, krypton and xenon are effectively O_2^- ions, then the Rice–Allnatt theory is again seen to provide adequate representation of the observations.

6.9 The Smoluchowski equation

The Fokker–Planck equation describing the phase evolution of the coarse-grained singlet distribution $\bar{f}_{(1)}(\boldsymbol{p}_1, \boldsymbol{q}_1, t)$ in the weak-coupling limit, (6.4.20).

$$\frac{\partial \bar{f}_{(1)}}{\partial t} + \frac{\boldsymbol{p}_1}{m}\frac{\partial \bar{f}_{(1)}}{\partial \boldsymbol{q}_1} = \beta_{(1)}\frac{\partial}{\partial \boldsymbol{p}_1}\left\{\left(\frac{\boldsymbol{p}_1}{m} - \boldsymbol{u}\right)\bar{f}_{(1)} + kT\frac{\partial \bar{f}_{(1)}}{\partial \boldsymbol{p}_1}\right\} \equiv M,$$

is in general difficult to solve. Even more so, of course, is the evolution of its two-body counterpart, $\bar{f}_{(2)}(\boldsymbol{p}_1, \boldsymbol{p}_2, \boldsymbol{q}_1, \boldsymbol{q}_2, t)$, briefly discussed in §6.6. We have seen, however, that for dense systems equilibration in momentum space tends to precede configurational equilibration. We might observe that in a dense realistic fluid it is primarily the quasi-Brownian dissipative motion (M-coupling) in the soft force field that is responsible for the momentum evolution whilst it is the short range strong coupling of the Chapman–Enskog type (J-coupling) that governs the configurational evolution, this being essentially a geometrical problem. Associated with these processes of course, are their corresponding soft and hard friction constants, the magnitude of which establishes the rate of approach to equilibrium. We have already seen that $\beta_S/\beta_H \sim 20$, and we might have anticipated that momentum evolution would precede evolution in configuration space. Indeed, if the general friction constant β is large enough (more precisely, if β_S is much larger than β_H) we may assume that momentum evolution is essentially complete, further equilibration occurring in configuration space alone. In this case the Fokker–Planck equation in phase space reduces to the Smoluchowski equation in configuration space. Whether the general magnitude of β is indeed large enough for this purpose we shall ignore for the moment. More precisely, β

must be sufficiently large for the condition $\beta\tau \gg 1$ to hold[28]. This reduction to the Smoluchowski equation was first made by Kramers[29] in association with the Brownian motion, and has been studied by Suddaby and Miles[30] more recently.

The Smoluchowski equation in singlet configuration space follows from (6.4.20) by integration of the singlet phase distribution $f_{(1)}(\boldsymbol{p}_1, \boldsymbol{q}_1, t)$ over the momentum $\boldsymbol{p}_1$, to yield an equation in configuration space of the form

$$\frac{\partial n_{(1)}}{\partial t}(\boldsymbol{q}_1, t) = -\frac{\partial \boldsymbol{j}_1}{\partial \boldsymbol{q}_1}, \tag{6.9.1}$$

where
$$n_{(1)}(\boldsymbol{q}_1, t) = \int f_{(1)}(\boldsymbol{p}_1, \boldsymbol{q}_1, t)\,\mathrm{d}\boldsymbol{p}_1 \tag{6.9.2}$$

and
$$\boldsymbol{j}_1 = \frac{\partial}{\partial \boldsymbol{q}_1}\left(\frac{kT}{m} n_{(1)}\right) - \frac{\langle \boldsymbol{F}_1 \rangle}{m\beta_{(1)}} n_{(1)}. \tag{6.9.3}$$

The equilibrium momentum distribution in the fluid is Maxwellian. We could, of course, have substituted this into the single-particle Fokker–Planck equation, thereby reducing it to the Smoluchowski equation. $\langle \boldsymbol{F}_1 \rangle$ is the mean force acting on particle 1, and may be written directly in terms of the potential of mean force, Ψ:

$$\langle \boldsymbol{F}_1 \rangle = -\nabla_1 \Psi(r) = \nabla_1 [kT \ln g_{(2)}(r)], \tag{6.9.4}$$

$g_{(2)}(r)$ being the equilibrium pair distribution function. We shall constantly find that the determination of transport coefficients, etc., require precise knowledge of the *equilibrium* distributions, and the current inadequacy of our knowledge of both the distribution functions and the effective pair interactions makes a quantitative assessment of the success of transport theory difficult.

No distinction is made here between the time-smoothed distribution $\bar{f}_{(1)}$, and $f_{(1)}$: this is justified provided the displacement from equilibrium is not too great.

The liquid cannot, of course, be treated as a system of Brownian particles since the representative particles are strongly coupled. This is most simply taken into account by considering a representative *pair* of molecules. The determination of the various transport coefficients associated with their corresponding fluxes and arising as a response to generalized forces in the liquid – thermal and velocity gradients – requires in principle the solution of the *two-body* Fokker–Planck equation, a particularly formidable task[31]. However, the reduction to a two-body Smoluchowski equation is readily effected[32], and the extension of (6.9.1) into the pair space of the particles yields

$$\frac{\partial n_{(2)}}{\partial t}(\boldsymbol{q}_1, \boldsymbol{q}_2, t) = -\frac{\partial \boldsymbol{j}_{1(2)}}{\partial \boldsymbol{q}_1} + \frac{\partial \boldsymbol{j}_{2(2)}}{\partial \boldsymbol{q}_2}, \tag{6.9.5}$$

where

$$\left.\begin{aligned} n_{(2)}(\boldsymbol{q}_1, \boldsymbol{q}_2) &= n_{(1)}(\boldsymbol{q}_1)\, n_{(1)}(\boldsymbol{q}_2)\, g_{(2)}(|\boldsymbol{q}_1 - \boldsymbol{q}_2|), & (a)\\ \boldsymbol{j}_{i(2)} &= \frac{N(N-1)}{m}\iint \boldsymbol{p}_i f_{(2)}(\boldsymbol{p}_1, \boldsymbol{p}_2, \boldsymbol{q}_1, \boldsymbol{q}_2)\, \mathrm{d}\boldsymbol{p}_1\, \mathrm{d}\boldsymbol{p}_2. & (b) \end{aligned}\right\} \tag{6.9.6}$$

$\boldsymbol{j}_{1(2)}$ and $\boldsymbol{j}_{2(2)}$ in (6.9.5), defined by (6.9.6*b*), are seen to represent the two components of momentum flux in the liquid, and as such represents an equation of continuity in the liquid. Equation (6.9.5) in terms of the three unknowns $n_{(2)}(\boldsymbol{q}_1, \boldsymbol{q}_2, t)$, $\boldsymbol{j}_{1(2)}(\boldsymbol{q}_1, \boldsymbol{q}_2, t)$ and $\boldsymbol{j}_{2(2)}(\boldsymbol{q}_1, \boldsymbol{q}_2, t)$ represents the most general statement of the configurational projection of the Liouville equation for times not too near $t = 0$, without specifying the particle interactions. Before we are able to express the two-body Smoluchowski equation in purely configurational form we must first specify the momentum currents $\boldsymbol{j}_{1(2)}$, $\boldsymbol{j}_{2(2)}$. It is more convenient, for reasons which will shortly become apparent, to work in terms of the relative coordinates $\boldsymbol{R} = \frac{1}{2}(\boldsymbol{q}_1 + \boldsymbol{q}_2)$, $\boldsymbol{r} = \frac{1}{2}(\boldsymbol{q}_1 - \boldsymbol{q}_2)$ locating the centre of gravity of the pair, and their separation, respectively. To determine the *relative* momentum currents $\mathscr{J}_1(\boldsymbol{R}, \boldsymbol{r}, t)$ and $\mathscr{J}_2(\boldsymbol{R}, \boldsymbol{r}, t)$ we take moments of the two-body Fokker–Planck equation by multiplying throughout by $\boldsymbol{\Pi} = \boldsymbol{p}_1 + \boldsymbol{p}_2$ and $\boldsymbol{\pi} = \boldsymbol{p}_1 - \boldsymbol{p}_2$ and then integrating over all values of $\boldsymbol{\Pi}$ and $\boldsymbol{\pi}$. The relative momentum currents thus obtained then have the form

$$\mathscr{J}_1 = \frac{1}{\beta_{(2)}}\left\{\frac{\partial}{\partial \boldsymbol{R}}\left(\frac{kT}{m} n_{(2)}\right) + \frac{\partial \mathscr{J}_1}{\partial t}\right\}, \tag{6.9.7}$$

$$\mathscr{J}_2 = \frac{1}{\beta_{(2)}}\left\{\frac{\partial}{\partial \boldsymbol{r}}\left(\frac{kT}{m} n_{(2)}\right) - \frac{\langle \boldsymbol{F}_{(2)}\rangle}{m\beta_{(2)}} n_{(2)} + \frac{\partial \mathscr{J}_2}{\partial t}\right\}, \tag{6.9.8}$$

where $\langle \boldsymbol{F}_{(2)} \rangle$ is the mean force acting on the relative pair. Some lengthy but straightforward analysis is required in arriving at the above expressions for $\mathscr{J}_1$ and $\mathscr{J}_2$. Kirkwood has shown that these expressions are correct to within terms of order $1/\beta_{(2)}^2$, which are negligible provided $\beta_{(2)}$ is sufficiently large.

At this stage we may further suppose that for mechanical equilibrium the time rate of change of the momentum fluxes $\dot{\mathscr{J}}_1$, $\dot{\mathscr{J}}_2$ is zero since the Smoluchowski equation assumes the momentum evolution to be essentially complete. Insertion of the resulting expressions for the relative momentum currents in the *relative* form of (6.9.5), i.e.

$$\frac{\partial n_{(2)}}{\partial t} + \frac{\partial \mathscr{J}_1}{\partial \boldsymbol{R}} + \frac{\partial \mathscr{J}_2}{\partial \boldsymbol{r}} = 0 \tag{6.9.9}$$

yields the final Smoluchowski equation in the configurational subspace of molecular pairs for a system of uniform temperature and velocity:

$$\frac{\partial n_{(2)}}{\partial t} = -\left\{\frac{\partial}{\partial \boldsymbol{R}} \cdot \left[\frac{\partial}{\partial \boldsymbol{R}}\left(\frac{kT}{m\beta_{(2)}} n_{(2)}\right)\right] + \frac{\partial}{\partial \boldsymbol{r}} \cdot \left[\frac{\partial}{\partial \boldsymbol{r}}\left(\frac{kT}{m\beta_{(2)}} n_{(2)}\right) - \frac{\langle \boldsymbol{F}_{(2)} \rangle}{m\beta_{(2)}} n_{(2)}\right]\right\}. \tag{6.9.10}$$

For a system subject to no external fields wherein the evolution is independent of the location of the centre of gravity of the pair, i.e. $\partial n_{(2)}/\partial \boldsymbol{R} = 0$:

$$\frac{\partial n_{(2)}}{\partial t} = -\operatorname{div}\left\{\frac{kT}{m\beta_{(2)}} \operatorname{grad} n_{(2)} + n_{(2)} \frac{\operatorname{grad} \Psi}{m\beta_{(2)}}\right\} \tag{6.9.11}$$

in vector notation. The replacement $\langle \boldsymbol{F}_{(2)} \rangle = -\operatorname{grad}(\Psi/kT)$ has been made for the mean force on the pair. For thermodynamic equilibrium the left hand side of (6.9.11) vanishes. It is then readily verified that the expression in the curly brackets (the flow vector $\mathscr{J}_2$) also vanishes if $n_{(2)} = \rho_0 \exp(-\Psi/kT)$, the ordinary relation between the equilibrium distribution and the potential of average force.

In order to calculate the soft force contribution to the energy and momentum fluxes we need first to determine the perturbed pair distribution (6.9.6*a*) in terms of which the liquid responds to the generalized force. The quantity $(kT/m\beta_{(2)})$ may be interpreted as a diffusion coefficient whereupon (6.9.10) and (6.9.11) describe the diffusive transport of energy, momentum, etc. More immediately, the Smoluchowski equation is seen to have the general form describing the diffusion of a Brownian particle provided the identification $D = kT/m\beta_{(2)}$ is made, representing the coefficient of diffusion of a molecular pair.

We have already encountered a determination of the perturbed distribution (6.9.6 *a*) in the analysis of Croxton and Ferrier[34], §4.4. In that case the distortion of the pair distribution arose as a response to the field gradient at the liquid surface in their treatment of the surface tension. Moreover, the distribution was determined not on the basis of the Smoluchowski equation, but instead as a variational minimization of the Hamiltonian at the surface. The surface tension γ was then immediately identified in terms of the coefficients of a harmonic mode expansion of the angular distortion of the spherically symmetric bulk pair distribution arising in the vicinity of the liquid–vapour interface. This approach, treating the action of external forces by the inclusion of a potential in the Hamiltonian of the fluid was introduced initially by Kubo[35]. This situation must, however, be distinguished from the essentially different case of a perturbation of an already established local equilibrium. In this case the transport process does not arise simply as a mechanical response and cannot be approached directly through the system Hamiltonian. No rigorous determination of the transport coefficients in this second situation has as yet been given.

Suddaby and Miles[36] in their treatment of the Brownian motion have extended (6.9.10) to obtain a fourth-order Smoluchowski equation correct to order $1/\beta^2_{(2)}$. These equations are prohibitively difficult to implement in the determination of the transport coefficients and will only become of importance when the second-order theory has been thoroughly worked[80]. We shall not, for this reason, reproduce their equation here.

We have seen above by direct substitution that an equilibrium distribution of the form

$$n_{(2)}(\boldsymbol{q}_1, \boldsymbol{q}_2, \infty) = \rho_0 \exp(-\Psi/kT) = \rho_0 g^0_{(2)}(r) \qquad (6.9.12)$$

(where ^{0} signifies the equilibrium value) satisfies the Smoluchowski equation for an isotropic fluid of spherical particles. Under steady (time-independent) non-equilibrium conditions the Smoluchowski equation (6.9.11) will contain other flux components relating to the momentum and energy currents which arise from the application of velocity and temperature gradients, respectively. Moreover the pair distribution will distort in response to the steady applied forces, adopting the time-independent form (for small deviations from equilibrium):

$$n_{(2)}(\boldsymbol{r}) = \rho_0 \, g^0_{(2)}(r) \, [1 + w(\boldsymbol{r})], \tag{6.9.13}$$

where $w(\boldsymbol{r})$ is to be determined from the Smoluchowski equation solved subject to the appropriate boundary conditions. This we shall consider in the next two sections.

6.10 The transport coefficients: viscous shear flow[38]

In the following determinations of the coefficients of shear and bulk viscosity and the thermal conductivity, a general approximation will be made based upon the assumption that in the fluid it is the soft component of the interaction which is primarily responsible for the approach to equilibrium. Thus, for example, the coefficient of shear viscosity has two components,

$$\eta = \eta_K + \eta_V,$$

where the first applies to a dilute gas, whilst the second is the dominant contribution for dense gases and liquids. Similarly, for the thermal conductivity,

$$\lambda = \lambda_K + \lambda_V.$$

We shall in this and the following section discuss only the dominant potential contribution designated by the subscript V – the kinetic component following directly from elementary kinetic theory. We might have anticipated that this approximation from the relative magnitudes of $\beta^S_{(2)}$ and $\beta^H_{(2)}$; however, as Rice[56] observes, in making this approximation we must be quite sure as to which friction coefficient we refer since numerical estimates depend sensitively upon our choice of $\beta_{(2)}$. At the end of §6.11 we present Rice's rationalization of the theory in terms of the Rice–Allnatt approach.

As we have seen, (6.9.11), if temperature and velocity are uniform the pair distribution is determined by the Smoluchowski equation

$$\frac{\partial n_{(2)}}{\partial t} = -\operatorname{div}\left\{\frac{kT}{m\beta_{(2)}}\operatorname{grad} n_{(2)} + \frac{n_{(2)}}{m\beta_{(2)}}\operatorname{grad}\Psi\right\}, \qquad (6.10.1)$$

the expression in the curly bracket being identified as the flow vector, $\mathscr{J}_2$ (cf. 6.9.8). In thermal equilibrium,

$$n_{(2)} = \rho_0 g^0_{(2)}, \quad \mathscr{J}_2 = \dot{\mathscr{J}}_2 = 0.$$

For a steady non-uniform state we assume the general distribution (6.9.13), and for the specification of viscous flow we let the fluid velocity $\boldsymbol{c}$ depend upon coordinates in an arbitrary way, subject to the condition $\operatorname{div}\boldsymbol{c} = 0$. The bulk flow of the liquid may be regarded as exerting a differential dragging effect on each of the molecules of the pair, given by $m\beta_{(2)}\boldsymbol{c}$, and therefore supplements the mean force term. Neglecting terms of magnitude $\boldsymbol{c}w$, since both are assumed small, the steady flow is then given by direct substitution from (6.10.1):

$$\mathscr{J}_2 = \left\{\frac{kT}{m\beta_{(2)}}\operatorname{grad}\{\rho_0 g^0_{(2)}(r)[1+w(\boldsymbol{r})]\}\right.$$

$$\left. + \frac{1}{m\beta_{(2)}}(\operatorname{grad}\Psi - m\beta_{(2)}\boldsymbol{c})\,\rho_0\, g^0_{(2)}(r)\,[1+w(\boldsymbol{r})]\right\}$$

$$= \left\{-[1+w(\boldsymbol{r})]\frac{kT}{m\beta_{(2)}}\rho_0 . \langle \boldsymbol{F}_{(2)}\rangle + \rho_0 g^0_{(2)}(r)\frac{kT}{m\beta_{(2)}}\operatorname{grad} w(\boldsymbol{r})\right.$$

$$\left. + \frac{kT}{m\beta_{(2)}}\rho_0\langle \boldsymbol{F}_{(2)}\rangle - \frac{kT}{m\beta_{(2)}}\rho_0\langle \boldsymbol{F}_{(2)}\rangle . w(\boldsymbol{r}) - \boldsymbol{c}\rho_0 g^0_{(2)}(r)\right\},$$

where $\operatorname{grad} g^0_{(2)}(r) = -g^0_{(2)}(r)\operatorname{grad}(-\Psi/kT) = +\dfrac{\langle \boldsymbol{F}_{(2)}\rangle}{kT} g^0_{(2)}(r)$,

i.e.
$$\mathscr{J}_2 = \rho_0 g^0_{(2)}(r)\left\{\frac{kT}{m\beta_{(2)}}\operatorname{grad} w(\boldsymbol{r}) - \boldsymbol{c}\right\}. \qquad (6.10.2)$$

Recalling the identity

$$\operatorname{div}(a\boldsymbol{A}) = \boldsymbol{A}.\operatorname{grad} a + a\operatorname{div}\boldsymbol{A}, \quad \text{and} \quad \operatorname{div}\boldsymbol{c} = 0^{(39)}$$

the function $w(\boldsymbol{r})$ is determined from the time-independent Smoluchowski equation

$$0 = \operatorname{div}\mathscr{J}_2,$$

i.e.
$$\nabla^2 w(\boldsymbol{r}) - \frac{\langle \boldsymbol{F}_{(2)}\rangle}{kT} . \operatorname{grad} w(\boldsymbol{r}) = -\frac{\beta_{(2)} m}{kT} \boldsymbol{c} . \langle \boldsymbol{F}_{(2)}\rangle \qquad (6.10.3)$$

subject to the boundary condition

$$\operatorname{grad} w(\boldsymbol{r}) = 0 \quad \text{at} \quad r = \infty$$

so that at infinity the flow is, from (6.10.2), $\rho_0 g^0_{(2)}(r) . \boldsymbol{c}$. For the purpose of calculating the shear viscosity we assume the flow is laminar and Newtonian with a rate of shear $\frac{1}{2}\alpha$ having Cartesian velocity components $c_x = c_z = 0$; $c_y = \alpha x$. The right hand side of (6.10.3) then becomes

$$-\frac{\alpha \beta_{(2)} m}{(kT)^2} . \frac{\partial \Psi}{\partial r} . r \sin^2\theta \sin\phi \cos\phi.$$

Assuming the same angular dependence for the left hand side we have the form

$$w(\boldsymbol{r}) = \frac{\alpha\beta_{(2)} m}{(kT)^2} u(r) \sin^2\theta \sin\phi\cos\phi = \frac{\alpha\beta_{(2)} m}{(kT)^2} u(r) P^2_2(\cos\theta), \quad (6.10.4)$$

i.e.
$$n_{(2)}(\boldsymbol{r}) = \rho_0 g^0_{(2)}(r) \left[P^0_0(\cos\theta) + \frac{\alpha\beta_{(2)} m}{(kT)^2} u(r) P^2_2(\cos\theta) \right],$$

where $u(r)$ is an as yet unknown function of the separation and represents the purely *radial* distortion of the equilibrium pair distribution. Substitution of (6.10.4) into (6.10.3) immediately yields the second-order differential equation determining $u(r)$:

$$\frac{d^2 u}{dr^2} + \left(\frac{2}{r} - \frac{1}{kT}\frac{d\Psi}{dr}\right)\frac{du}{dr} - \frac{6}{r^2} = -\frac{m\beta_{(2)}\alpha}{(kT)^2}\frac{d\Psi}{dr} r. \qquad (6.10.5)$$

This equation has been obtained by both Kirkwood[40] and Eisenschitz[38], and its solution in

$$\eta = \frac{2\pi\rho_0^2}{15} \beta_{(2)} \int_0^\infty \frac{d\Phi(r)}{dr} g^0_{(2)}(r) u(r) r^3 \, dr, \qquad (6.10.6)$$

should yield a value of the coefficient of viscosity, η, presuming the value of the two-body friction constant $\beta_{(2)}$ to be known.

On the basis of the assumption that the fluid is incompressible, we arrived at the time-independent Smoluchowski equation for the radial distortion, $u(r)$ (equation (6.10.5)). Had we incorporated possible compressibility effects (see ref. 39) we should have then to take into account dissipative effects arising from compressions and rarefactions in the fluid, even when it is at rest. Associated with this process is an internal or bulk viscosity, about which there is still some controversy. Insertion of the distribution

$$n_{(2)} = \rho_0 g^0_{(2)}[1 + w(\mathbf{r})]$$

under these circumstances would have led to a second radial equation, obtained by Kirkwood, Buff and Green[40], for the spheric component:

$$\frac{d^2 s}{dr^2} + \left(\frac{2}{r} - \frac{1}{kT}\frac{d\Psi}{dr}\right)\frac{ds}{dr} = -\frac{r}{kT}\frac{d\Psi}{dr}. \tag{6.10.7}$$

This is to be solved subject to the boundary conditions

$$\frac{ds}{dr} = 0 \quad \text{as} \quad r \to \infty;$$

$$r^2\frac{ds}{dr} = 0 \quad \text{as} \quad r \to 0.$$

Study of these equations makes it unlikely that $s(r)$ is strongly temperature dependent – the final expression for the bulk viscosity is given as

$$\zeta = \frac{\pi\rho_0^2}{9kT}\beta_{(2)}\int_0^\infty \frac{d\Phi}{dr} g^0_{(2)}(r)\, s(r)\, r^3\, dr. \tag{6.10.8}$$

This expression for the bulk viscosity is particularly important since continuum hydrodynamics is unable to provide detailed information about it.

There are, however, still quite formidable difficulties yet to be overcome. Equation (6.10.5) being of second order requires two boundary conditions to determine $u(r)$ uniquely. There is general agreement on one of these, viz.

$$u(r) \to 0 \quad \text{as} \quad r \to \infty. \tag{6.10.9}$$

Kirkwood, Buff and Green[40] take for the second boundary condition what has become known as the 'weak' condition

$$u(r) \to 0 \quad \text{as} \quad r \to 0. \tag{6.10.10}$$

This assumption ensures that $u(r)$ is everywhere finite. However, Eisenschitz[38] objects to this condition on the grounds that it leads to a degree of anisotropy not observed experimentally. Eisenschitz dispenses with the condition at the origin entirely and instead adopts a second 'strong' condition in the infinite radial limit:

$$r^3 u(r) = 0 \quad \text{as} \quad r \to \infty. \tag{6.10.11}$$

This condition requires that $u(r)$ falls off much more strongly for large r than does that of Kirkwood, although the second Eisenschitz boundary condition does become infinite at the origin. Eisenschitz points out, however, that the calculated viscosity will still remain finite: in practice the finite size of the particle provides an obvious cut-off, and does not allow the application of a condition at $r = 0$. Divergence at the origin therefore does not arise, and the calculated viscosity on the basis of both the Kirkwood, Buff and Green, and the Eisenschitz boundary conditions does indeed remain finite, although the alternative conditions lead to markedly different radial solutions.

The ambiguity in the boundary conditions is an unsatisfactory aspect of the present theory, and further work is needed to resolve this situation. Certain qualitative features, however, enable us to make some sort of distinction. The Eisenschitz solution to (6.10.5) proves to be particularly sensitive to the potential function, and since the particle potential is strongly temperature dependent, we might anticipate that $u(r)$ determined subject to the 'strong' boundary conditions (6.10.9) and (6.10.11) will, upon insertion in (6.10.6), yield a coefficient of shear viscosity which is strongly temperature dependent. The temperature dependence of the 'weak' Kirkwood function proves to be much less strong, and given that the pair friction constant $\beta_{(2)}$ is virtually temperature independent (see §6.7), we might reasonably conclude that the 'strong' boundary conditions are to be preferred bearing in mind the observed sensitivity of shear viscosity upon temperature. Since the radial distortion $u(r)$ is zero for a flowing gas, Suddaby and Hales[127] observe that $u(r)$ should decrease as the intermolecular force between a representative pair of molecules is turned off. These authors show that the 'strong' boundary condition results in solutions to the equation for $u(r)$ which satisfy this condition, whilst the 'weak'

Kirkwood condition does not: they therefore conclude that it is the Eisenschitz condition which is appropriate.

Eisenschitz and Suddaby[37] in their development of the fourth-order Smoluchowski equation find that a unique solution appears to exist subject only to the application of the strong boundary condition – the weak condition gives rise to some ambiguity in the form of solution. Moreover, Eisenschitz and Cole[41] find that the strong boundary conditions may be applied to systems of particles interacting with more general, asymmetrical (e.g. dipole) pair potentials. However, we should remember that Eisenschitz rejected the 'weak' Kirkwood condition on the grounds that the predicted anisotropy was at variance with observation. More recently Champion[42] has reported measurements of a small anisotropy in the flowing liquid, and it does seem that the Eisenschitz condition cannot be accepted without reservation. The divergence of $u(r)$ near the origin appears implausible on physical grounds, and this has been attributed to a failure in the Smoluchowski equation in this region. The Smoluchowski equation, is, of course, only a satisfactory approximation to the Fokker–Planck equation if the friction constant $\beta_{(2)}$ is large enough. This is certainly the case for the Brownian motion, but the uncertainty in $\beta_{(2)}$ in liquid theory makes the replacement of the Fokker–Planck equation by that of Smoluchowski more dubious.

Orton[43] has experimentally determined the coefficient of shear viscosity for liquid argon at 89 °K to be 2.39×10^{-3} poise. From this value the pair friction constant $\beta_{(2)}$ is found to be $4.24 \times 10^{11}\, s^{-1}$ and as such provides a basis for comparison of the various theoretical estimates. The first calculation of the coefficient of shear viscosity was given by Kirkwood, Buff and Green[40]. These authors adapted an analytic pair function to coincide with the first peak of an experimental determination. Their pair potential was principally LJ (6–12), but was replaced by a hard sphere approximation at small separations. It should be noted that the pair interaction and the assumed radial distribution function are essentially incompatible in the most sensitive region of the integrand. The friction constant was not calculated on the basis of a molecular model of decorrelation of the two-body force autocorrelation. Kirkwood *et al.*

TABLE 6.10.1. *Comparison of experimental and calculated shear viscosities of liquid argon at 89 °K*

	$\eta\times 10^3$ (poise)	$\beta_{(2)}\times 10^{-11}$ (s^{-1})	Boundary condition
Orton (expt)[43]	2.39	4.24	
Kirkwood *et al.*[40]	1.27	73	Weak
Zwanzig *et al.*[44]	0.84	0.35	Weak
Eisenschitz[38]	1.91	3.5	Strong

obtained $\eta = 1.27\times 10^{-3}$ poise, $\beta_{(2)} = 7.3\times 10^{12}\,s^{-1}$. The calculated viscosity is about half that measured by Orton, and the friction constant is almost certainly too large (see table 6.10.1). In a later calculation, Zwanzig, Kirkwood, Stripp and Oppenheim[44] obtained a more satisfactory value of the friction coefficient, but their final estimate of the coefficient of shear viscosity was reduced to about one third of the experimental value: $\eta = 0.84\times 10^{-3}$ poise,

$$\beta_{(2)} = 0.35\times 10^{11}\,s^{-1}.$$

Zwanzig *et al.* also 'optimized' the radial distribution function with respect to the thermodynamic functions, in particular the pressure (note the similarity between the viscosity and pressure integrals), and this amounted to the scaling $g_{(2)}(r)\to g_{(2)}(1.026r)$. An estimate of the bulk viscosity was also obtained: $\zeta = 0.42\times 10^{-3}$ poise. Both these estimates were, of course, determined subject to the weak boundary condition.

The radial distortion $u(r)$ determined by Eisenschitz leads, in conjunction with an estimate of $\beta_{(2)}$ obtained by Eisenschitz and Wilford[45], to the most satisfactory numerical estimate of the shear viscosity. Unfortunately apparent numerical accuracy alone is insufficient to distinguish between the two boundary conditions: the viscosity integral contains the imperfectly known functions $g_{(2)}(r)$ and $\Phi(r)$, not to mention the pair friction constant. However, there are two further important experimental facts relating to the variation of viscosity with pressure at constant temperature, and the variation of viscosity with temperature at constant density. We see from fig. 6.10.1 that liquid argon exhibits a virtually linear variation

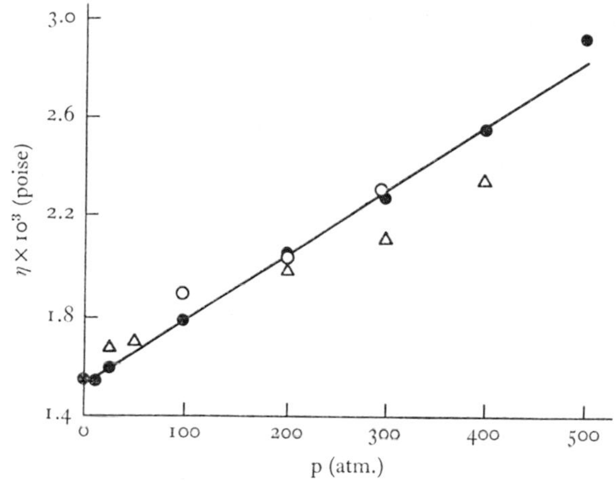

Fig. 6.10.1. The variation of the coefficient of shear viscosity of liquid argon with pressure at 101.8 °K: ● Lowry, Rice and Gray[46]; ○ Zhdanova[47]; △ Scott[48]. (Redrawn by permission from Rice and Gray, *The Statistical Mechanics of Simple Liquids*, (Wiley).)

of η with pressure at constant temperature, the coefficient ranging from $\sim 1.7 \times 10^{-3}$ poise at 50 atm., 101.8 °K, to $\sim 2.8 \times 10^{-3}$ poise at 500 atm. The effect of temperature at constant density appears to be rather weak, this being at variance with the Arrhenius temperature dependence proposed by Andrade[49] and indirectly supported by the strong boundary condition of Eisenschitz. It therefore appears that one of the earlier distinctions in favour of the Eisenschitz theory on the basis of its strong temperature dependence must be reconsidered in the light of the fact that only about one quarter of the 'temperature effect' may be directly attributed to this cause, the rest being identified as a density effect[50]. Indeed, we have already seen the susceptibility of the computed thermodynamic functions to a shift in the radial distribution function. Moreover, Zwanzig *et al.* have specifically invoked a scaling factor of only 1.026 for this very purpose. Clearly it is too early to attempt to draw a distinction between the Kirkwood and Eisenschitz conditions: this aspect of the problem must remain an important focus of attention both theoretically and experimentally.

In as far as the soft component of the friction constant contributes to the uncertainty of the calculated viscosity, it is of interest to consider the high temperature high density limit when $T \gg \epsilon/k$, and $\beta^{\mathrm{H}}_{(2)} \gg \beta^{\mathrm{S}}_{(2)}$. Under these circumstances, of course, the weak Fokker–Planck operator M is replaced by its Chapman–Enskog counterpart, J. Moreover, the hard friction constant is unambiguously calculable, and it is of interest therefore to estimate the shear viscosity in this thermodynamic limit. $\beta^{\mathrm{S}}_{(2)}$ cannot be easily estimated since no data on self-diffusion exist at high temperature and density. Rice and Gray[51] make such a comparison between the Enskog theory and experiment with good agreement for the shear viscosity of argon at 273 and 348 °K. The progressive discrepancy at high densities may presumably be attributed to the inadequacy of the low density estimate of $\beta^{\mathrm{H}}_{(2)}$.

6.11 The transport coefficients: thermal conductivity[38]

For specifying the thermal conduction we now assume the liquid is at rest, and that T is an arbitrary time-independent function of the coordinates subject only to the Fourier–Laplace condition

$$\nabla^2 T = \lambda^{-1}\dot{T} = 0$$

i.e. the usual heat flow equation holds. From (6.10.1) the energy current is given as

$$\mathscr{J}_2 = \left\{\frac{k}{m\beta_{(2)}}\{T\,\mathrm{grad}\,\rho_0 g^0_{(2)}[1+w(\boldsymbol{r})]+\rho_0 g^0_{(2)}[1+w(\boldsymbol{r})]\,\mathrm{grad}\,T\}\right.$$
$$\left.+\frac{\rho_0 g^0_{(2)}}{m\beta_{(2)}}[1+w(\boldsymbol{r})]\,\mathrm{grad}\,(kT\Psi)\right\}. \quad (6.11.1)$$

Neglecting terms of magnitude $w\,\mathrm{grad}\,T$, $\mathrm{grad}^2\,T$, etc., since for small deviations from equilibrium these quantities are small, the flow is given by

$$\mathscr{J}_2 = \frac{k}{m\beta_{(2)}}\rho_0 g^0_{(2)}(r)\,[T_0\,\mathrm{grad}\,w(\boldsymbol{r})+(1+\Psi)\,\mathrm{grad}\,T], \quad (6.11.2)$$

where we have used the relation

$$\mathrm{grad}\,g^0_{(2)}(r) = -g^0_{(2)}(r)\,\mathrm{grad}\left(-\frac{\Psi}{kT}\right).$$

T_0 is the temperature at the origin of coordinates. The temperature-dependent correction function to the equilibrium distribution $w(\boldsymbol{r})$ is then given by the time-independent Smoluchowski equation

$$0 = \operatorname{div} \mathscr{J}_2,$$

i.e.
$$\nabla^2 w(\boldsymbol{r}) - \frac{\langle F \rangle}{kT} \cdot \operatorname{grad} w(\boldsymbol{r}) = \frac{\Psi}{(kT)^2 T_0} \langle F \rangle \operatorname{grad} T. \quad (6.11.3)$$

Equation (6.11.3) is solved subject to the boundary condition

$$T_0 \operatorname{grad} w + \operatorname{grad} T = 0$$

so that, from (6.11.2), the flow is zero at infinity. For the purpose of calculating the thermal conductivity we suppose the temperature is distributed according to the linear relation

$$T = T_0(1 + \xi z),$$

so that the gradient is constant:

$$\nabla_z T = \zeta T_0; \quad \nabla_x T = \nabla_y T = 0$$

Under these circumstances the right hand side of (6.11.3) is

$$\frac{\xi \Psi}{(kT)^2} \frac{\partial \Psi}{\partial r} \cos \theta$$

and we therefore assume a form $w(\boldsymbol{r}) = v(r) \cos \theta$ which when inserted in (6.11.3) yields the differential relation

$$\frac{d^2 v}{dr^2} + \left(\frac{2}{r} - \frac{1}{kT}\frac{d\Psi}{dr}\right)\frac{dv}{dr} - \frac{2v}{r^2} = \frac{\xi \Psi}{(kT)^2}\frac{d\Psi}{dr} \quad (6.11.4)$$

with the boundary conditions $dv/dr = -\xi$, $v/r = -\xi$ $(r = \infty)$. This equation was first obtained by Eisenschitz[38] and again, in essentially the same form, by Zwanzig, Kirkwood, Oppenheim and Alder[44]. In this case the boundary conditions are not so controversial. Equation (6.11.4) is solved for $v(r)$, and substituted in the expression for the thermal conductivity:

$$\lambda_V = \frac{\pi \rho_0^2}{2kT\beta_{(2)}} \int_0^\infty \int_0^\pi \int_0^{2\pi} g_{(2)}^0(r) \left\{ \left(r\frac{d\Phi}{dr} - \Psi(r) \right) \left(\frac{dv}{dr} + \frac{\Psi(r)}{kT} \right) \cos^2 \theta \right.$$
$$\left. + \left(\frac{r}{2}\frac{d\Phi}{dr} - \Psi(r) \right) \left(\frac{v}{r} + \frac{\Psi(r)}{kT} \right) \sin^2 \theta \right\} \sin \theta \, dr \, d\theta \, d\phi, \quad (6.11.5)$$

where, in the total expression for the thermal conductivity

$$\lambda = \lambda_K + \lambda_V,$$

the kinetic component λ_K is considered negligible with respect to the configurational term, λ_V, in a dense fluid. Measurements of the thermal conductivity of liquid argon have been reported by Uhlir[(52)], by Ziebland and Burton[(53)] and by Keyes[(54)]. Measurements over extensive ranges of temperature and pressure have been reported by Ikenberry and Rice[(55)]. The general agreement between the measurements is good. The value calculated by Zwanzig, Kirkwood, Stripp and Oppenheim[(44)] of

$$\lambda_V = 2.4 \times 10^{-4} \,\mathrm{cal\, g^{-1}\, s^{-1}\, deg^{-1}}$$

was determined with a distribution function scaled so as to give the correct equilibrium pressure at 90 °K. This estimate is surprisingly near the experimental value[(55)] of

$$\lambda = 2.96 \times 10^{-4} \,\mathrm{cal\, g^{-1}\, s^{-1}\, deg^{-1}}.$$

However, Zwanzig *et al.* demonstrated that this was a largely fortuitous result: scaling of the distribution function by the smallest amount led to large variations in the estimate of the thermal conductivity. Once again we observe that the calculation of the transport coefficients depends ultimately upon an adequate knowledge of the equilibrium distributions and pair interaction.

The most extensive comparison of theory and experiment has been made for the Rice–Allnatt liquid. Rice[(56)] observes that the expression (6.11.5) for the configurational component of the thermal conductivity based ultimately on the Fokker–Planck–Smoluchowski approximation is unable to account for the large energy and momentum transfers which occur during strongly repulsive encounters. Contributions from the soft part of the intermolecular potential are adequately dealt with in the Eisenschitz and Kirkwood theories, i.e. small momentum and energy transfers are correctly treated. However, in as far as the configurational component of the thermal conductivity λ_V is adequate, the appropriate friction constant must be used – that is, the soft component ($\beta_{(2)}^{S}$) only. Subtraction of the rigid core contribution leads to appreciable changes in the effective friction constant. In this sense impulsive exchanges cannot be neglected and an approach of the Rice–Allnatt type is required which incorporates both aspects of the dissipative process if the formulation is not to exclude completely the possibility of such exchanges.

It would probably be fair to say that the Rice–Allnatt approach to the transport coefficients is qualitatively reasonably sound, although not too much emphasis should yet be placed on the apparent quantitative success of the theory. Significant quantitative estimates of the practical utility of the theory must await fully reliable distribution and pair interaction functions; only then will it be possible to assess how we might refine our theories. Rice and Gray[51] investigate the high temperature, high density limit for the thermal conductivity, wherein the hard friction constant which is unambiguously calculable dominates the dissipative process. In this case the Fokker–Planck operator reduces to one of the Enskog type, and as for the shear viscosity, leads to quite good agreement with the experimental values for liquid argon at 273 and 348 °K in the density range 0–600ρ/ρ_0 (ρ_0 = molar density).

In the meantime we observe[57] that the coefficient of shear viscosity, as given by (6.10.6), is directly proportional to the friction constant, whilst the thermal conductivity, given by (6.11.5), is inversely proportional to β. It would therefore appear a useful test of the theory to measure the β-independent quantity $\eta\lambda$ (or $\xi\lambda$, although data are scarce). Uncertainty in the $\eta\lambda$-product is then reduced to the uncertainty in the functions $g^0_{(2)}(r)$ and $\Phi(r)$. A meaningful theoretical estimate of the product is within the scope of the theory, and on the basis of the estimates of Zwangig *et al.* we have $\eta\lambda_V = 2.02\times10^{-7}$ against the experimental value of 4×10^{-7} – a factor of 2 greater. This is no cause for alarm, however, since the great sensitivity of the integrals to the scale of the radial functions permits little more than an order of magnitude estimate.

Finally, we must question the adequacy of the Smoluchowski equation as an approximation to the Fokker–Planck equation for the discussion of time-independent non-equilibrum processes. We have previously only specified that the friction constant, or more precisely the product $\beta\tau$, be 'sufficiently large'. Momentum evolution is then assumed to be complete, whereupon the Smoluchowski equation describes the configurational anisotropy arising in response to a generalized force. How large is 'sufficiently large'? As yet we have no definite answer[80]; Suddaby and Miles[36] have suggested that it might be as large as $4 \times 10^{12}\,s^{-1}$ whereupon a fourth-order Smoluchowski equation such as these authors have proposed becomes almost mandatory.

6.12 Linear response theory – the Kubo relations

In this section we shall consider the response of a *uniform* system to an external perturbation. This treatment is important because most non-equilibrium states are set up by perturbing a system that is initially in thermal equilibrium. The final non-equilibrium state will be stationary or time dependent, depending of course upon the time dependence of the perturbation. The type of perturbation we are considering can be expressed in purely mechanical terms either as a modification of the Hamiltonian or as a modification of the boundary conditions. There is also the class of transport processes which arise in response to non-uniformities within the system itself. At the present time it seems that all flow processes, except those arising from thermal gradients, can be set up by purely mechanical perturbations. Although a number of methods have been developed to handle thermal conduction and mass diffusion, it does appear that the application of linear response theory is not altogether satisfactory in these cases, and here we shall restrict ourselves exclusively to the former class of process.

The essential features of the linear response formalism, stimulated perhaps by the Kirkwood expression[32] of the friction constant as a diffusion coefficient in momentum space, were first proposed by M. S. Green[58]. Of the former class of processes mentioned above, Kubo[59] obtained rigorous expressions for the transport coefficients as time integrals over the autocorrelation of the appropriate flux. In the latter class of transport processes no rigorous derivation of the transport coefficients has yet been given, and the problem may be directly ascribed to the difficulty of expressing the non-mechanical perturbation of the system Hamiltonian.

The phenomenological relations of linear response theory are experimentally evident[60], provided the deviations from equilibrium are not too great. Thus we have in physics many such examples of the linear response of an open system giving rise to a stationary state which is not, by virtue of the dissipative processes, in thermodynamic equilibrium. Perhaps the most obvious examples are those of viscous and ohmic dissipation wherein we have a linear relation between the generalized flux, $\boldsymbol{J}$, and the generalized driving force, $\boldsymbol{X}$. The coefficient of proportionality, L, is a time-independent phenomenological transport coefficient, independent of the driving force. The phenomenological theory is largely confined to linear responses, but can be extended to non-linear situations. The microscopic theory will also be restricted to linear responses, although it too could easily be extended.

A generalized flux $\boldsymbol{J}_i$ arising in response to the applied force $\boldsymbol{X}_i$ rarely, if ever, develops at the complete exclusion of other secondary currents. Thus, in general a given flux is associated with more than one driving force, and cross-phenomena arise, so that for time-independent, or at least very slowly varying driving forces, we write

$$\boldsymbol{J}_i = \sum_j L_{ij}\boldsymbol{X}_j; \tag{6.12.1}$$

or in matrix notation $\{\boldsymbol{J}\} - \{L\}\{\boldsymbol{X}\}$.

The most general form of linear relation between currents and driving forces is[61]

$$\boldsymbol{J}_i(\boldsymbol{q},t) = \sum_j \iint L_{ij}(\boldsymbol{q}-\boldsymbol{q}',t-t')\boldsymbol{X}_j(\boldsymbol{q}',t)\,\mathrm{d}\boldsymbol{q}'\,\mathrm{d}t'. \tag{6.12.2}$$

Examples of these cross-phenomena are the electric field which produces both a charge current and a thermal current; a thermal gradient will cause both a thermal current and a mass current (Soret effect) and, conversely, a concentration gradient will give rise to a temperature gradient (the Dufour effect). We are therefore obliged to deal with the matrix of phenomenological coefficients $\{L\}$, (6.12.1), wherein the diagonal elements L_{ii} refer to conjugate fluxes and forces, and the off-diagonal elements L_{ij} refer to the cross-phenomena. On the basis of microscopic reversibility

Onsager[62] has shown that for a 'proper' choice of fluxes and forces, the matrix of phenomenological transport coefficients is symmetric:

$$L_{ij} = L_{ji} \tag{6.12.3}$$

and as such expresses the reciprocity between linked irreversible effects. This symmetry condition will not apply for all systems and those involving rotational or magnetic effects are notable exceptions. The question arises as to what constitutes a proper choice of fluxes and forces, for if these are incorrectly chosen the reciprocal relation (6.12.3) no longer holds. Conventionally the approach is as follows: we assume the time-independent state to be one of maximum entropy and minimum entropy production, subject to the application of external constraints such as electric fields or shearing forces. Thus if a set of s functions A_i representing various parameters of the system, energy, composition, momentum, etc., adopt their mean equilibrium values $\langle A_i \rangle$, then we may express the equilibrium (maximum) entropy S_0, as

$$S_0 = S_0(\langle A_1 \rangle, \langle A_2 \rangle, \ldots, \langle A_S \rangle).$$

We further suppose that for a system not too far displaced from equilibrium, the non-equilibrium entropy $S < S_0$ is given as

$$S = S(A_1, A_2, A_3, \ldots, A_S)$$

although we must be careful in dealing with non-equilibrium functions of the system, A_i. The fluctuations are assumed small, whereupon S may be Taylor expanded about S_0 in terms of the deviation from equilibrium

$$a_i = A_i - \langle A_i \rangle, \tag{6.12.4}$$

$$S = S_0 - \frac{1}{2} \sum_{i,j=1}^{S} \left(\frac{\partial^2 S}{\partial A_i \partial A_j} \right)_{A=\langle A \rangle} a_i a_j. \tag{6.12.5}$$

We now define the flux $\boldsymbol{J}_i$ by

$$\boldsymbol{J}_i = \dot{a}_i \tag{6.12.6}$$

and differentiation of (6.12.5) with respect to time yields[68]

$$\frac{dS}{dt} = \sum_{i=1}^{S} \boldsymbol{J}_i \boldsymbol{X}_i, \tag{6.12.7}$$

where $$X_i = \frac{\partial}{\partial a_i}(S - S_0). \quad (6.12.8)$$

It is clear from this that the maximization of the entropy, subject to the external constraints, provides the driving force for the approach to the steady state. Thus, as the system approaches a time-independent non-equilibrium distribution, so $\dot{S}$ and $\sum_{i=1}^{S} J_i X_i \to 0$. Unlike our previous discussions of irreversible phenomena, the relations (6.12.1), (6.12.2) are essentially macroscopic. Whether these expressions can be substantiated at a microscopic level remains to be seen, and we shall discuss this shortly.

The tasks of linear response theory are first to show that relations of the form (6.12.1) may be established from a microscopic standpoint, secondly to obtain exact closed expressions for the transport coefficients L_{ij} and finally to establish the formal properties of the phenomenological coefficients L_{ij}. We shall content ourselves here to consider only the second aspect – the expression of the transport coefficients in molecular terms. Kubo[59,61] considers the general problem of the response of a system to a mechanical perturbation. In that analysis the microscopic expression of L_{ij} is automatically obtained and moreover, the final linear relation between the generalized fluxes and forces is indeed of the form (6.12.1).

Here we merely consider the application of the linear regression relation (6.12.1)

$$J_i = \frac{da_i}{dt} = \sum_j L_{ij} X_j$$

to the non-equilibrium of a uniform system. Onsager proposed that the above relation should apply to the regression of fluctuations as well as external perturbations. As it stands, (6.12.4) implies an exponential regression with consequential difficulties at $t = 0$, for then the response would appear instantaneously when in actual fact there must be a transient period during which the flux has first to build up[63]. It is these transients which are involved in the $[\dot{\psi}(t)]_{t=0} = 0$ form of the correlation functions.

In particular we are concerned with the coarse-grained derivative[64]

$$\frac{d\bar{a}_i}{dt} = \frac{1}{\tau}[\langle a_i(t+\tau)\rangle - a_i(t)], \quad (6.12.9)$$

where $a_i(t)$ denotes the initial value of a_i, and $\langle a_i(t+\tau)\rangle$ represents the mean value of a_i at a time τ later, subject to the condition that the initial values of a_i were $a_i(t)$. τ is taken to be sufficiently long to smooth over transients but nevertheless short on a macroscopic scale. The forces $\boldsymbol{X}_i$ are taken to be a function of the $\{a(t)\}$ so that we have

$$\frac{1}{\tau}[\langle a_i(t+\tau)\rangle - a_i(t)] = \sum_j L_{ij}\boldsymbol{X}_j(\{a(t)\}). \qquad (6.12.10)$$

Mutliplying throughout by $a_l(0)$ and averaging over all $\{a(t)\}$,

$$\frac{1}{\tau}[\langle a_i(t)\, a_l(0)\rangle - \langle a_i(0)\, a_l(0)\rangle] = \sum_j L_{ij}\langle \boldsymbol{X}_j(\{a\})\, a_l(0)\rangle. \qquad (6.12.11)$$

Now, since $$a_i(\tau) = \int_0^\tau \dot{a}_i(t')\,\mathrm{d}t' + a_i(0)$$

(6.12.7) may be written

$$\int_0^\tau \langle \dot{a}_i(t')\, a_l(0)\rangle\,\mathrm{d}t' = \sum_j L_{ij}\langle \boldsymbol{X}_j(\{a\})\, a_l(0)\rangle, \qquad (6.12.12)$$

where we have observed that $\langle \dot{a}_i(0)\, a_l(0)\rangle = 0$. Invariance of $\langle \dot{a}_i(t')\, a_l(0)\rangle$ to a displacement of the origin of time t enables us to write

$$\begin{aligned}\langle \dot{a}_i(t')\, a_l(0)\rangle &= -\langle a_i(t')\, \dot{a}_l(0)\rangle \\ &= -\int_0^{t'} \langle \dot{a}_i(t'')\, \dot{a}_l(0)\rangle\,\mathrm{d}t'' \qquad (6.12.13)\end{aligned}$$

whilst the right hand side of (6.12.12) may be written

$$\langle \boldsymbol{X}_j(\{a\})\, a_l(0)\rangle = \delta_{jl}$$

from (6.12.8). Thus, from (6.12.12) and (6.12.13) we have finally

$$\left.\begin{aligned}L_{il} &= \frac{1}{\tau}\int_0^\tau\int_0^{t'} \langle \dot{a}_i(t'')\, \dot{a}_l(0)\rangle\,\mathrm{d}t''\,\mathrm{d}t' \\ &= \int_0^\tau \langle \dot{a}_i(t'')\, a_l(0)\rangle\left(1-\frac{t''}{\tau}\right)\mathrm{d}t''.\end{aligned}\right\} \qquad (6.12.14)$$

This last expression may be further reduced if $\tau \gg t''$ whereupon

$$L_{il} = \int_0^\tau \langle \dot{a}_i(t'')\, \dot{a}_l(0)\rangle\,\mathrm{d}t'',$$

i.e. if the correlation drops to zero well within the coarse graining time, τ. A similar expression has been obtained by Hashitsume[65].

As we mentioned at the beginning of this section, we are principally concerned here with uniform systems, that is, those whose intensive variables such as temperature and density are constant throughout the body of the liquid. Thermal conduction and mass diffusion are typical processes occurring in a *non-uniform* system. Montroll[66] has successfully dealt with the latter process, but so far no similar method has been devised for thermal processes. An open system can be driven by external agencies into a non-uniform state that can be maintained indefinitely: if this can be described directly in purely mechanical terms, then we are dealing with the response of the system to a mechanical driving force. This problem has already been dealt with. A discussion of the theory of non-uniform systems is given by Chester[67].

Nakano[69] has shown that the transport coefficients in a uniform non-equilibrium system may be determined from a variational principle. Furthermore, he has shown that if the rate of production of entropy is defined as in (6.12.7)[68], then the variational principle for the transport coefficients enables one to establish a principle of the production of maximum entropy. Chester[67] observes that this result goes over to that of Kohler[70] for a weakly coupled system.

The transport coefficients thus appear as time integrals over the autocorrelation of the appropriate flux. Thus, for example, the autocorrelation function for the diffusion coefficient in an isotropic system is given as

$$D = \int_0^\infty \langle v_{x_i}(0)\, v_{x_i}(t)\rangle_0 \,\mathrm{d}t, \tag{6.12.15}$$

where $v_{x_i}(t)$ is the x-component of the velocity of the ith atom at time t, and the subscript 0 denotes thermal averaging over the equilibrium ensemble. The value of the mean square fluctuation of velocity is

$$\langle v_{x_i}^2(0)\rangle_0 = \frac{kT}{m}, \tag{6.12.16}$$

where m is the atomic mass: a collisionless gas would, of course, maintain this initial value of its velocity autocorrelation and the diffusion constant would be infinite. In fact atomic coupling serves to decorrelate the velocity vectors in time in some complicated manner and a finite diffusion constant is obtained. The precise specification of the velocity autocorrelation and its related quantity, the force autocorrelation, is of outstanding importance and we shall consider various models in the next section (§6.13). If, for example, we assume that the interaction between a given atom and the rest of the system serves to randomize the motions, then the velocity autocorrelation decays exponentially with a decay time τ_D. We then have

$$\langle v_{x_i}(0)\, v_{x_i}(t)\rangle_0 \simeq \langle v_{x_i}^2(0)\rangle_0 \exp(-|t|/\tau_D) \qquad (6.12.17)$$

which, from (6.12.15), (6.12.16) yields the relaxation constant

$$\tau_D = mD/kT.$$

This value of the decay time may be obtained from the Langevin diffusion equation[71].

We now quote relations obtained from the fluctuation-dissipation formalism when applied to momentum transfer. The coefficients of viscosity are expressed in terms of the stress tensor, $\boldsymbol{\sigma}$. The shear viscosity η, for example, is given as

$$\eta = \frac{\rho}{kT}\int_0^\infty \langle \sigma^{xy}(0)\,\sigma^{xy}(t)\rangle_0\, dt, \qquad (6.12.18)$$

where

$$\boldsymbol{\sigma} = \sum_i \left[\frac{1}{m}\boldsymbol{p}_i\boldsymbol{p}_i - \tfrac{1}{2}\sum_j \boldsymbol{r}_{ij}\nabla_{ij}\Phi(r_{ij})\right] + \sum_i \boldsymbol{r}_i\boldsymbol{X}_i. \qquad (6.12.19)$$

In defining the momentum flux, or stress tensor $\boldsymbol{\sigma}$, we find that the external forces $\boldsymbol{X}_i$ on molecule i are important. The relaxation time of $\sum_i \boldsymbol{r}_i\boldsymbol{X}_i$ is assumed to be so short that it may be replaced by its average value, $-pV$, which is determined from the virial theorem. Thus, (6.12.19) may be written

$$\boldsymbol{\sigma} = \sum_i \left(\frac{1}{m}\boldsymbol{p}_i\boldsymbol{p}_i - \tfrac{1}{2}\sum_j \boldsymbol{r}_{ij}\nabla_{ij}\Phi(r_{ij})\right) - pV\mathbf{1}. \qquad (6.12.20)$$

The form of the stress tensor used by Helfand[72] is equivalent to that proposed by Irving and Kirkwood[73], and is seen to have both kinetic and potential components leading to a two-component shear viscosity coefficient $\eta = \eta_K + \eta_V$ (cf. §6.10). The mean square fluctuation of stress is given by

$$\langle \sigma^{xy}(0)^2\rangle_0 = \frac{kT\rho G}{m}, \qquad (6.12.21)$$

where G is the rigidity modulus[74]. If the stress correlation is assumed to decay exponentially, (6.12.18) and (6.12.21) yield a relaxation time

$$\tau_\eta = \eta/G \tag{6.12.22}$$

which is known as the Maxwell relaxation time.

The bulk viscosity ζ, not arising from a flow process, would be expected to depend upon only the diagonal elements of $\boldsymbol{\sigma}$:

$$\zeta = \frac{\rho}{kT}\int_0^\infty \langle \sigma^{xx}(0)\,\sigma^{xx}(t)\rangle_0\,\mathrm{d}t. \tag{6.12.23}$$

The off-diagonal elements vanish for a fluid in thermodynamic equilibrium of course, and moreover, depend upon the velocity gradient according to the rule $\sigma^{xy} = (v^{xy}+v^{yx})/2$, etc. The kinetic contributions to the diagonal components of the stress tensor amount to pV, and in consequence ζ is purely a function of the interaction (cf. (6.10.8)). If there is a single relaxation time for the stress a result analogous to (6.12.22) is obtained. In general, however, the stress-correlation functions may be expected to show a complicated structure reflecting the actual modes of motion in the fluid.

Although the stress tensor (6.12.20) applies equally to gases and liquids the relative importance of the two contributions to the momentum flow differs greatly in the two cases, the kinetic mechanism predominating in the former. Assuming an exponential decay of the kinetic component of the stress-tensor autocorrelation, and identifying the relaxation time τ_η as $\hat{l}/\bar{c}$, where $\hat{l}$ is the mean free path and $\bar{c}$ as the mean atomic velocity, the kinetic relation

$$\eta = K\rho\bar{c}\hat{l} \tag{6.12.24}$$

is immediately obtained from (6.12.18). K is a numerical constant of the order of unity. It is apparent from this that there is presumably a simple relationship between the coefficients of shear viscosity and diffusion[49]. Further simple reasoning relates the mean free path to the atomic diameter σ, and we have

$$\lambda = K'm/\rho\sigma^2,$$

where K' is again of the order unity. The kinetic component of the shear viscosity is therefore

$$\eta = KK'm\bar{c}/\sigma^2 \tag{6.12.25}$$

for a dilute hard sphere gas. Thus it appears that the coefficient of shear viscosity is independent of pressure and proportional to the square root of the absolute temperature (i.e. $\bar{c}$). For hard spheres the above relationship is reasonably well obeyed, although even at moderately high densities discrepancies arise on account of the inadequacy of the assumed mode of decay of the stress tensor autocorrelation. For more realistic interactions the failure is even more pronounced. More rigorous calculations due principally to Boltzmann, Enskog and Chapman allowed Lennard-Jones to deduce the parameters of the atomic interaction from the measured viscosity coefficient of dilute gases.

For thermal conductivity λ, the autocorrelation may be shown to be[67, 75]

$$\lambda = \frac{\rho}{3kT^2}\int_0^\infty \langle \boldsymbol{J}(0).\boldsymbol{J}(t)\rangle \, dt, \tag{6.12.26}$$

where $\boldsymbol{J}$ is the heat current tensor:

$$\boldsymbol{J} = \sum_{i=1}^{N} \left\{ \frac{\boldsymbol{p}_i^2}{2m}\mathbf{1} + \frac{1}{2}\sum_{j\neq i=1}^{N} (\Phi(r_{ij})\mathbf{1} - \boldsymbol{r}_{ij}\nabla_{ij}\Phi(r_{ij}))\right\} . \frac{\boldsymbol{p}_i}{m}. \tag{6.12.27}$$

This definition differs slightly from that of Helfand[72]. The mean square fluctuation of the heat current appears to have no simple physical meaning – see a discussion of this point by Schofield[76]. The heat flux tensor contains both kinetic ($\boldsymbol{J}_K$) and potential ($\boldsymbol{J}_V$) components arising from the net transport of kinetic and potential energy across an arbitrary reference plane. In dilute gases $\boldsymbol{J}_K$ makes the dominant contribution, whereas in liquids $\boldsymbol{J}_V$ dominates the total heat flux. If thermal equilibrium is set up rapidly compared with the time for $\boldsymbol{J}$ to dissipate, the heat conduction current will tend to decay as $\delta(t)$. In this case fluctuations will diffuse with a diffusion coefficient given by

$$D_\lambda = \lambda/C_p\rho. \tag{6.12.28}$$

For a perfect gas D_λ is identical to the particle diffusion coefficient, D.

It is clear that there is a close relation here between linear response theory and the Boltzmann equation, provided we restrict ourselves to elastic interactions and an exponential decay of the autocorrelation. For a more complete discussion of this point see the papers listed under ref. 77.

As we have observed above, the decay of the autocorrelation will not have the simple exponential form for systems at high density: instead they will show considerable structure which arises from the vibrational and other modes of atomic motion. To investigate the form of the autocorrelation experimentally (and, for that matter, theoretically) it is convenient to form the spectral density of the autocorrelation, $\psi(t)$, as

$$\tilde{\psi}(\omega) = \frac{2}{\pi}\int_0^\infty \psi(t)\cos(\omega t)\,\mathrm{d}t. \tag{6.12.29}$$

It is evident from (6.12.29) that at $\omega = 0$ we recover the transport coefficient

$$\tilde{\psi}(0) = 2L_{ii}/\pi$$

for the ith conjugates. This, however, does not tell us much about the entire spectral density. Radiation scattering methods are required to give experimental data at arbitrary ω, and these are discussed extensively by Egelstaff[78]. In general, the analytic description of the time evolution of the autocorrelation of a dynamical variable is difficult: attention has in the main been directed at the development of the velocity autocorrelation, which we shall now consider.

6.13 The velocity autocorrelation

In this section we shall consider the various attempts which have been made to describe the evolution of the velocity autocorrelation. The exact formulation is essentially a restatement of the general N-body problem, and it is therefore not possible at present to give a rigorous calculation of the velocity autocorrelation, and therefore a model of the liquid is generally employed. Rahman's molecular dynamics investigations of the microscopic behaviour of liquid argon have provided an extremely valuable insight into the nature of the microkinetics. In consequence we have a reasonably accurate

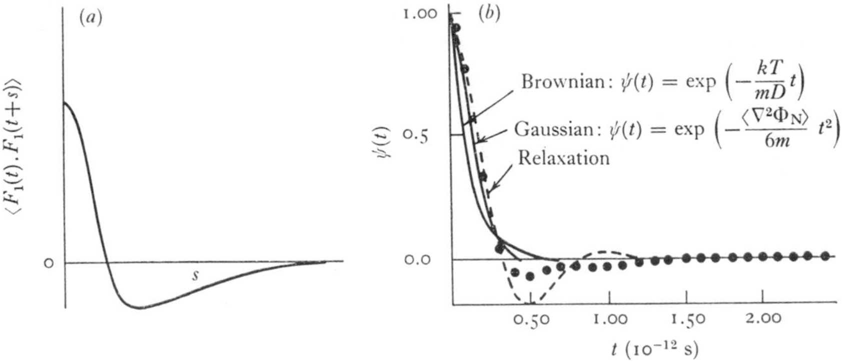

Fig. 6.13.1. (*a*) Schematic force autocorrelation. The negative region represents the statistical effect of atomic backscattering. (*b*) Comparison of the molecular dynamic velocity autocorrelation of Rahman for liquid argon at 94.4 °K with the Gaussian, Brownian and relaxation (equation (6.13.10)) models. Only the relaxation model is able to reproduce the negative region of the autocorrelation. (Redrawn by permission from Rice and Gray, *The Statistical Mechanics of Simple Liquids* (Wiley).)

qualitative understanding of the atomic motions. It was apparent from our investigation of the single-particle friction constant as a time integral over the force autocorrelation that a simple Gaussian decay was inadequate at liquid densities. It is evident that the spatial and dynamical coupling between a system of particles at high density can only serve to establish a correlation between adjacent dynamical events: only for a system sufficiently dilute that negligible backscattering of a particle occurs within its cage of neighbours can we anticipate a purely Gaussian behaviour. Otherwise, we must expect an initially Gaussian decay of the velocity autocorrelation during which time the particle executes a quasi-Brownian motion in the fluctuating mean force field of the cage interior. Thereafter a large-momentum impulsive exchange of the representative particle with the cell boundary occurs. Otherwise, we must expect an initially Gaussian decay of the velocity autocorrelation during which time the particle executes a quasi-Brownian motion in the fluctuating mean force field of its neighbours. In the course of this trajectory the momentum exchange $\Delta \boldsymbol{p}_i$ is understood to be small in comparison to its intrinsic mean momentum $\boldsymbol{p}_i$. Thereafter, a large quasi-elastic impulsive momentum exchange of the representative particle with the cell boundary occurs, engendering strong quasi-elastic backscattering of the itinerant particle (fig. 6.13.1 (*a*)). It is in this period that the departure from a Gaussian is most pronounced: the subsequent quasi-Brownian motion serves to completely decorrelate the momentum vector with respect to its initial value. We may alternatively consider that the decorrelation is essentially Gaussian, periodically punctuated by large-momentum quasi-elastic impulsive exchanges, the local structure establishing the temporal extent of the vibratory or 'rattling' component. On this scheme we imagine the undamped cosine form of the autocorrelation appropriate to a harmonic oscillator to be modulated by a Gaussian envelope. Here we understand the motion of an atom to consist of a thermal vibratory cloud whose centre of oscillation satisfies the Langevin equation and executes a Brownian motion. Sears[79] essentially adopted this standpoint in his treatment of a fluid as a system of diffusing Einstein oscillators. We should observe however, that no clear-cut distinction may be made between the high frequency vibratory modes and the low frequency diffusive modes: there is coupling between the two components, and this is a current difficulty of the theory. As the local structure relaxes with decreasing density, so the vibratory component damps out and we eventually recover the purely Gaussian decay appropriate to the stochastic evolution of a phase variable which is known to occur in dilute systems. We recall Rahman's (§5.9) molecular dynamics investigation of the departure from Gaussian form of the single-particle motion. It was found that for short times, times of the order $\bar{d}/\sqrt{T}$, where $\bar{d}$ is the mean atomic spacing, the evolution was essentially Gaussian. Thereafter the development of non-Gaussian terms, 10–20 per cent of the Gaussian term, was observed. Eventually, after about 10^{-11} s the Gaussian form was recovered and a mean square displacement having a time dependence appropriate to that of a Brownian particle was obtained. This result is entirely in accord with the qualitative microscopic description presented above.

We recall the equation of continuity in momentum space (cf. (6.9.9))

$$\frac{\partial n_{(2)}}{\partial t} + \nabla_{12}\cdot\mathscr{J}_1 + \nabla\cdot\mathscr{J}_2 = 0 \qquad (6.13.1)$$

which expresses the evolution of the divergence of the two momentum fluxes $\mathscr{J}_1$ and $\mathscr{J}_2$. These momentum currents are determined in terms of the moments

$$\left.\begin{aligned}\mathscr{J}_1 &= \frac{1}{2m}\iint \boldsymbol{\pi} f_{(2)}(\boldsymbol{p}_1, \boldsymbol{p}_2, \boldsymbol{q}_1, \boldsymbol{q}_2, t)\, \mathrm{d}\boldsymbol{\Pi}\, \mathrm{d}\boldsymbol{\pi},\\ \mathscr{J}_2 &= \frac{1}{2m}\iint \boldsymbol{\Pi} f_{(2)}(\boldsymbol{p}_1, \boldsymbol{p}_2, \boldsymbol{q}_1, \boldsymbol{q}_2, t)\, \mathrm{d}\boldsymbol{\Pi}\, \mathrm{d}\boldsymbol{\pi},\end{aligned}\right\} \tag{6.13.2}$$

where $\boldsymbol{\pi} = \boldsymbol{p}_1 - \boldsymbol{p}_2$, $\boldsymbol{\Pi} = \boldsymbol{p}_1 + \boldsymbol{p}_2$. $\mathscr{J}_1$ represents the relative flux and $\mathscr{J}_2$ the centre of mass flux.

We now consider a simple model due to Gray in terms of which we are able to determine the qualitative features of the velocity autocorrelation. Gray replaces the centre of mass term in (6.13.1), $\nabla . \mathscr{J}_2$, by a relaxation term $\beta(n_{(2)} - n^0_{(2)})$, where $n^0_{(2)}$ is the local equilibrium value. Since the transport of mass, momentum and energy is dependent on the relative motion of the pair and not on the motion of the centre of mass this approximation appears satisfactory. If by the relative momentum flux we understand

$$\mathscr{J}_1 = -\frac{n_{(2)}}{2m}\langle \boldsymbol{p} \rangle \tag{6.13.3}$$

(cf. (6.13.2)), then (6.13.1) may be written

$$\frac{\partial n_{(2)}}{\partial t} + \beta(n_{(2)} - n^0_{(2)}) - \frac{1}{2m}\langle \boldsymbol{p} \rangle . \nabla_{12} n^0_{(2)}, \tag{6.13.4}$$

where, to a first approximation, the pair distribution function appearing in (6.13.3) has been replaced by its local equilibrium value, $n^0_{(2)}$.

If we now multiply (6.13.4) throughout by $\nabla\Phi(r)$ and integrate over $\boldsymbol{r}$ we obtain

$$\frac{\mathrm{d}}{\mathrm{d}t}\int \nabla\Phi(r)\, n_{(2)}\, \mathrm{d}\boldsymbol{r} + \int \beta(n_{(2)} - n^0_{(2)})\, \nabla\Phi(r)\, \mathrm{d}\boldsymbol{r} - \frac{1}{2m}\int \langle \boldsymbol{p} \rangle\, \nabla\Phi(r) . \nabla n^0_{(2)}\, \mathrm{d}\boldsymbol{r} = 0. \tag{6.13.5}$$

Since $$\langle \boldsymbol{F}_1 \rangle = \frac{\mathrm{d}}{\mathrm{d}s}\langle \boldsymbol{p} \rangle,$$

the first term becomes $\langle \ddot{p} \rangle$. The term in $\beta n^0_{(2)}$ gives the equilibrium average force, which is zero, whilst the last term becomes $-\beta\gamma\langle \boldsymbol{p}\rangle$ from the definition of the friction constant and where γ is a simple positive constant. Thus we may rewrite (6.13.5) in the form of a differential equation in $\langle p \rangle$:

$$\langle \ddot{p} \rangle + \beta\langle \dot{p} \rangle + \gamma\beta\langle p \rangle = 0. \tag{6.13.6}$$

It now remains to establish a relationship between $\langle p \rangle$ and $\psi(t)$ the velocity autocorrelation, and to solve (6.13.6). The normalized autocorrelation is related to the conditionally averaged momentum by

$$\psi(t) = \frac{\langle \boldsymbol{p}(0).\boldsymbol{p}(t)\rangle}{\langle \boldsymbol{p}(0).\boldsymbol{p}(0)\rangle} = \frac{\langle \boldsymbol{p}(0).\boldsymbol{p}(t)\rangle}{mkT}$$

whereupon (6.13.6) becomes

$$\ddot{\psi} + \beta\dot{\psi} + \gamma\beta\psi = 0 \tag{6.13.7}$$

which must be solved subject to the boundary conditions $\psi(0) = 1$, $\dot{\psi}(0) = 0$. The solution is

$$\psi(t) = \frac{1}{\theta_+ - \theta_-}\{\theta_+ \exp(\theta_- t) - \theta_- \exp(\theta_+ t)\}, \tag{6.13.8}$$

where θ_+ and θ_- are the roots of the auxilliary equation

$$\theta_\mp = -\frac{\beta}{2} \mp \left(\frac{\beta^2}{4} - \gamma\beta\right)^{\frac{1}{2}}. \tag{6.13.9}$$

When $\beta^2 > 4\gamma\beta$ the correlation function $\psi(t)$ is a monotonic function of time, whilst in the contrary case it is an oscillatory decaying function of time. If $\beta \gg \gamma$ then the roots of the auxilliary equation (6.13.9) become

$$\theta_+ \simeq -\gamma, \quad \theta_- \simeq -\beta$$

so that

$$\psi(t) = \left(1 - \frac{\gamma}{\beta}\right)^{-1}\left[\exp(-\gamma t) - \frac{\gamma}{\beta}\exp(-\beta t)\right]$$

$$\simeq \exp(-\gamma t). \tag{6.13.10}$$

Provided we make the identification $\gamma = kT/mD$, (6.13.10) is seen to relate to the velocity autocorrelation of a Brownian particle, and $\gamma = \tau_D^{-1}$ the diffusive relaxation time (cf. (6.12.17)). On the other hand, if $4\gamma\beta > \beta^2$, i.e. $\gamma \gg \beta$, so that structural relaxation is much slower than momentum relaxation, then we obtain a damped oscillatory velocity autocorrelation characteristic of a dense fluid.

We compare the various estimates of the velocity autocorrelation in fig. 6.13.1 (*b*). Gray's analysis is seen to yield an autocorrelation in qualitative agreement with the molecular dynamics result of Rahman for liquid argon. From (6.13.9) we are able to obtain the condition for the development of a negative region on this relaxation model – it is simply that $4\gamma > \beta$, or alternatively that structural relaxation proceeds more slowly than momentum relaxation. If we take the limit $\gamma \gg \beta$ we obtain the strongly oscillatory velocity autocorrelation appropriate to a solid wherein the structural relaxation is virtually zero (i.e. $D_{\text{solid}} = (kT/m\gamma) \sim 0$). On the other hand Nelson[126], using a molecular dynamic approach, finds that the velocity autocorrelation of a low density square-well fluid is of the form (6.13.10). Thus (6.13.8) appears to concur broadly with the qualitative discussion at the beginning of this section. In particular we note that the decay is essentially Brownian, and depending upon the configurational relaxation of the system, contains a vibratory component. This is clearly demonstrated in the spectral density $\tilde{\psi}(\omega)$ of the velocity autocorrelation, (6.12.29), for liquid sodium[83], shown in fig. 6.13.2. The low frequency diffusive modes are seen to be quite distinct from the high frequency vibratory modes more or less centred on $\omega_0 = 2 \times 10^{13}\,\text{s}^{-1}$ which is virtually identical to the vibrational frequency in the solid. This is to be expected, of course, since the isochoric specific heat is almost unchanged on melting – the specific heat at constant volume per particle being $3.1k$ (371 °K) and $3.4k$ (373 °K) in the solid and liquid phases respectively. C_v is slightly greater in the liquid phase on account of the diffusive modes. Sears exploits the distinct vibratory and diffusive components of the spectral density to propose a kinetic model wherein atoms are assumed to vibrate as Einstein oscillators with characteristic frequency ω_0. Diffusive modes are incorporated in terms of a centre of atomic oscillation which is itself executing a Brownian motion. The two components are not independent as Sears assumes them to be, of course. The vibratory and diffusive modes are coupled by virtue of the condition that $\int_0^\infty \tilde{\psi}(\omega)\,d\omega$ be equal to the total number of degrees of freedom, as in the case of the Debye solid.

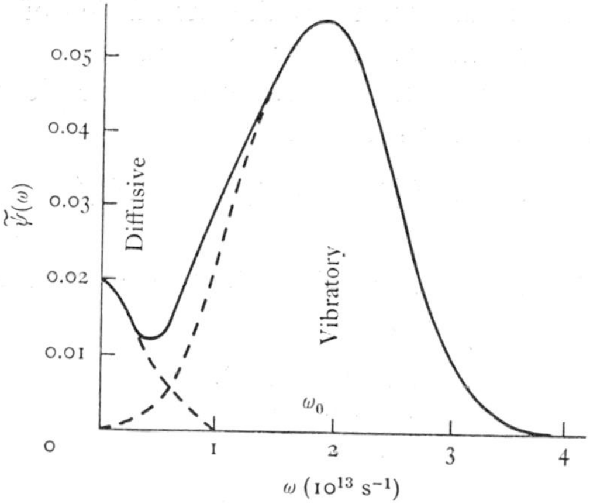

Fig. 6.13.2. Spectral density of liquid sodium at 373 °K, determined from the neutron scattering experiments of Egelstaff. A qualitative subdivision into vibratory and diffusive components is indicated. (Redrawn by permission from Egelstaff, *Introduction to the Liquid State* (Academic Press).)

Sears locates the centre of oscillation at some mean position $\boldsymbol{R}(t)$, where

$$\boldsymbol{R}(t) = \frac{1}{\tau}\int_{t-\tau/2}^{t+\tau/2} \boldsymbol{r}(t')\,\mathrm{d}t'. \tag{6.13.11}$$

τ is the period of oscillation (i.e. $2\pi/\omega_0$), and $\boldsymbol{r}$ denotes the instantaneous position of the atom whose velocity autocorrelation is to be determined. We need now to locate the position of the atom with respect to its centre of oscillation. We define the following displacements:

(i) Instantaneous displacement of atom from its centre of vibration

$$= \boldsymbol{r}(t) - \boldsymbol{R}(t) = \boldsymbol{b}(t);$$

(ii) Position of atom j relative to centre of oscillation

$$= \boldsymbol{r}_j(t) - \boldsymbol{R}(t) = \boldsymbol{a}_j(t).$$

We may immediately write down the instantaneous total force $\mathscr{F}(t)$ acting on the representative particle:

$$\mathscr{F}(t) = m\boldsymbol{A}(t) = m\omega_0^2\boldsymbol{b}(t) - \boldsymbol{K}(t)\,\boldsymbol{b}(t) + \dots, \tag{6.13.12}$$

where $\boldsymbol{A}(t)$ is the fluctuating driving force ($\langle \boldsymbol{A}\rangle = 0$), the second term is the harmonic restoring force, proportional to the displacement

$\boldsymbol{b}(t)$, and the third term represents the modulation of the restoring force by the thermal motions of atoms near the central atom. Higher terms would include anharmonic contributions. Symmetry requirements enable us to write

$$\langle \boldsymbol{A}_\alpha(t)\rangle = 0; \quad \langle \boldsymbol{A}_\alpha(t).\boldsymbol{A}_\beta(t)\rangle = \delta_{\alpha\beta}\langle \boldsymbol{A}_\alpha^2(t)\rangle; \quad \langle \boldsymbol{K}_{\alpha\beta}(t)\rangle = 0, \tag{6.13.13}$$

where α, β are Cartesian indices. It is straightforward to show that, in terms of the pair potential,

$$\omega_0^2 = \frac{1}{m}\sum_j \left\langle \frac{\partial^2\Phi}{\partial a_{\alpha j}^2}(a_j)\right\rangle$$

$$= \frac{\rho}{m}\int_v g_{(2)}(a)\frac{\partial^2\Phi}{\partial a_\alpha^2}(a)\,\mathrm{d}\boldsymbol{a} \tag{6.13.14}$$

and

$$\boldsymbol{A}_\alpha(t) = \frac{1}{m}\sum_j \frac{\partial\Phi}{\partial a_j}(a_j).\frac{a_{\alpha j}}{a_j} \tag{6.13.15}$$

so that

$$\langle \boldsymbol{A}_\alpha^2(t)\rangle = \frac{\rho}{3m^2}\int_v \left(\frac{\partial\Phi}{\partial a}(a)\right)^2 g_{(2)}(a)\,\mathrm{d}\boldsymbol{a}$$
$$+\frac{\rho^2}{3m^2}\int_v\int_v \frac{\partial\Phi}{\partial a_1}(a_1)\frac{\partial\Phi}{\partial a_2}(a_2)g_{(2)}(a_1, a_2)\frac{\boldsymbol{a}_1.\boldsymbol{a}_2}{a_1 a_2}\,\mathrm{d}\boldsymbol{a}_1\,\mathrm{d}\boldsymbol{a}_2. \tag{6.13.16}$$

Sears then makes the approximation that the forces $\boldsymbol{A}(t)$ and $\boldsymbol{K}(t)$ are not prescribed functions of time, but are random functions of time. This ignores the fact that by influencing the motion of its neighbours, the central atom affects the time dependence of these functions, and hence influences its own motion. This reactive effect of the central atom on itself must not be ignored since it is responsible for the dissipation of the thermal vibrational energy of the central atom to the surrounding liquid. Thus, if $\boldsymbol{A}(t)$ and $\boldsymbol{K}(t)$ are to be treated as stochastic variables it is necessary to introduce some additional dissipative term into the expression for $\mathscr{F}(t)$. The simplest choice is to take this dissipative term proportional to the velocity of the central atom, $\mu\dot{\boldsymbol{r}}(t)$. With a suitable redefinition of the parameter we may incorporate the dissipative thermal modulation term $\boldsymbol{K}(t)\,\boldsymbol{b}(t)$ to obtain finally

$$m\ddot{\boldsymbol{r}}(t) = m\boldsymbol{A}(t) - m\omega_0^2[\boldsymbol{r}(t)-\boldsymbol{R}(t)] + \mu m\dot{\boldsymbol{r}}(t). \tag{6.13.17}$$

Equation (6.13.17) represents the equation of motion of a damped Einstein oscillator. The motion of the centre of oscillation is taken to satisfy the Langevin diffusion equation (6.4.33) in the absence of an external field:

$$m\ddot{\boldsymbol{R}}(t)+m\gamma\dot{\boldsymbol{R}}(t) = \boldsymbol{B}(t). \qquad (6.13.18)$$

$\boldsymbol{B}(t)$ is the fluctuating driving force, ($\langle\boldsymbol{B}(t)\rangle = 0$), γ is another arbitrary friction constant, and $\boldsymbol{R}(t)$ locates the instantaneous centre of oscillation. It is interesting at this stage to make a brief digression, and determine the velocity autocorrelation of a Brownian particle, or alternatively the purely diffusive component of the autocorrelation $\psi(t)$ from the Langevin equation (6.13.18). Multiplying throughout by $\exp(\gamma t)$ and integrating we obtain

$$\exp(\gamma t)\,\boldsymbol{v}(t)-\boldsymbol{v}(0) = \frac{1}{m}\int_0^t \exp(\gamma t')\,\boldsymbol{B}(t')\,\mathrm{d}t'. \qquad (6.13.19)$$

To obtain the velocity autocorrelation we now have to average over all atoms having the same initial velocity, and then multiply by $\boldsymbol{v}(0)$ and thermalize. Thus,

$$\exp(\gamma t)\,\langle\boldsymbol{v}(t).\boldsymbol{v}(0)\rangle - \langle v^2(0)\rangle = 0. \qquad (6.13.20)$$

The right hand side is zero provided we consider times sufficiently long that $\langle\boldsymbol{B}(0).\boldsymbol{B}(t)\rangle = 0$. We immediately obtain

$$\psi(t) = \frac{kT}{m}\mathrm{e}^{-\gamma t} \qquad (6.13.21)$$

since $\langle v^2(0)\rangle = kT/m$ (equation (6.12.16)). The friction constant is immediately given as

$$D = \int_0^\infty \psi(t)\,\mathrm{d}t = kT/m\gamma,$$

i.e.

$$\gamma = kT/mD. \qquad (6.13.22)$$

The Langevin autocorrelation is shown in fig. 6.13.1 and by comparison with Rahman's results is seen to give a rough description of the initial decay of $\psi(t)$ as anticipated in the opening discussion. There is, however, an essential disagreement at the origin when the Brownian autocorrelation predicts a cusp, whereas in fact $\dot{\psi}(0) = 0$.[63] The spectral density is immediately given in terms of the Lorentzian:

$$\tilde{\psi}(\omega) = \frac{1}{\pi}\frac{kT}{m}\frac{\gamma}{\omega^2+\gamma^2} \qquad (6.13.23)$$

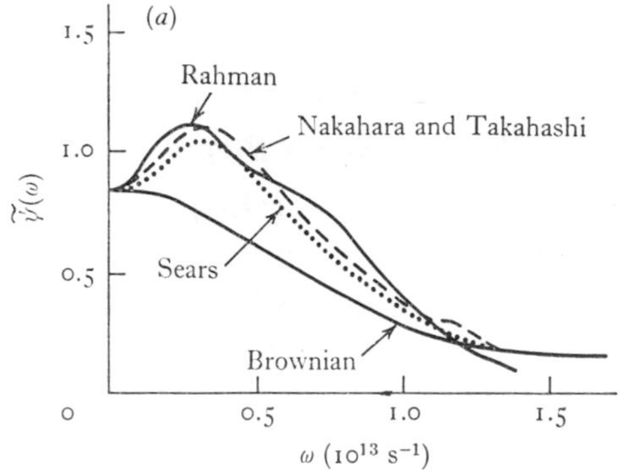

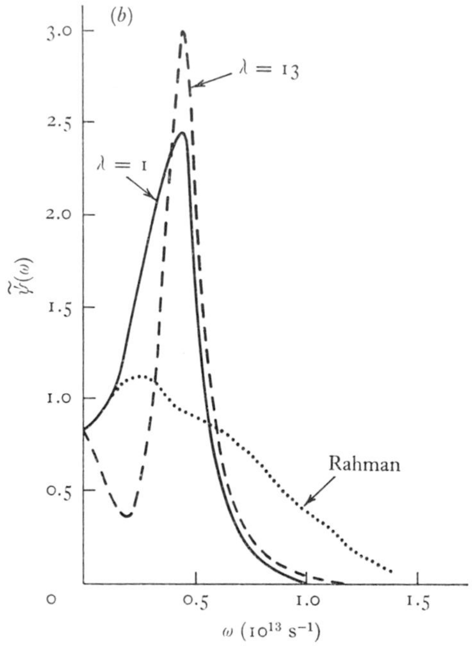

Fig. 6.13.3. (*a*) Comparison of various models of the phonon spectrum of liquid argon at ~ 90 °K with the molecular dynamics results of Rahman. Both the Sears and the Nakahara and Takahashi models qualitatively reproduce the vibratory peak: a simple Brownian model is evidently quite inadequate in this region of the spectrum. (*b*) The parameter λ in the Nakahara–Takahashi model of the spectral distribution represents the ratio of the fluctuating driving force of the Einstein oscillator to the fluctuating Brownian force. Large values of λ therefore emphasize the vibratory aspect of the spectrum (cf. fig. 6.13.2), and the actual magnitude of this parameter will be sensitive to the form of the minimum in $\Phi(r)$ or, more percisely, $\Psi(r)$. (Redrawn by permission from Nakahara and Takahashi, *Proc. Phys. Soc.* **89**, 747 (1966).)

and is shown in fig. 6.13.3(*a*). Whilst there is qualitative agreement with Rahman's result[81] only in the broadest sense, the spectral density given by (6.13.23) can be rejected on the grounds that it predicts an infinite second moment:

$$\int_0^\infty \omega^2 \tilde{\psi}(\omega)\, \mathrm{d}\omega. \qquad (6.13.24)$$

$\tilde{\psi}(\omega)$ needs to fall off faster than ω^{-2} to ensure convergence of the integral. This need not surprise us since we have not as yet attempted to describe the high frequency form of the spectral density. Kubo[82] has shown that this difficulty may be overcome by extending the definition of the friction constant to make it frequency dependent, and in this case the fluctuation-dissipation theorem is satisfied if

$$\tilde{\gamma}(\omega) = \frac{1}{3mkT}\int_0^\infty \langle \boldsymbol{B}(t)\,.\,\boldsymbol{B}(t+s)\rangle\, \mathrm{e}^{-\mathrm{i}\omega s}\, \mathrm{d}s, \qquad (6.13.25)$$

where (6.7.8) represents the zero frequency limit of this expression. Furthermore, if we assume that the decay of the stochastic force autocorrelation is exponential with a time constant γ', then

$$\tilde{\gamma}(\omega) = \frac{kT}{mD}\frac{\gamma'}{\gamma'+\mathrm{i}\omega}. \qquad (6.13.26)$$

Equation (6.13.23) then becomes

$$\tilde{\psi}(\omega) = \frac{D}{\pi}\frac{\gamma^2\gamma'^2}{(\omega^2+\gamma'^2)\,\omega^2+\gamma^2\gamma'^2} \qquad (6.13.27)$$

which has a finite second moment given approximately by $kT\gamma\gamma'/m$, where $\gamma = kT/mD$. It may be shown that the higher moments of (6.13.27) are also finite[84].

We now continue with Sears' analysis of the velocity autocorrelation of diffusing Einstein oscillators. To define the model completely the stochastic variables $\boldsymbol{A}(t)$, $\boldsymbol{B}(t)$ must be specified. In a theory of Brownian motion one would take these as purely random functions of time[85]. This, however, is not justified in the present situation since there is an upper limit on the frequency content of these functions of the order of ω_0. The simplest type of random process having a finite correlation time is the one-dimensional

Gaussian–Markov process for which the correlation function is a simple exponential[86], which, taking into account (6.13.13), enables us to write

$$\langle \boldsymbol{A}_\alpha(t).\boldsymbol{A}_\beta(t')\rangle = \delta_{\alpha\beta}\langle \boldsymbol{A}_\alpha^2\rangle \exp(-\mu'|t'-t|), \qquad (6.13.28)$$

$$\left.\begin{aligned}\langle \boldsymbol{B}_\alpha(t).\boldsymbol{B}_\beta(t')\rangle &= \delta_{\alpha\beta}\langle \boldsymbol{B}_\alpha^2\rangle \exp(-\gamma'|t'-t|),\\ \langle \boldsymbol{A}_\alpha(t).\boldsymbol{B}_\beta(t')\rangle &= 0.\end{aligned}\right\} \qquad (6.13.29)$$

The solution of (6.13.17) and (6.13.18) is elementary, and Sears finds the spectral density

$$\tilde{\psi}(\omega) = \frac{D}{\pi}\frac{\gamma'\gamma^2}{(\omega^2-\omega_0^2)^2+\mu^2\omega^2}\left\{\frac{\gamma'\omega_0^4}{(\omega^2+\gamma'^2)(\omega^2+\gamma^2)}+\frac{\lambda\mu'\omega^2}{\omega^2+\mu'^2}\right\}, \qquad (6.13.30)$$

where D is the macroscopic diffusion coefficient and $\lambda = \langle \boldsymbol{A}_\alpha^2\rangle/\langle \boldsymbol{B}_\alpha^2\rangle$. Thus, the itinerant oscillator model is characterized by seven parameters $\langle \boldsymbol{A}_\alpha^2\rangle$, $\langle \boldsymbol{B}_\alpha^2\rangle$, μ, μ', γ, γ' and ω_0. If the pair potential and pair correlation function are known, ω_0 and $\langle \boldsymbol{A}_\alpha^2\rangle$ may be determined from (6.13.14) and (6.13.15).

An obvious objection which can be raised against the starting equations of Sears is that using (6.13.17) and (6.13.18) one does not obtain (6.13.11).[87] Since Sears does not use (6.13.11) explicitly one might argue that this equation is redundant. In that case it is not clear what the precise meaning of $\boldsymbol{R}(t)$ should be. Nakahara and Takahashi[88] observe that (6.13.17) and (6.13.18) are inconsistent with the processes (6.13.28) and (6.13.29). Since it is assumed that the stochastic variables $\boldsymbol{A}(t)$, $\boldsymbol{B}(t)$ do not have a white spectrum, but are cut off beyond $\sim \omega_0$, it follows from the fluctuation-dissipation theorem that μ and γ are no longer constants but functions of time given by relations similar to (6.13.25). Nakahara and Takahashi then obtain

$$\ddot{\boldsymbol{r}}(t)+\int_0^t \mu(t-t')\,\dot{\boldsymbol{r}}(t')\,\mathrm{d}t' + \omega_0^2\{\boldsymbol{r}(t)-\boldsymbol{R}'(t)\} = \boldsymbol{A}(t) \qquad (6.13.31)$$

and

$$\ddot{\boldsymbol{R}}(t)+\int_0^t \gamma(t-t')\,\dot{\boldsymbol{R}}(t')\,\mathrm{d}t' = \boldsymbol{B}(t). \qquad (6.13.32)$$

Here the function $\mu(t)$ represents a retarded effect of the frictional force. Analogous relationships to (6.13.26) are obtained for $\tilde{\mu}(\omega)$

and $\tilde{\gamma}(\omega)$ as functions of (μ', ω), (γ', ω) respectively. Nakahara and Takahashi finally obtain for the spectral density

$$\tilde{\psi}(\omega) = \tilde{\psi}_1(\omega) + \tilde{\psi}_2(\omega), \tag{6.13.33}$$

where

$$\tilde{\psi}_1(\omega) = \frac{(kT)^2}{\pi}\mu'\langle A^2\rangle \frac{\omega^2}{(\omega^2-\omega_0^2)^2(\omega^2+\mu'^2)+\langle A^2\rangle^2\omega^2},$$

$$\tilde{\psi}_2(\omega) = \frac{(kT)^2}{\pi}\gamma'\omega_0^4 \frac{\langle B^2\rangle}{(\omega^2+\gamma'^2)\,\omega^2+\langle B^2\rangle^2} \times \frac{\omega^2+\mu'^2}{(\omega^2-\omega_0^2)^2(\omega^2+\mu'^2)+\langle A^2\rangle^2\omega^2}.$$

The values of ω_0 and $\langle A^2\rangle$ can be calculated from (6.13.14) and (6.13.16): for a Lennard-Jones system Sears obtains

$$\omega_0 = 0.42 \times 10^{13}\,\mathrm{s}^{-1}$$

$$\langle A^2\rangle = 3 \times 10^{33}\,\mathrm{cm}^2\,\mathrm{s}^{-4} \tag{6.13.34}$$

for liquid argon at 90 °K.

In the Nakahara–Takahashi treatment there are only five parameters, $\langle A^2\rangle$, $\langle B^2\rangle$ (or λ), μ', γ' and ω_0 instead of the seven in Sears's case. The only undetermined parameter is $\lambda = \langle A^2\rangle/\langle B^2\rangle$. In fig. 6.13.3(*b*) we compare the spectral density curve (6.13.33) for two values of the parameter λ with Rahman's distribution for liquid argon at 90 °K. The discrepancy is seen to be very considerable, even for $\lambda = 1$. The significance of the parameter λ should be emphasized: it represents the ratio of the fluctuating driving force of the Einstein oscillator to the fluctuating Brownian driving force. Thus large values of λ tend to emphasize the vibratory aspect of the frequency distribution, as is apparent from fig. 6.13.3(*b*). The lifetime of a phonon of frequency ω is proportional to $\tilde{\psi}(\omega)$, and if a system is to have pronounced vibratory components this tells us something of the nature of the pair potential. Evidently the sharper well in liquid sodium leads to a better definition of the oscillator frequency and a value of $\lambda \sim 13$ seems more appropriate (see fig. 6.13.2). The excessively sharp maximum of $\tilde{\psi}(\omega)$ is almost undoubtedly due to the assumption of a single oscillator frequency. What is required here is a Debye-like phonon spectrum coupled with low frequency Brownian modes.

Nakahara and Takahashi found that only by using a value of $\langle A^2 \rangle$ ten times greater than the Lennard-Jones estimate (6.13.34) is a reasonable fit obtained (fig. 6.13.3). A number of serious problems remain, however. The experimental value of $\langle A^2 \rangle$ exceeds the theoretical value estimated by Sears (6.13.34) by a factor of ten. The effect of three-body forces and anharmonic effects are neglected in the calculation of $\langle A^2 \rangle$; their inclusion evidently decreases the value of $\langle A^2 \rangle$. Moreover there are problems concerning the value of γ, the friction constant for the Brownian motion appearing in (6.13.18). On the simple Brownian model $\gamma \propto 1/m$, (6.13.22). However, from the physical significance of this term γ should be proportional to $1/m^*$, where m^* is an *effective mass*; $m^* > m$. The effective mass of the Brownian particle arises from its dragging of neighbouring particles with it, thereby impeding its motion. This is conveniently incorporated in terms of an effective mass of the Brownian particle, but does engender some uncertainty in the value of γ. The values of the various parameters used by Sears and Nakahara and Takahashi are compared in table 6.13.1. There is also some uncertainty in the application of the fluctuation-dissipation theorem as we shall now consider.

If we make the simplification that the friction constant effecting the stochastic forces is the same in both the Einstein and Langevin equations, i.e. $\mu' = \gamma'$, then (6.13.35) simplifies at once to give[84]

$$\tilde{\psi}(\omega) = \frac{D}{\pi}\left\{\frac{\gamma^2\gamma'^2}{(\omega^2+\gamma'^2)\,\omega^2+\gamma^2\gamma'^2}+\frac{\gamma\gamma'}{\omega_0^2}\frac{\lambda\omega^2}{\omega^2+\gamma'^2}\right\}$$
$$\times\left\{\left(\frac{\omega^2-\omega_0^2}{\omega_0^2}\right)^2+\frac{\omega^2\lambda^2}{\omega^2+\gamma'^2}\right\}^{-1}$$

which may be compared with the Brownian spectral density (6.13.27), and Sears's solution (6.13.30). The components in the equation above of Nakahara and Takahashi may be immediately identified. The first factor in brackets on the right hand side essentially represents the Brownian envelope which, for realistic interactions at liquid densities, dominates the momentum evolution. The second factor, modulated by the first, describes the vibratory component having a pronounced maximum at ω_0. The sharpness of this maximum depends upon the magnitude of λ:

TABLE 6.13.1. *Itinerant oscillator: parameters arising in $\tilde{\psi}(\omega)$ (90 °K)*

	Sears	Nakahara and Takahashi	
λ	13	13	1
ω_0 (10^{13} s^{-1})	0.4	0.42	0.42
$\langle A^2\rangle$ (10^{34} cm^2 s^{-4})	0.3	3	3
$\langle B^2\rangle$ (10^{34} cm^2 s^{-4})	2.5	0.23	3
μ' (10^{13} s^{-1})	2.4	2.3	2.1
γ' (10^{13} s^{-1})	0.5	0.16	2.1
$\mu(0)$† (10^{13} s^{-1})	μ 0.7	0.7	0.8
$\gamma(0)$† (10^{13} s^{-1})	γ 0.5	0.8	0.8

† The quantities $\mu(0)$ and $\gamma(0)$ are the values of $\mu(\omega)$, $\gamma(\omega)$ at $\omega = 0$, respectively. These values are cited for comparison with the corresponding parameters in Sears's calculations.

the greater λ the greater the 'selectivity', and this is apparent from fig. 6.13.3(*b*).

Damle, Sjölander and Singwi(89) have pointed out that the application of the fluctuation-dissipation by Nakahara and Takahashi is incorrect: when properly treated the *structure* of Sears's original equations, (6.13.17), (6.13.18), is modified to read

$$\ddot{\boldsymbol{r}}(t)+\int_0^t \mu(t-t')\,\dot{\boldsymbol{r}}(t')\,\mathrm{d}t'+\frac{\alpha^2}{m}[\boldsymbol{r}(t)-\boldsymbol{R}(t)] = \boldsymbol{A}(t) \quad (6.13.35)$$

and

$$\ddot{\boldsymbol{R}}(t)+\int_0^t \gamma(t-t')\,\dot{\boldsymbol{R}}(t')\,\mathrm{d}t'-\frac{\alpha}{m^*}[\boldsymbol{r}(t)-\boldsymbol{R}(t)] = \boldsymbol{B}(t), \quad (6.13.36)$$

where m denotes the mass of the atom under consideration, and the surrounding atoms have been replaced by a fictitious centre whose coordinate is $\boldsymbol{R}(t)$ and whose mass is m^*. These equations are of the damped stochastic type and are identical to those of Nakahara and Takahashi(88) but for the presence of a restoring term in (6.13.36), which follows from Newton's third law, and is also a consequence of the fluctuation-dissipation theorem, as we shall see. Defining $\boldsymbol{r}(0) = \boldsymbol{R}(0)$, the above equations may be rewritten

$$\ddot{\boldsymbol{r}}(t)+\int_0^t [\mu(t-t')+\omega_0^2]\,\dot{\boldsymbol{r}}(t')\,\mathrm{d}t'-\int_0^t \omega_0^2\dot{\boldsymbol{R}}(t')\,\mathrm{d}t' = \boldsymbol{A}(t), \quad (6.13.37)$$

$$\ddot{\boldsymbol{R}}(t)+\int_0^t [\gamma(t-t')+\omega_1^2]\,\dot{\boldsymbol{R}}(t')\,\mathrm{d}t'-\int_0^t \omega_1^2\dot{\boldsymbol{r}}(t')\,\mathrm{d}t' = \boldsymbol{B}(t), \quad (6.13.38)$$

where $\omega_0^2 = \alpha^2/m$ and $\omega_1^2 = \alpha^2/m^*$.

Equations (6.13.37) and (6.13.38) may now be written as a single matrix equation

$$\dot{\boldsymbol{V}}(t)+\int_0^t \Gamma(t-t')\,\boldsymbol{V}(t')\,\mathrm{d}t' = \boldsymbol{F}(t), \quad (6.13.39)$$

where

$$\left.\begin{aligned} \boldsymbol{V}(t) &= \begin{bmatrix}\dot{\boldsymbol{r}}(t)\\ \dot{\boldsymbol{R}}(t)\end{bmatrix}, \quad \Gamma(t) = \begin{bmatrix}\mu(t)+\omega_0^2 & -\omega_0^2\\ -\omega_1^2 & \gamma(t)+\omega_1^2\end{bmatrix}\\ \text{and} \quad \boldsymbol{F}(t) &= \begin{bmatrix}\boldsymbol{A}(t)\\ \boldsymbol{B}(t)\end{bmatrix}. \end{aligned}\right\} \quad (6.13.40)$$

Kubo[90] has shown that

$$\phi(p) = [p\mathbf{1}+\Gamma(p)]^{-1}\,\phi^0 \quad (6.13.41)$$

and

$$\Gamma(p)\,\phi^0 = F(p), \quad (6.13.42)$$

where $\phi_{ij}(p)$, $\Gamma_{ij}(p)$ and $F_{ij}(p)$ are, respectively, the Laplace transforms of the matrix elements

$$\langle \boldsymbol{V}_i(t)\,.\,\boldsymbol{V}_j(0)\rangle, \quad \Gamma_{ij}(t) \quad \text{and} \quad \langle \boldsymbol{F}_i(t)\,.\,\boldsymbol{F}_j(0)\rangle,$$

and

$$\phi_{ij}^0 = \langle \boldsymbol{V}_i(0)\,.\,\boldsymbol{V}_j(0)\rangle. \quad (6.13.43)$$

It is evident that $\dot{\boldsymbol{R}}(0)$ and $\dot{\boldsymbol{r}}(0)$ are quite uncorrelated, and so we have (cf. (6.12.16))

$$\phi^0 = \begin{bmatrix}3kT/m & 0\\ 0 & 3kT/m^*\end{bmatrix}. \quad (6.13.44)$$

The diagonal terms of (6.13.42) are, therefore,

$$\frac{3kT}{m}\left[\mu(p)+\frac{\omega_0^2}{p}\right] = A(p), \quad (6.13.45)$$

$$\frac{3kT}{m^*}\left[\gamma(p)+\frac{\omega_1^2}{p}\right] = B(p), \quad (6.13.46)$$

and these are the fluctuation-dissipation relations. $\mu(p)$, $\gamma(p)$, $A(p)$ and $B(p)$ are the Laplace transforms of $\mu(t)$, $\gamma(t)$, $\langle \boldsymbol{A}(t)\,.\,\boldsymbol{A}(0)\rangle$ and $\langle \boldsymbol{B}(t)\,.\,\boldsymbol{B}(0)\rangle$ respectively. It should be noted that $\langle \boldsymbol{A}(t)\,.\,\boldsymbol{B}(0)\rangle \neq 0$.

Damle *et al.* then split each of the stochastic forces $\boldsymbol{A}(t)$ and $\boldsymbol{B}(t)$ into two components:

$$\boldsymbol{A}(t) = \boldsymbol{A}'(t)+(\alpha/m)(3kT)^{\frac{1}{2}}\boldsymbol{A}'', \tag{6.13.47}$$

$$\boldsymbol{B}(t) = \boldsymbol{B}'(t)+(\alpha/m^*)(3kT)^{\frac{1}{2}}\boldsymbol{B}'', \tag{6.13.48}$$

where $\boldsymbol{A}'(t)$ and $\boldsymbol{B}'(t)$ correspond to the two friction terms in (6.13.45), (6.13.46); $(\alpha/m)(3kT)^{\frac{1}{2}}\boldsymbol{A}''$ and $(\alpha/m^*)(3kT)^{\frac{1}{2}}\boldsymbol{B}''$ corresponding to the two restoring forces. Now, $\boldsymbol{A}''$ is a stochastic force such that $\langle \boldsymbol{A}''\rangle = 0$, $\langle \boldsymbol{A}''.\boldsymbol{A}''\rangle = 1$ and is uncorrelated to $\boldsymbol{A}'$ and $\boldsymbol{B}'$. Then the fluctuation-dissipation relations (6.13.45), (6.13.46) take the form

$$\left.\begin{aligned}\frac{3kT}{m}\mu(p) &= A'(p),\\ \frac{3kT}{m^*}\gamma(p) &= B'(p),\end{aligned}\right\} \tag{6.13.49}$$

where $A'(p)$ and $B'(p)$ are the Laplace transforms of

$$\langle \boldsymbol{A}'(t).\boldsymbol{A}'(0)\rangle, \quad \langle \boldsymbol{B}'(t).\boldsymbol{B}'(0)\rangle$$

respectively. Equations (6.13.41) and (6.13.44) give

$$\phi_{11}(p) = \frac{3kT}{m}\left[p+\mu(p)+\frac{\omega_0^2\{\gamma(p)+p\}}{p\{\gamma(p)+p\}+\omega_1^2}\right]^{-1}, \tag{6.13.50}$$

where $\gamma(p)$ and $\mu(p)$ have yet to be determined, and

$$\begin{aligned}\phi_{11}(p) &= \int_0^\infty e^{-pt}\,\phi_{11}(t)\,dt\\ &= \int_0^\infty e^{-pt}\,\langle \dot{\boldsymbol{r}}(t).\dot{\boldsymbol{r}}(0)\rangle\,dt.\end{aligned} \tag{6.13.51}$$

The spectral function of the normalized velocity autocorrelation is given by

$$\psi(\omega) = \frac{2m}{3\pi kT}\int_0^\infty \langle \dot{\boldsymbol{r}}(t).\dot{\boldsymbol{r}}(0)\rangle\cos\omega t\,dt. \tag{6.13.52}$$

Damle, Sjölander and Singwi[89] consider two models for the generally complicated decay of the correlation of the fluctuating force terms $\langle \boldsymbol{A}'(t).\boldsymbol{A}'(0)\rangle$, $\langle \boldsymbol{B}'(t).\boldsymbol{B}'(0)\rangle$. The two models

TABLE 6.13.2. *Itinerant oscillator: parameters arising in* $\tilde{\psi}(\omega)$ (94.4 °K)

	Damle, Sjölander and Singwi	
	Exponential	Gaussian
ω_0^2 (10^{24} s^{-2})	1.5	1.5
ω_1^2 (10^{24} s^{-2})	2.1	2.1
$m\langle A'^2\rangle/3kT\pi$ (10^{24} s^{-2})	71.5	57.7
γ', c (10^{13} s^{-1})	γ' 1.14	c 0.78
$\gamma(0)$ (10^{13} s^{-1})	0.25	0.31

considered incorporate an exponential and a Gaussian decay. For the exponential correlation we have

$$\left.\begin{aligned}\langle \boldsymbol{A}'_\alpha(t)\,.\,\boldsymbol{A}'_\beta(0)\rangle &= \delta_{\alpha\beta}\langle A'^2_\alpha\rangle \exp(-\mu'|t|),\\ \langle \boldsymbol{B}'_\alpha(t)\,.\,\boldsymbol{B}'_\beta(0)\rangle &= \delta_{\alpha\beta}\langle B'^2_\alpha\rangle \exp(-\gamma'|t|),\end{aligned}\right\} \tag{6.13.53}$$

$$\langle \boldsymbol{A}'_\alpha(t)\,.\,\boldsymbol{B}'_\beta(t')\rangle = 0 \tag{6.13.54}$$

(cf. Sears's equations (6.13.28), (6.13.29)) whereupon from (6.13.49), and (6.13.50):

$$\mu(p) = \frac{m}{3kT\pi}\frac{\langle A'^2_\alpha\rangle}{\mu'+p}, \tag{6.13.55}$$

$$\gamma(p) = \frac{m^*}{3kT\pi}\frac{\langle B'^2_\alpha\rangle}{\gamma'+p}. \tag{6.13.56}$$

$\langle A'^2_\alpha\rangle$ may be calculated from Sears's original equations as may ω_0. The values of the various parameters in the exponential and Gaussian models of Damle *et al.* are given in table 6.13.2. In the case of a Gaussian decay of the force autocorrelation, we have

$$\langle \boldsymbol{A}'_\alpha(t)\,.\,\boldsymbol{A}'_\alpha(0)\rangle = \delta_{\alpha\beta}\langle A^2_\alpha\rangle\, e^{-ct^2}, \tag{6.13.57}$$

where c is a constant. It may then be shown that

$$\mu(p) = \frac{m\langle A'^2_\alpha\rangle}{3kT\pi}\left(\frac{\pi}{4c}\right)^{\frac{1}{2}} \exp\left(\frac{p^2}{4c}\right) \operatorname{erf}\left(\frac{p}{(4c)^{\frac{1}{2}}}\right). \tag{6.13.58}$$

The final spectral densities of Damle, Sjölander and Singwi are very similar to those of Sears, and Nakahara and Takahashi, although as Damle *et al.* observe, this does not prove the correctness of the model, even though they are all in essential agreement with Rahman's curve[81].

As we mentioned above, the starting equation for $\boldsymbol{R}(t)$ does not, in the cases of Sears, and Nakahara and Takahashi, contain the restoring force term. That this term should be there can be seen from (6.13.42), where

$$\begin{bmatrix} \Gamma_{11}(p)\,\phi_{11}^0 & \Gamma_{12}(p)\,\phi_{22}^0 \\ \Gamma_{21}(p)\,\phi_{11}^0 & \Gamma_{22}(p)\,\phi_{22}^0 \end{bmatrix} = \begin{bmatrix} F_{11}(p) & F_{12}(p) \\ F_{21}(p) & F_{22}(p) \end{bmatrix}. \tag{6.13.59}$$

It has been assumed here that ϕ^0 is a diagonal matrix. Since $F_{12}(p) = F_{21}(p)$, we have

$$\Gamma_{12}(p)\,\phi_{22}^0 = \Gamma_{21}(p)\,\phi_{11}^0 \tag{6.13.60}$$

or

$$\frac{kT}{m^*}\Gamma_{12}(p) = \frac{kT}{m}\Gamma_{21}(p). \tag{6.13.61}$$

The presence of a restoring term in (6.13.37) corresponds to a finite value of $\Gamma_{12}(p)$, which demands the presence of a similar term in (6.13.38) corresponding to $\Gamma_{21}(p)$. It might be said that m^* could be very large, whereupon the restoring term would vanish, but then so would $F_{22}(p)$, which is contrary to what was assumed by Sears[79] and Nakahara and Takahashi[88]. In fact, Damle *et al.* find $m^* = 0.7m$ for liquid argon, which is very much smaller than one would intuitively expect. The explanation of this surprising effective mass is not yet clear since it relates to the fictitious centre of mass of the neighbouring atoms and is in some way linked to the dynamical effect of the surroundings.

Slow-neutron scattering experiments[91] and machine computations[81] have both served to confirm the details of the self-motion of atoms in simple classical liquids in terms of a vibratory and diffusive component. Several models, discussed above, have been proposed to account for this kind of motion, but by their very nature, inevitably involve parameters in a rather *ad hoc* manner which cannot, therefore, be easily related to microscopic quantities. Several authors have attempted to establish a relation for the self-motion of atoms in a simple classical liquid, in particular the velocity autocorrelation, based essentially on the Liouville equation and involving a knowledge of no more than the interaction potential and the static pair distribution function, $g_{(2)}(r)$. Nijboer and Rahman[92] have shown that the representation of the velocity autocorrelation in terms of its time expansion, (6.13.62), is unsatisfactory on account of its slow convergence:

$$\psi(t) = \psi(0) - \psi^{(2)}(0)\frac{t^2}{2!} + \psi^{(4)}(0)\frac{t^4}{4!} - \dots$$
$$= \sum_{n=0}^{\infty} (-1)^n \psi^{(2n)}(0)\, t^{2n}. \qquad (6.13.62)$$

$\psi(t)$ has been expressed in terms of a Maclaurin expansion about $\psi(0)$, with the requirement that it be an even function of time, although we have already seen that $\dot{\psi}(0) = 0$ (§6.12). The bracketed superscripts represent repeated differentiation with respect to time. The first two and the general coefficient are given below. Higher order terms, $\psi^{(4)}(0)\dots$, are given by Nijboer and Rahman explicitly:

$$\left.\begin{aligned}
\psi(0) &= \langle \dot{x}(0)\,.\,\dot{x}(0)\rangle = kT/m,\\
\psi^{(2)}(0) &= \langle \ddot{x}(0)\,.\,\ddot{x}(0)\rangle = \frac{1}{m^2}\left\langle\left(\frac{\partial\Phi}{\partial x}\right)^2\right\rangle = \frac{kT}{m^2}\left\langle\frac{\partial^2\Phi}{\partial x^2}\right\rangle\\
&= \frac{kT\rho}{3m^2}\int g_{(2)}(r)\,\nabla^2\Phi(r)\,\mathrm{d}\boldsymbol{r},\\
&\;\;\vdots\\
\psi^{(2n)} &= \left\langle\frac{\partial^{n-1}x(0)}{\partial t^{n-1}}\,.\,\frac{\partial^{n-1}x(0)}{\partial t^{n-1}}\right\rangle.
\end{aligned}\right\} \qquad (6.13.63)$$

$\nabla^2\Phi$ may be determined experimentally[19] (see §6.7) by means of isotope separation; Rahman has obtained a rough estimate of $\psi^{(4)}$ from a computer experiment. The resulting small-time expansion of the normalized velocity autocorrelation for liquid argon at 85.8 °K is then

$$\frac{\langle v(0)\,.\,v(t)\rangle}{\langle v(0)^2\rangle} = 1 - 22.5(t\times 10^{12})^2 + 277.0(t\times 10^{12})^4 - \dots. \qquad (6.13.64)$$

In fig. 6.13.4 we have plotted Rahman's molecular dynamic normalized velocity autocorrelation for liquid argon

$$(\rho = 1.407\,\mathrm{g\,cm^{-3}}, \quad 85.8\,°\mathrm{K}),$$

together with the successive approximations arising from the inclusion of the second and third terms of (6.13.64). The result is discouraging; inclusion of the t^4-term even prevents the function from developing its negative region. Nevertheless, Isbister and

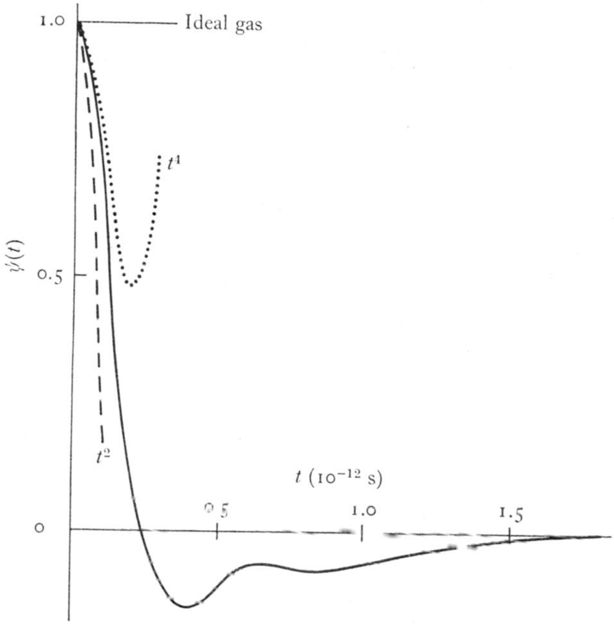

Fig. 6.13.4. The normalized velocity autocorrelation for liquid argon determined by Rahman. The effect of including terms of order t^2 and t^4 (equation (6.13.64)) to the ideal gas value ($\psi = 1.00$) seems to suggest that a time expansion at anything other than the smallest times is not a promising approach to $\psi(t)$. (Redrawn by permission from March, *Liquid Metals* (Pergamon).)

McQuarrie[127] have recently used simple parametrized analytic expansions for both the velocity autocorrelation function and its associated memory function. The parameters are fixed either by comparison with the Maclaurin expansion (6.13.62), through terms in t^4, or alternatively to assume that the long time decay of the velocity autocorrelation is exponential, whereupon terms to order t^2 only are retained. Values of the self-diffusion coefficient D in the range 4.26×10^{-5} and 3.32×10^{-5} (reduced units) are obtained, in comparison with Rahman's value of $D = 3.52 \times 10^{-5}$. Singwi and Tosi[93] discuss the evolution of the velocity autocorrelation, and establish its power spectrum, by a method which is equivalent to summing the time expansion (6.13.64) to infinite order, although only approximately. The method was first proposed by Tjon[94] in discussing spin autocorrelation, and consists in evaluating the velocity autocorrelation function from a linear integro-differential equation whose kernel has the significance of a memory function. The defining relation is (cf. (6.13.39))

$$\frac{d\psi(t)}{dt}+\int_0^t K(t-t')\,\psi(t')\,dt' = 0. \qquad (6.13.65)$$

The derivation of this equation direct from a simplified Liouville equation, appropriate to any normalized phase function, is based on a general formalism given by Zwanzig[95], and also used by Mori[96] in similar contexts. This same relation has been used by Berne, Boon and Rice[97].

It seems convenient to discuss the velocity autocorrelation in terms of the memory function, $K(t)$ since both vibratory and diffusive aspects may be relatively easily incorporated. Taking $K(t)$ to be a constant we obtain an oscillatory autocorrelation, whilst if we choose $K(t)$ to be a δ-function, we recover Langevin's equation which gives an exponential decay of $\psi(t)$. Machine computations by Rahman for liquid argon have indicated (fig. 6.13.5) that $K(t)$ has two important characteristic features: first that it drops by an order of magnitude very sharply – within 3×10^{-13} s – and then secondly, shows a much slower time dependence. It has been shown by various authors that these two characteristic features of the memory function computed by Rahman can be understood in terms of the static pair correlation function and the interatomic potential. From (6.13.65) we have immediately

$$\psi(p) = \frac{1}{p+K(p)}, \qquad (6.13.66)$$

where $K(p)$ is the Laplace transform of $K(t)$. Comparing (6.13.66) with the result of Damle, Singwi and Sjölander, (6.13.50), and remembering that $\phi_{11}(p) \equiv \psi(p)$, we identify the transform of the function as

$$K(p) = \mu(p)+\frac{\omega_0^2\{\gamma(p)+p\}}{p\{\gamma(p)+p\}+\omega_1^2}. \qquad (6.13.67)$$

If we now set μ and γ to be constants, inversion of (6.13.67) yields a memory kernel

$$K(t) = \mu\delta_{t=0}+\frac{\omega_0^2}{(a-b)}\{(a-\gamma)\,e^{-at}+(\gamma-b)\,e^{-bt}\}, \qquad (6.13.68)$$

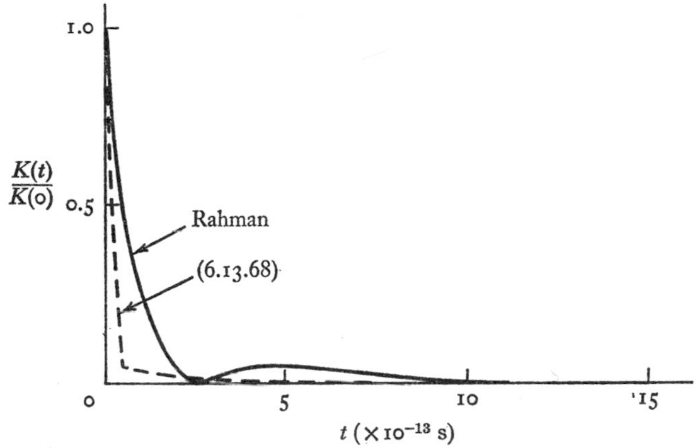

Fig. 6.13.5. The normalized memory function.

where
$$\left.\begin{aligned} a &= \tfrac{1}{2}[\gamma \pm \{\gamma^2 - 4\omega_1^2\}^{\frac{1}{2}}], \\ b &= 2\omega_1^2[\gamma \pm \{\gamma^2 - 4\omega_1^2\}^{\frac{1}{2}}]^{-1}. \end{aligned}\right\} \tag{6.13.69}$$

This expression is seen to be in qualitative agreement with Rahman's function[98] (fig. 6.13.5), with a δ-function at the origin followed by a long range exponential decay of the memory. The memory function $K(t)$ is essentially real, and conditions are therefore placed on a and b in (6.13.69) such that they are everywhere real and positive this being a physical requirement for the decay of the memory. Moreover, if a and b are small, the long range form of $K(t)$ is sensibly constant, whereupon the velocity autocorrelation is an oscillatory function of time corresponding to the case of a damped Einstein oscillator. Thus the two limiting cases are contained within (6.13.65) for appropriate choices of the memory function. The great advantage of the present approach, of course, is that one can guess a reasonable functional form for the kernel $K(t)$.

The above development is, however, at variance with the fluctuation-dissipation theorem. We should instead have retained the time-dependent friction constants, and inverted the entire expression, (6.13.67). Thus, taking the general time-dependent expressions (6.13.45), (6.13.46) for $\mu(p)$ and $\gamma(p)$ respectively, we should, on the basis of an exponential decorrelation of the stochastic forces, utilize expressions (6.13.55), (6.13.56), whilst if we assumed a Gaussian decay we should have used (6.13.58) in (6.13.67). The memory kernel and the corresponding velocity autocorrelation would then eventually be obtained after some lengthy, but trivial manipulation. In the search for analytic solutions several authors have taken advantage of the relatively simple form of the kernel to suggest trial functions on the basis of physical arguments and mathematical expediency. A simple and reasonable choice for the memory function in a liquid which embodies the two limiting cases for appropriate choices of the parameters is a Gaussian form:

$$K(t) = -A\,\mathrm{e}^{-Bt^2}. \qquad (6.13.70)$$

However, as Singwi and Tosi[93] observe, adoption of the exponentially decaying memory function is inconsistent with the requirement that the higher even moments of $\tilde{\psi}(\omega)$ be finite: only the zeroth moment exists. At large times, however, this kernel becomes correct. Singwi and Tosi find for the Gaussian memory function

$$A = \langle \omega^2 \rangle_{\mathrm{av}}; \quad B = \pi^{\frac{1}{2}} mD/2kT,$$

where $\langle \omega^2 \rangle_{\mathrm{av}}$ represents the second moment of $\tilde{\psi}(\omega)$, and D is the diffusion constant. Taking the molecular dynamics values of Rahman[99] for liquid argon – $\langle \omega^2 \rangle_{\mathrm{av}} = 50 \times 10^{24}\,\mathrm{s}^{-2}$, $D = 1.88 \times 10^{-5}$ $\mathrm{cm}^2\,\mathrm{s}^{-1}$ (85.5 °K), and $55 \times 10^{24}\,\mathrm{s}^{-2}$, $2.43 \times 10^{-5}\,\mathrm{cm}\ \mathrm{s}^{-1}$ (94.4 °K), respectively – these authors determine the spectral densities for the Gaussian kernel, (6.13.70). Their results are shown in fig. 6.13.6. The velocity autocorrelations are in qualitative agreement with Rahman's results, showing a pronounced negative region, but followed by a series of damped oscillations. This emphasis of the vibratory aspect of the self-motion is apparent in the spectral density curves. A decrease in the diffusion coefficient or in the second moment of the spectrum heightens and narrows the peak in $\tilde{\psi}(\omega)$, as we should expect qualitatively. Berne *et al.*[97] have obtained the frequency spectrum for an exponential kernel (note, however, the reservations made above), and these are also compared in fig. 6.13.6. This present approach clearly represents a substantial improvement in the description of the velocity autocorrelation over the Langevin model or the truncated

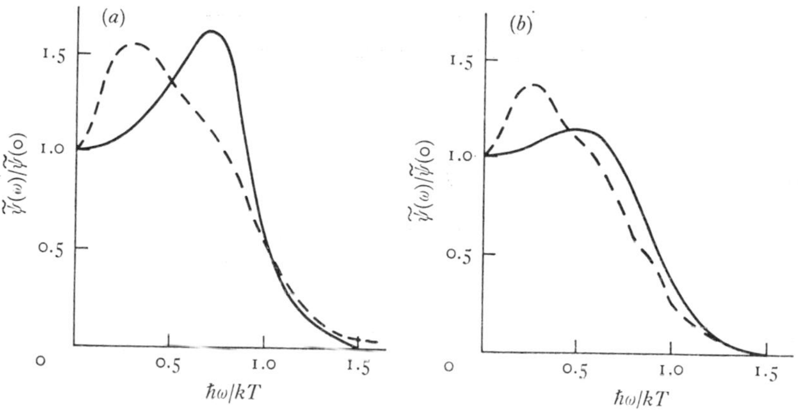

Fig. 6.13.6. The spectral density of Singwi and Tosi[93] on the basis of a Gaussian memory kernel. Their results are compared with Rahman's molecular dynamics determinations for liquid argon at (*a*) 85.5 and (*b*) 94.4 °K (broken line). (Redrawn by permission of Singwi and Tosi, *Phys. Rev.* **153**, 157 (1967).)

time expansion, (6.13.62). It is also clear however, that the use of a simple memory function oversimplifies the problem.

Several more sophisticated expressions of the memory kernel have been made[100] with varying degrees of success: ultimately, however, they all reduce to an analysis in terms of diffusive and vibratory modes of atomic motion. How these components are coupled remains to be determined, but it is probably true to say that the problem may be most directly attacked in terms of the memory function approach outlined above.

It is well known that the phonon spectra of perfect crystals contain non-analytic points, the van Hove singularities. A recent suggestion has been made by Gaskell and March[101] to the effect that the spectral density of a classical liquid exhibits a singularity at $\omega = 0$. Ernst *et al.*[102] have recently shown that the normalized velocity autocorrelation in the limit of long times is given as

$$\operatorname*{Lt}_{t\to\infty} \psi(t) \simeq \frac{2}{3\rho}\left\{4\pi\left(D+\frac{\eta}{\rho m}\right)t\right\}^{-\frac{3}{2}}, \qquad (6.13.71)$$

where η is the shear viscosity, ρ the number density and m the atomic mass. It may then be shown in a straightforward manner that at low frequencies $\tilde{\psi}(\omega)$ may be expanded as an even function of ω:

$$\tilde{\psi}(\omega) = \tilde{\psi}(0) + d_1\,\omega^{\frac{1}{2}} + d_2\,\omega + d_3\,\omega^{\frac{3}{2}} + d_4\,\omega^2 + \ldots, \quad (6.13.72)$$

where

$$\tilde{\psi}(0) = D/\pi$$

$$d_1 = -(2\pi)^{\frac{1}{2}}\frac{2}{3\rho}\left\{4\pi\left(D+\frac{\eta}{\rho m}\right)\right\}^{-\frac{3}{2}}\frac{kT}{m\pi}. \quad (6.13.73)$$

It is evident by taking the derivative of (6.13.72) that $\tilde{\psi}(\omega)$ develops a negatively infinite slope at $\omega = 0$, and Gaskell and March suggest that this might provide a qualitative explanation for the initial decrease in $\tilde{\psi}(\omega)$ from the value $\tilde{\psi}(0) = D/\pi$ observed experimentally by Egelstaff(103) and by Randolph(104) for liquid sodium (see fig. 6.13.2). This qualitative subdivision of the spectral density into diffusive and vibratory modes has not, as yet, been revealed by molecular dynamics calculations on this liquid metal. However, as we observed in our discussion of the Nakahara–Takahashi phonon spectrum (fig. 6.13.3(*b*)), for a value of the parameter $\lambda = 13$ this qualitative subdivision can be reproduced. Indeed, λ has the significance of the ratio of the mean square driving force for the Einstein oscillator to the mean square driving force for Langevin diffusion, and large values of λ indicate a subordinate role adopted by the Langevin forces. Under these circumstances D and η become small, enhancing the singularity given by (6.13.73). On the other hand, as Gaskell and March observe, in the limit $D \to 0$, $\eta \to \infty$ corresponding to a glass or solid, d_1 tends to zero as does D/π. Moreover, if the coefficients d_2 and d_3 behave similarly to d_1, and these authors believe they do, then we are left with a Debye-like term $d_4\,\omega^2$, corresponding to the propagation of acoustic modes in the crystal. Finally, it is pointed out that an incoherent neutron scattering measurement of $\underset{\omega\to 0}{\mathrm{Lt}}(\omega^{\frac{1}{2}}\,\partial\tilde{\psi}/\partial\omega)$ could, from (6.13.73) and (6.13.72), be used to determine the coefficient of shear viscosity, taking $\tilde{\psi}(0) = D/\pi$.

From the equation of motion of an atom in a liquid and an approximation involving the decoupling of a statistical average,

Gaskell *et al.*[105] have obtained the following expression for the memory function

$$K(t-\tau) = \frac{\rho}{3m}\iint g_{(2)}(r)\,\nabla^2\Phi(|\boldsymbol{r}-\boldsymbol{x}|)\,G'_{\mathrm{S}}(\boldsymbol{x},t-\tau)\,\mathrm{d}\boldsymbol{r}\,\mathrm{d}\boldsymbol{x}, \quad (6.13.74)$$

where $g_{(2)}(r)$ is the equilibrium radial distribution function, $\Phi(r)$ is the interatomic potential and

$$G'_{\mathrm{S}}(\boldsymbol{x},t-\tau) = \langle\delta[x-x_{ij}(t)+x_{ij}(\tau)]\rangle$$

is the probability that the change in separation of atoms i and j in time $t-\tau$ is x. If it is assumed that each particle migrates independently from its initial to its final position, G'_{S} has the same form, within the Gaussian approximation, as the self-correlation function, but with an effective mass m^*. Hence, approximately we have

$$G'_{\mathrm{S}}(\boldsymbol{x},t-\tau) = \{4\pi a(t-\tau)\}^{-\frac{3}{2}}\exp\left(\frac{-x^2}{4a(t-\tau)}\right), \quad (6.13.75)$$

where

$$a(t) = \frac{kT}{m^*}\int_0^t (t-\tau)\,\psi(\tau)\,\mathrm{d}\tau \quad (6.13.76)$$

and $m^* = m/2$ is the reduced mass in the relative motion of the two atoms. If we were dealing with a perfect gas, the normalized velocity autocorrelation is time independent whereupon (6.13.76) becomes kTt^2/m. As we observed above, the two distinct regions of the memory function are to be associated with the hard core repulsive and soft long range attractive regions of pair potential. Gaskell calculates the hard and soft components of the memory function,

$$K(t) = K_{\mathrm{H}}(t)+K_{\mathrm{S}}(t),$$

on the basis of (6.13.74) where $a(t)$ has been taken from some results given by Nijboer and Rahman based on a computer determination of the velocity autocorrelation function in liquid argon. At small times a linear trajectory approximation is made: $a(t)\to kTt^2/m$ (see above), whilst asymptotically we have $a(t)\to 2Dt+c$ where D is the diffusion constant and c is a constant. It is therefore appropriate to use the small time value for $a(t)$ in the determination of $K_{\mathrm{H}}(t)$ and the asymptotic value for $K_{\mathrm{S}}(t)$. There is, however, some ambiguity as to how a realistic potential is to be split up into its hard and soft components. One choice is to separate the regions at σ where $\Phi(\sigma) = 0$. This is investigated by Gaskell, but a second choice

$$\left.\begin{aligned} \Phi_{\mathrm{H}}(r) &= 4\epsilon\left\{\left(\frac{\sigma}{r}\right)^{12} - \left(\frac{\sigma}{r}\right)^{6}\right\} + 2\epsilon & r < \sigma \\ &= 2\epsilon\left(\frac{\sigma}{r}\right)^{12} & r \geqslant \sigma \\ \text{and}\qquad \Phi_{\mathrm{S}}(r) &= -2\epsilon & r > \sigma \\ &= 4\epsilon\left\{\frac{1}{2}\left(\frac{\sigma}{r}\right)^{12} - \left(\frac{\sigma}{r}\right)^{6}\right\} & r \geqslant \sigma \end{aligned}\right\} \qquad (6.13.77)$$

seems more appropriate and yields a more satisfactory memory kernel. The results from this choice applied to (6.13.74) are shown in fig. 6.13.7(*a*). The former choice of hard and soft components yields a much narrower peak at the origin, and the fall in $K_H(t)$ is much too rapid. One aspect as yet not reproduced in any of the analyses of the memory function to date is the minimum at around 4×10^{-13} s followed by a maximum at about 6.5×10^{-13} s (see fig. 6.13.5). It does appear, however, that these features have considerable bearing upon the spectral density $\tilde{\psi}(\omega)$. In fig. 6.13.7(*b*) Gaskell's normalized spectral density $\tilde{\psi}(\omega)/\tilde{\psi}(0)$ is compared to Rahman's molecular dynamics result[106] obtained under the same thermodynamic conditions but using the Buckingham exp-6 potential. The major discrepancy is seen to be in the difference in height in the two peaks, a feature ascribed by Björkmann to the absence of the oscillation in $K(t)$ around 5×10^{-13} s (fig. 6.13.5). The diffusion constant is given as

$$\begin{aligned} D &= (\pi kT/m)\,\tilde{\psi}(0) \\ &= (kT/m)\left\{\int_0^\infty K(t)\,\mathrm{d}t\right\}^{-1} \\ &= 1.95 \times 10^{-5}\,\mathrm{cm}^2\,\mathrm{s}^{-1}. \end{aligned}$$

6.14 General theory of interacting particles[107]

The Liouville equation describing the N-body phase evolution

$$\frac{\mathrm{d}f_{(N)}}{\mathrm{d}t} = \frac{\partial f_{(N)}}{\partial t} + \sum_{j=1}^{N}\left\{\frac{\boldsymbol{p}_j}{m}\frac{\partial f_{(N)}}{\partial \boldsymbol{q}_j} + \boldsymbol{F}_j\frac{\partial f_{(N)}}{\partial \boldsymbol{p}_j}\right\} = 0 \qquad (6.14.1)$$

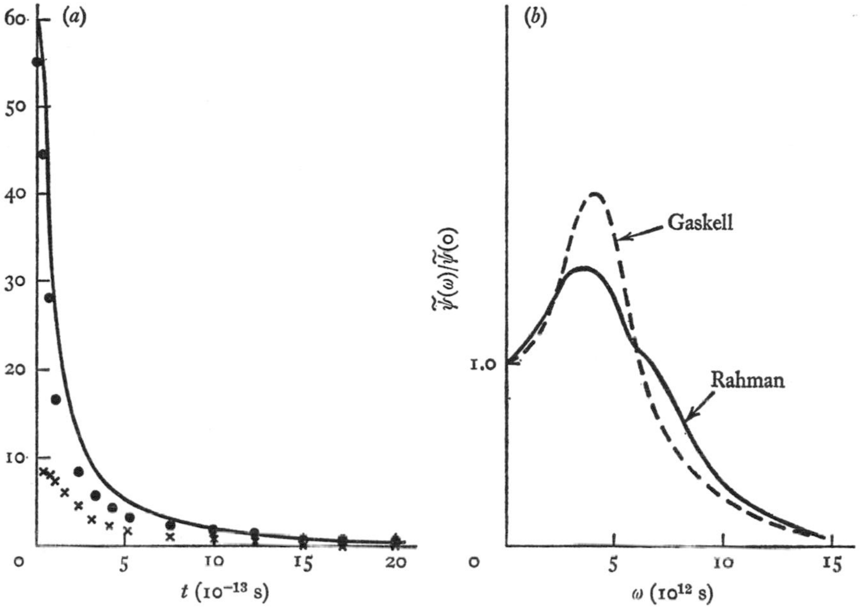

Fig. 6.13.7. (a) The form of Gaskell's memory function on the basis of soft (×) and hard (●) interactions, together with (——) the combined function

$$K(t) = K_H(t) + K_S(t).$$

(b) The normalized spectral density of Gaskell compared with Rahman's molecular dynamic computations using a Buckingham exp-6 potential. (Redrawn by permission from Gaskell, *J. Phys. Chem.* **4**, 1466 (1971); Gaskell and Barker, *J. Phys. Chem.* **5**, 353 (1972).)

may be written in terms of the linear Hermitian operator $\mathscr{L}$, where if we set

$$\mathscr{L} = -\mathrm{i}\sum_{j=1}^{N}\left(\frac{\boldsymbol{p}_j}{m}\frac{\partial}{\partial \boldsymbol{q}_j}+\boldsymbol{F}_j\frac{\partial}{\partial \boldsymbol{p}_j}\right); \quad \mathrm{i} = \sqrt{-1} \tag{6.14.2}$$

then the Liouville equation takes the form

$$\mathrm{i}\frac{\partial f_{(N)}}{\partial t} = \mathscr{L} f_{(N)} \tag{6.14.3}$$

subject to the boundary condition that $f_{(N)}$ vanishes outside the volume V and for indefinitely large values of the particle momenta. In this development we understand the system Hamiltonian to be a two-component function

$$\mathscr{H} = \mathscr{H}_0 + \lambda\mathscr{H}_i,$$

where $\mathscr{H}_0$ refers to the kinetic non-interacting degrees of freedom whilst $\mathscr{H}_i$ refers to the configurationally-dependent components. The parameter λ is a dimensionless quantity characterizing the strength of the interaction which we shall find useful in classifying our later developments. To this decomposition of the Hamiltonian there clearly corresponds a similar decomposition of the Liouville operator

$$\mathscr{L} = \mathscr{L}_0 + \delta\mathscr{L},$$

the second component constituting a perturbation of the purely kinetic development. $\lambda = 0$ corresponds, of course, to the unperturbed situation. In the usual presentation of mechanics the essential quantities are the coordinates and momenta – their rates of change being given by Hamilton's canonical equations. Here, however, the phase distribution develops naturally in terms of collective coordinates wherein the conventional mechanical coordinates are replaced by their Fourier indices or wave vectors, ρ_k representing the amplitude of the kth vector. We shall be concerned here then with the evolution of the spectral density, which incidentally, is assumed continuous: the implicit assumption therefore being made is that $V \to \infty$, subject of course to N/V remaining constant. It turns out in fact that the very existence of irreversibility, the explanation of which must be considered one of the primary objectives of the general theory, is closely related to a continuous spectrum of wave vectors.

The Liouville equation (6.14.3) has the formal solution

$$f_{(N)}(\boldsymbol{p}_N, \boldsymbol{q}_N, t) = \exp(-i\mathscr{L}t) f_{(N)}(\boldsymbol{p}_N, \boldsymbol{q}_N, 0), \qquad (6.14.4)$$

where $\exp(-i\mathscr{L}t)$, designated the *propagator*, links the phase distributions at times 0 and t. Tolman[108] has observed that in this form, the Liouville equation (6.14.3) has the same form and boundary conditions as the Schrödinger equation, and is presumably susceptible to the same mathematical techniques. The interesting possibility of discussing the phase evolution in the same terms as applied to the time-dependent Schrödinger equation was first raised by Kirkwood[109], and forms the basis of the extraordinarily elegant theory of non-equilibrium which has subsequently been developed by Prigogine and his school[110].

An alternative expression of the general formal solution to the Liouville equation in operator form is obtained in terms of the Laplace transform as follows. By definition

$$\hat{f}_{(N)}(k) = \int_0^\infty f_{(N)}(t)\exp(-kt)\,dt, \qquad (6.14.5)$$

where $\hat{f}_{(N)}(k)$ represents the Laplace transform of the phase distribution. The Laplace transform of the Liouville equation is very simply shown to be

$$ik\hat{f}_{(N)}(k) - i f_{(N)}(0) = \mathscr{L}\hat{f}_{(N)}(k), \qquad (6.14.6)$$

where the fact that $\mathscr{L}$ is time independent has been used. Rearrangement yields immediately

$$\hat{f}_{(N)}(k) = -i(\mathscr{L} - ik)^{-1} f_{(N)}(0) \qquad (6.14.7)$$

and this is essentially the Laplace transform of the propagator equation, (6.14.4). The formal solution is given in terms of the resolvent as the inverse of (6.14.7):

$$f_{(N)}(t) = -\frac{1}{2\pi i}\int_{\infty+ic}^{-\infty+ic} (\mathscr{L} - z)^{-1}\exp(-izt) f_{(N)}(0)\,dz, \qquad (6.14.8)$$

where $z = ik$ is the complex variable. Equation (6.14.8) is in fact an example of the Bromwich integral[111], familiar in the theory of the complex variable, and as such represents the formal inversion of a Laplace transform. This resolvent formalism has been developed principally by Résibois[112] and allows one to handle the time dependence of the diagrams discussed later in this section in a particularly convenient way. This, as it turned out, was an essential step in the derivation of a general kinetic equation and a general H-theorem. We shall consider the resolvent formalism later; suffice it to say for the present that in either development the Fourier expansion of the phase function the space-*independent* part, corresponding to the distribution of velocities, may be separated from the space-*dependent* part. This is quite analogous to the degenerate Bose gas wherein we may separate the ground state from the excited states. In the present theory then, the homogeneous, space-independent component plays the role of the ground state whilst the correlations and inhomogeneities adopt the role of excitations in quantum theory. Moreover, the eigenvalue structure of the Liouville equation enables us to write

$$\mathscr{L}\phi_k = \alpha_k \phi_k, \tag{6.14.9}$$

where ϕ_k and α_k are eigenfunctions and eigenvalues of the Liouville operator, although no physical significance is as yet ascribed to these quantities. Since the Liouville equation is linear, we may expand the phase distribution in terms of the complete set of eigenfunctions:

$$f_{(N)}(t) = \sum_k a_k(t)\,\phi_k(\boldsymbol{p}_N, \boldsymbol{q}_N). \tag{6.14.10}$$

The coefficients a_k are functions of time and involve the phase, representing the temporal evolution amongst the eigenfunction set $\phi_{\{k\}}$. Inserting this expression of the distribution into the Liouville equation (6.14.3), and using (6.14.9), we immediately obtain

$$a_k(t) = c_k \exp(-\mathrm{i}\alpha_k t)$$

whereupon (6.14.10) becomes

$$f_{(N)}(t) = \sum_k c_k \exp(-\mathrm{i}\alpha_k t)\,\phi_k(\boldsymbol{p}_N, \boldsymbol{q}_N). \tag{6.14.11}$$

c_k is a numerical coefficient dependent upon $f_{(N)}(0)$.

To help fix our ideas we shall consider a system of N non-interacting particles moving in a cubic box of side L. In this case the Liouville operator simplifies to its purely kinetic form

$$\mathscr{L}_0 = -\mathrm{i}\sum_{j=1}^{N}\left(\frac{\boldsymbol{p}_j}{m}\frac{\partial}{\partial \boldsymbol{q}_j}\right) \tag{6.14.12}$$

whereupon we easily obtain

$$\phi_k = \frac{1}{L^{\frac{3}{2}}}\exp\{\mathrm{i}(\boldsymbol{k}_j\cdot\boldsymbol{q}_j)\}, \tag{6.14.13}$$

where $\boldsymbol{k}_j = 2\pi\boldsymbol{n}_j/L$, $\boldsymbol{n}_j$ being a vector whose components are integers. If now we extend the discussion to include interacting particles we must use the full Liouville operator $\mathscr{L}$, defined in (6.14.2). Provided the interactions are not too strong it is reasonable to use the original set of basis functions determined in the free particle case, (6.14.13). Under these circumstances we may write

$$f_{(N)}(\boldsymbol{p}_N, \boldsymbol{q}_N, t) = \left(\frac{2\pi}{L}\right)^{3N} \sum_{\{k\}} \rho_{\{k\}}(\boldsymbol{p}_N, t) \exp\left[\mathrm{i}\sum_j \left\{\boldsymbol{k}_j \cdot \left(\boldsymbol{q}_j - \frac{\boldsymbol{p}_j}{m}t\right)\right\}\right] \tag{6.14.14}$$

with
$$\rho_{\{k\}} = \left(\frac{L}{4\pi^2}\right)^{3N/2} a_{\{k\}}.$$

The notation $\{\boldsymbol{k}\}$ is taken to include all wave vectors corresponding to a specific $\boldsymbol{n}$, having total value $\boldsymbol{k}$. Equation (6.14.14) represents no more than a re-expression of the phase distribution in Fourier space. The time dependence of the amplitude $\rho_{\{k\}}$ now arises on account of the scattering to and from the state $\boldsymbol{k}$. The evolution in phase is now understood as an evolution of the spectral density and the determination of $\rho_{\{k\}}(\boldsymbol{p}_N, t)$ is now our primary concern. This approach to the Liouville equation seeks to replace the interaction between particles surrounded by a force field by an equivalent interaction between wave fields. A field is, in effect, a system having an infinite number of degrees of freedom, and hence field-theoretic perturbation problems have many features in common with the many-body problem in the limit $N \to \infty$. The perturbation technique developed below was very much inspired by the methods of quantum field theory.

Substituting (6.14.14) into the Liouville equation written in the form

$$\mathrm{i}\frac{\partial f_{(N)}}{\partial t} = (\mathscr{L}_0 + \lambda\delta\mathscr{L}) f_{(N)}$$

we obtain

$$\mathrm{i}\frac{\partial}{\partial t}\sum_{\{k\}} \rho_{\{k\}} \exp\left[\mathrm{i}\sum_j \left\{\boldsymbol{k}_j \cdot \left(\boldsymbol{q}_j - \frac{\boldsymbol{p}_j}{m}t\right)\right\}\right]$$
$$= (\mathscr{L}_0 + \lambda\delta\mathscr{L}) \sum_{\{k\}} \rho_{\{k\}} \exp\left[\mathrm{i}\sum_j \boldsymbol{k}_j \cdot \left(\boldsymbol{q}_j - \frac{\boldsymbol{p}_j}{m}t\right)\right]. \tag{6.14.15}$$

The definition of $\mathscr{L}_0$, (6.14.12), enables us to reduce (6.14.15) to the *interaction representation* as follows:

$$\mathrm{i}\sum_{\{k\}} \left(\frac{\partial \rho_{\{k\}}}{\partial t}\right) \exp\left[\mathrm{i}\sum_j \left\{\boldsymbol{k}_j \cdot \left(\boldsymbol{q}_j - \frac{\boldsymbol{p}_j}{m}t\right)\right\}\right]$$
$$= \lambda\delta\mathscr{L} \sum_{\{k\}} \rho_{\{k\}} \exp\left[\mathrm{i}\sum_j \left\{\boldsymbol{k}_j \cdot \left(\boldsymbol{q}_j - \frac{\boldsymbol{p}_j}{m}t\right)\right\}\right]. \tag{6.14.16}$$

Multiplication throughout by $\exp[\mathrm{i}\sum_l \boldsymbol{k}_l.\boldsymbol{q}_l]$ and integration over all $\boldsymbol{q}$ yields

$$\mathrm{i}\left(\frac{\partial\rho_{\{k\}}}{\partial t}\right) = \lambda\sum_{\{k'\}}\exp\left[\mathrm{i}\Sigma\boldsymbol{k}_j.\frac{\boldsymbol{p}_j}{m}t\right]\langle\{\boldsymbol{k}\}|\delta\mathscr{L}|\{\boldsymbol{k}'\}\rangle$$

$$\times\exp\left[-\mathrm{i}\Sigma\boldsymbol{k}'_j.\frac{\boldsymbol{p}_j}{m}t\right]\rho_{\{k'\}}(\boldsymbol{p}_N,t), \quad (6.14.17)$$

where account has been taken of the orthogonality properties of the eigenfunctions, and where we have used the matrix notation

$$\langle\{\boldsymbol{k}\}|\delta\mathscr{L}|\{\boldsymbol{k}'\}\rangle = \frac{1}{(2\pi)^N}\int\dots\int\exp[-\mathrm{i}\Sigma\boldsymbol{k}_j.\boldsymbol{q}_j]\,\delta\mathscr{L}$$

$$\times\exp[\mathrm{i}\Sigma\boldsymbol{k}'_j.\boldsymbol{q}_j]\,\mathrm{d}\boldsymbol{q}_1\dots\mathrm{d}\boldsymbol{q}_N; \quad (6.14.18)$$

that is, in the more familiar terminology of quantum mechanics, the transition probability $\boldsymbol{k}'\rightarrow\boldsymbol{k}$ between the two states is given by averaging the free particle eigenfunctions over the interaction Hamiltonian. Thus, quantum mechanically we should have instead of (6.14.17)

$$\mathrm{i}\left(\frac{\mathrm{d}c_k}{\mathrm{d}t}\right) = \sum_{k'}\exp[\mathrm{i}E_k t/\hbar]\langle\boldsymbol{k}|\mathscr{H}_i|\boldsymbol{k}'\rangle\exp[\mathrm{i}E_{k'}t/\hbar]\,c_{k'}, \quad (6.14.19)$$

where $c_{k'}$ are the corresponding Fourier coefficients, although in the quantum mechanical case the amplitudes $c_{k'}$ are purely functions of time. Equation (6.14.19) is immediately understood as relating to the population of state $\boldsymbol{k}$ subject to the interaction whose Hamiltonian is $\mathscr{H}_i$. The evolution of the kth amplitude will therefore explicitly depend upon the number of particles in the initial state $c_{k'}\exp[\mathrm{i}E_{k'}t/\hbar]$, the occupation probability of the final state $\exp[\mathrm{i}E_k t/\hbar]$, and the transition probability $\langle\boldsymbol{k}|\mathscr{H}_i|\boldsymbol{k}'\rangle$, the whole being summed over all initial states.

As we might have anticipated, unperturbed terms of the type $\langle\boldsymbol{k}|\mathscr{L}_0|\boldsymbol{k}'\rangle$ vanish on the grounds of orthogonality of the eigenfunctions, and we shall occasionally write the above matrix element in Kroenecker delta form: $\langle\boldsymbol{k}|\mathscr{L}_0|\boldsymbol{k}'\rangle\,\delta^{\mathrm{Kr}}_{k,k'}$ – the element vanishing except when the argument $(\boldsymbol{k}-\boldsymbol{k}')$ vanishes, i.e. when $\boldsymbol{k}=\boldsymbol{k}'$. Such transitions are termed *diagonal* in the sense that they return the system to its initial state. $\delta\mathscr{L}$, on the other hand, is then *off-diagonal*, and we understand $\delta\mathscr{L}$ to induce transitions from states $\{\boldsymbol{k}'\}$ to $\{\boldsymbol{k}\}$. This approach, then, forms the basis of the entire development: we shall pursue the evolution of the amplitudes $\rho_{\{k\}}(\boldsymbol{p}_N, t)$ arising from scattering to and from the state $\boldsymbol{k}$ under the action of the perturbation $\delta\mathscr{L}$.

The coefficient $\rho_{\{0\}}$ has an especially simple and important physical meaning. We determine the Fourier coefficient in the usual way as

$$\rho_{\{0\}}(\boldsymbol{p}_N, t) = \frac{1}{(2\pi)^{N/2}}\int\cdots\int f_{(N)}(\boldsymbol{p}_N, \boldsymbol{q}_N, t)\,\mathrm{d}\boldsymbol{q}_1\ldots\mathrm{d}\boldsymbol{q}_N \quad (6.14.20)$$

whereupon it is immediately evident that this coefficient describes the energy evolution of the system, containing only the kinetic components. We shall now attempt a perturbation expansion of (6.14.17) in terms of the parameter λ. First, however, consider the following illustration:

$$\frac{\partial y}{\partial t}(x,t) = \lambda\mathscr{K}(x,t)\,y(x,t) \quad (6.14.21)$$

in which $\mathscr{K}$ is a time-dependent operator acting on y and λ is some parameter. If the initial value of y is

$$y(x,0) = y_0(x) \quad (6.14.22)$$

then we may write (6.14.21) as

$$y(x,t) = y_0(x) + \lambda\int_0^t \mathscr{K}(x,t_1)\,y_0(x)\,\mathrm{d}t_1 \quad (6.14.23)$$

which serves as a basis for a new iteration:

$$y(x,t) = y_0(x) + \lambda\int_0^t \mathscr{K}(x,t_1)\,y_0\,\mathrm{d}t_1$$

$$+\lambda^2\int_0^t\int_0^{t_1} \mathscr{K}(x,t_1)\,\mathscr{K}(x,t_2)\,y_0\,\mathrm{d}t_2\,\mathrm{d}t_1. \quad (6.14.24)$$

Differentiation establishes the correctness of this power expansion of $y(x, t)$ in λ. Applying this iteration scheme to (6.14.17) we have

$$\rho_{\{0\}}(t) = \rho_{\{0\}}(0) + \frac{\lambda}{i}\sum_{\{k'\}}\int_0^t \langle\{0\}|\,\delta\mathscr{L}\,|\{\boldsymbol{k}'\}\rangle \exp\left[-i\Sigma \boldsymbol{k}_j.\frac{\boldsymbol{p}_j}{m}.t_1\right]\rho_{\{k'\}}(0)\,dt_1$$
$$+\left(\frac{\lambda}{i}\right)^2 \sum_{\{k'\}\{k''\}}\int_0^t\int_0^{t_1}\langle\{0\}\,|\delta\mathscr{L}\,|\,\{\boldsymbol{k}'\}\rangle \exp\left[-i\Sigma\boldsymbol{k}'_j.\frac{\boldsymbol{p}_j}{m}t_1\right]$$
$$\times\exp\left[-i\Sigma\boldsymbol{k}'_j.\frac{\boldsymbol{p}_j}{m}t_2\right]\langle\{\boldsymbol{k}'\}\,|\delta\mathscr{L}\,|\,\{\boldsymbol{k}''\}\rangle$$
$$\times\exp\left[-i\Sigma\boldsymbol{k}''_j.\frac{\boldsymbol{p}_j}{m}t_2\right]\rho_{\{k''\}}(0) + \left(\frac{\lambda}{i}\right)^3\dots \qquad (6.14.25)$$

Whilst it is not explicitly apparent here, it may quite simply be shown that the Fourier expansion of $\delta\mathscr{L}$ in $\boldsymbol{q}$ (or more precisely in $|\boldsymbol{q}_i - \boldsymbol{q}_j|$ for a pairwise decomposable interaction) directly determines which matrix elements $\langle \boldsymbol{k}\,|\delta\mathscr{L}\,|\,\boldsymbol{k}'\rangle$ will appear. Thus, to a term proportional to $\exp(\pm i\boldsymbol{m}\boldsymbol{q})$ in $\mathscr{H}_i$ (or equivalently, in the Fourier transform of the pair potential, $\tilde{\Phi}$) there will exist a corresponding non-vanishing matrix element $\langle \boldsymbol{k}\,|\delta\mathscr{L}\,|\,\boldsymbol{k}\pm\boldsymbol{m}\rangle$. The physical significance of these non-diagonal elements is quite clear: transitions of the type $\boldsymbol{k} \neq \boldsymbol{k}'$ represent inelastic interactions between particles while the diagonal transitions correspond to elastic exchanges during which, however, transient 'excitations' may occur. For weak interactions we may take it that $\boldsymbol{m}$ adopts the values ± 1 only, so that

$$\langle \boldsymbol{k}\,|\delta\mathscr{L}\,|\,\boldsymbol{k}'\rangle = 0 \quad \text{for} \quad \boldsymbol{k}' \neq \begin{cases}\boldsymbol{k}+1,\\ \boldsymbol{k}-1.\end{cases}$$

It would be quite impossible, of course, to attempt to write down the complete expansion (6.14.25). Instead we have to isolate the important terms and demonstrate that $\rho_{\{0\}}(t)$ satisfies a much simpler 'master equation' than the original Liouville equation, and moreover, we have to show that this equation provides us with all the information we are interested in. The condition $\boldsymbol{m} = \pm 1$ restricts the transitions which may occur; for example it is not possible to effect the transition $\rho_{\{0\}}(0)$ to $\rho_{\{0\}}(t)$ in an odd number of steps, i.e

$$\langle\{0\}\,|\delta\mathscr{L}\,|\,\{\boldsymbol{k}''\}\rangle\langle\{\boldsymbol{k}''\}\,|\delta\mathscr{L}\,|\,\{\boldsymbol{k}'\}\rangle\langle\{\boldsymbol{k}'\}\,|\delta\mathscr{L}\,|\,\{0\}\rangle = 0$$

since the first two single-step transitions cannot be compensated by the third. This means that only even-number transitions need be retained in (6.14.25). We also need an initial condition on the $\rho_{\{k\}}(0)$ which appear throughout (6.14.25) in defining the initial density of states. Here we adopt the (unnecessarily restrictive) condition

$$\rho_{\{k\}}(0) = 0 \quad \{\boldsymbol{k}\} \neq 0 \tag{6.14.26}$$

often termed the random phase assumption. Since $\rho_{\{0\}}(0)$ is the only non-vanishing initial amplitude only transitions $\{0\} \rightarrow \{0\}$ may occur. Equation (6.14.25) therefore reduces to

$$\begin{aligned}\rho_{\{0\}}(t) = \rho_{\{0\}}(0) &+ \left(\frac{\lambda}{i}\right)^2 \sum_{\{k'\}} \int_0^t \int_0^{t_1} \langle\{0\}|\delta\mathscr{L}|\{\boldsymbol{k}'\}\rangle \\ &\times \exp\left[-i\Sigma \boldsymbol{k}'_j . \frac{\boldsymbol{p}_j}{m}(t_1 - t_2)\right] \langle\{\boldsymbol{k}'\}|\delta\mathscr{L}|\{0\}\rangle \rho_{\{0\}}(0)\, dt_1\, dt_2 \\ &+ \left(\frac{\lambda}{i}\right)^4 \sum_{\{k'\}\{k''\}\{k'''\}} \int_0^t \int_0^{t_1} \int_0^{t_2} \int_0^{t_3} \langle\{0\}|\delta\mathscr{L}|\{\boldsymbol{k}'\}\rangle \\ &\times \exp\left[-i\Sigma \boldsymbol{k}'_j . \frac{\boldsymbol{p}_j}{m}(t_1 - t_2)\right] \langle\{\boldsymbol{k}'\}|\delta\mathscr{L}|\{\boldsymbol{k}''\}\rangle \\ &\times \exp\left[-i\Sigma \boldsymbol{k}''_j . \frac{\boldsymbol{p}_j}{m}(t_2 - t_3)\right] \langle\{\boldsymbol{k}''\}|\delta\mathscr{L}|\{\boldsymbol{k}'''\}\rangle \\ &\times \exp\left[-i\Sigma \boldsymbol{k}'''_j . \frac{\boldsymbol{p}_j}{m}(t_3 - t_4)\right] \langle\{\boldsymbol{k}'''\}|\delta\mathscr{L}|\{0\}\rangle \\ &\times \rho_{\{0\}}(0)\, dt_1\, dt_2\, dt_3\, dt_4 + \left(\frac{\lambda}{i}\right)^6 \ldots\end{aligned} \tag{6.14.27}$$

We have now to decide which terms in (6.14.27) we wish to retain. It may be shown that the relaxation time varies with the interaction parameter as

$$\tau \sim \frac{1}{\lambda^2}$$

which is dimensionally incorrect but qualitatively acceptable: a non-interacting system will evidently require an infinite time to reach equilibrium. Further, if we anticipate an evolution in time of the form $\exp(-t/\tau)$ then it is evident that the general term in (6.14.27) we would wish to retain will be of order $(\lambda^2 t)^n$, and it is these components we shall attempt to extract from the series (6.14.27).

The terms appearing in (6.14.27) are conveniently discussed in terms of diagrams. Each matrix element of the non-interacting type $\langle\{\boldsymbol{k}\}\,|\mathscr{L}_0|\,\{\boldsymbol{k}'\}\rangle$ *must* have $\boldsymbol{k}' = \boldsymbol{k}$ if it is not to vanish, and may be represented by a set of lines from right to left equal in number to the number of non-vanishing $\boldsymbol{k}$-vectors in the set $\{\boldsymbol{k}\}$. Individual lines in a diagram are labelled by the particle that is represented by the $\boldsymbol{k}$-vector. This construction is closely allied to that of quantum field theory with the additional complication of distinguishability in the classical case. In the case of the non-diagonal elements $\langle\{\boldsymbol{k}\}\,|\delta\mathscr{L}|\,\{\boldsymbol{k}'\}\rangle$, the number of wave vectors increases or decreases corresponding to the *creation* or *destruction* of a correlation, respectively. In these cases the creation or destruction process is represented by a vertex where the lines representing the wave vectors cross in a characteristic way. Because we assume pairwise interaction of the particles ij, we have the wave vectors $\boldsymbol{k}_i'$, $\boldsymbol{k}_j'$ entering on the right and the modified vectors $\boldsymbol{k}_i$, $\boldsymbol{k}_j$ leaving on the left.

Thus, the second-order term in (6.14.27) corresponds to the *cyclic diagram* (6.14.28), where the magnitude of $\boldsymbol{k}'$ is restricted by the Fourier components appearing in $\tilde{\Phi}$, and by the parameter λ. The cycle appearing in (6.14.28) represents a diagonal fragment in that it refers to the transition $\{0\} \to \{0\}$, occasionally termed a *vacuum–vacuum transition*, again from the corresponding quantum field terminology. Conservation of wave vectors is implied.

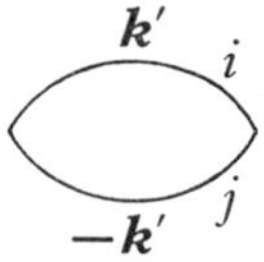

(6.14.28)

We now consider the fourth-order diagrams, (6.14.27), and here two possibilities arise. Either the four vertices are separated in time as two successive cycles, (6.14.29),

(6.14.29)

otherwise they overlap in some way, (6.14.30):

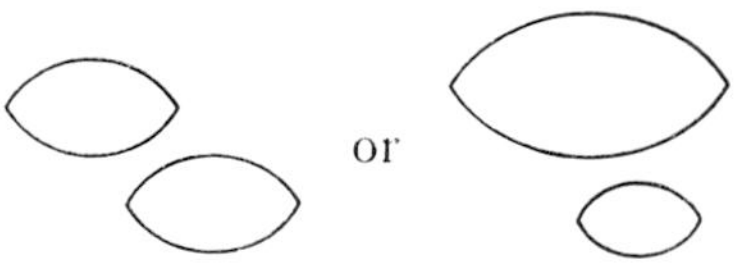

(a) (b)

(6.14.30)

In (6.14.29) we show a succession of two diagonal fragments corresponding to two independent collisions. Each of the diagrams (*a*) and (*b*) in (6.14.30) constitutes a *single* diagonal fragment, since none of the intermediate states $\{\boldsymbol{k}\}$ is identical with the initial state. Prigogine has shown that diagrams of type shown in (6.14.29) give contributions of order $(\lambda^2 t)^2$, whilst those in (6.14.30) give a contribution $\sim \lambda^4 t$: these latter may evidently be neglected in the weak-coupling (small λ) limit. We may now construct the general term in which we are interested as

$$\left(\frac{\lambda}{\mathrm{i}}\right)^{2n} \sum_{\{k\}_1} \dots \sum_{\{k\}_n} \int_0^t \int_0^{t_1} \dots \int_0^{t_{2n-1}} \langle 0 \,|\delta\mathscr{L}|\, \{\boldsymbol{k}\}_1\rangle \langle \{\boldsymbol{k}\}_1 \,|\delta\mathscr{L}|\, 0\rangle$$
$$\times \langle 0 \,|\delta\mathscr{L}|\, \{\boldsymbol{k}\}_2\rangle \langle \{\boldsymbol{k}\}_2 \,|\delta\mathscr{L}|\, 0\rangle \dots \langle 0 \,|\delta\mathscr{L}|\, \{\boldsymbol{k}\}_n\rangle \langle \{\boldsymbol{k}\}_n \,|\delta\mathscr{L}|\, 0\rangle$$
$$\times \exp \Sigma \left\{ \mathrm{i}\frac{\boldsymbol{p}_n}{m}(t_{2n-1}-t_n) \right\} \mathrm{d}t_1 \dots \mathrm{d}t_{2n}. \qquad (6.14.31)$$

After performing the time integrations, Prigogine and Brout show that the series (6.14.27), the general term of which is (6.14.31), may be written finally as

$$\rho_{\{0\}}(t) = \rho_{\{0\}}(0) + \lambda^2 t O_0 \rho_{\{0\}}(0) + \frac{\lambda^4 t^2}{2!} O_0^2 \rho_{\{0\}}(0)$$
$$+ \dots + \frac{(\lambda^2 t)^n}{n!} O_0^n \rho_{\{0\}}(0) + \dots \qquad (6.14.32)$$
$$= \rho_{\{0\}}(0) \exp [\lambda^2 t O_0], \qquad (6.14.33)$$

where O_0 is an operator related to the sum over cyclic elements

$$\langle 0 | \quad \rangle \langle \quad | 0 \rangle$$

having initial and final states zero. Differentiating (6.14.33) with respect to time we obtain the master equation describing the evolution of the amplitude $\rho_{\{0\}}$

$$\partial \rho_{\{0\}}(t)/\partial t = \lambda^2 O_0 \rho_{\{0\}}(t). \qquad (6.14.34)$$

It is immediately evident that (6.14.34) describes a Markovian process: the evolution of the Fourier amplitude $\dot{\rho}_{\{0\}}(t)$ is proportional to its current value and, moreover, is independent of its previous history. Equation (6.14.34) describing the distribution of energy may be easily shown to be entirely equivalent to the Fokker–Planck equation relating to the evolution in phase space. It will be recalled that in our treatment of the Fokker–Planck equation we indicated the existence of transition probabilities $\overset{p+\Delta p}{\underset{p}{W}}$; these are rather directly recovered from (6.14.34), and we direct the interested reader to Prigogine's monograph[107].

What happens if we allow the interactions to become arbitrarily strong, that is remove the restriction on λ, but still suppose that the concentration c if the system is low? For low concentrations the relaxation time will be inversely proportional to c:

$$\tau \sim \frac{1}{c}$$

and we now wish to retain terms of order $\tau c \sim 1$ regardless of the value of λ. In the weak-coupling limit previously discussed we observed that the general non-vanishing matrix element

$$\langle \boldsymbol{k} | \delta\mathscr{L} | \boldsymbol{k} \pm \boldsymbol{m} \rangle$$

arises from the component $\exp(\pm i\boldsymbol{m}\boldsymbol{q})$ of the Fourier-transformed pair interaction $\tilde{\Phi}$. For strong interactions the restriction on $\boldsymbol{m}$ no longer applies and two-body transitions having three vertices (hence of order λ^3) of the type

$$\langle\{0\}|\delta\mathscr{L}|\{\boldsymbol{k}''\}\rangle\langle\{\boldsymbol{k}''\}|\delta\mathscr{L}|\{\boldsymbol{k}'\}\rangle \times \langle\{\boldsymbol{k}'\}|\delta\mathscr{L}|\{0\}\rangle \equiv$$

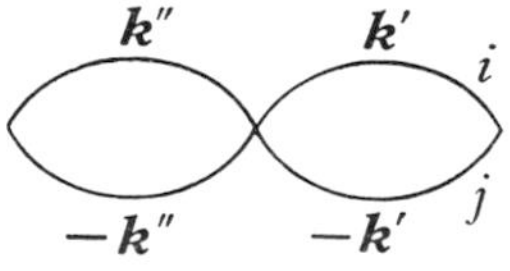

occur. Three-body interactions will also enter to order λ^2 shown in (6.14.35):

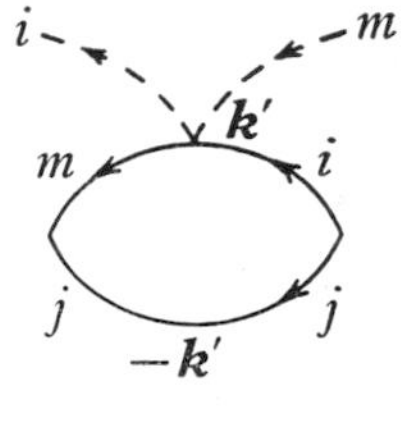

(6.14.35)

Here two interacting particles i and j initially having $\{0\}$ are excited to the states $\boldsymbol{k}'_i - \boldsymbol{k}'_j$. Before the $\{0\} \rightarrow \{0\}$ transition is completed, however, a third particle m having wave vector zero interacts with i as indicated. m adopts the wave vector $\boldsymbol{k}'_p$ whilst i adopts the wave vector zero. The wave vectors $\boldsymbol{k}'_p$ and $\boldsymbol{k}'_j$ then recombine to yield the final diagonal state. This amounts to the transition $\{0\} \rightarrow \{0\}$ effected in the field of a third particle. The three-body diagram of the type shown in (6.14.36) is generally used instead of that of (6.14.35):

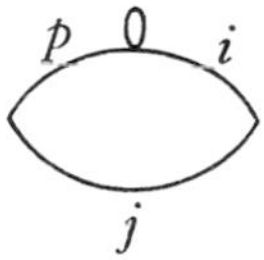

(6.14.36)

As we might anticipate, at low densities the development of many-body diagrams is negligible in comparison with the two-body diagrams discussed earlier. We therefore neglect the three- and four-body diagrams of order λ^4 shown in (6.14.37).

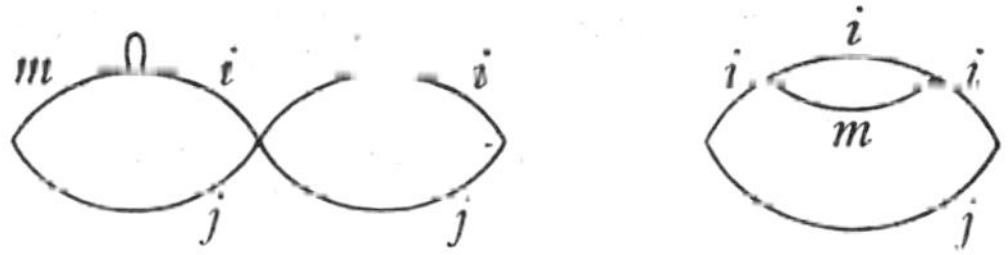

Three-body diagrams of order λ^4

(6.14.37)

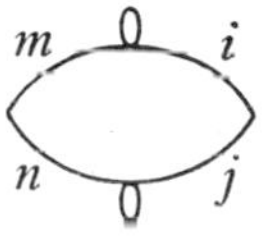

Four-body diagram of order λ^4

We have already discussed two-body diagrams of order λ^4

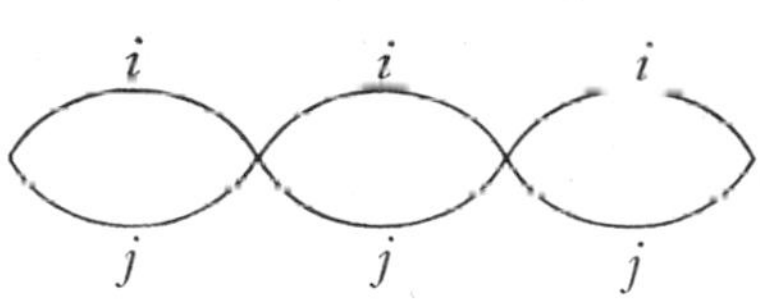

as we have the non-overlapping *independent* two-body interactions:

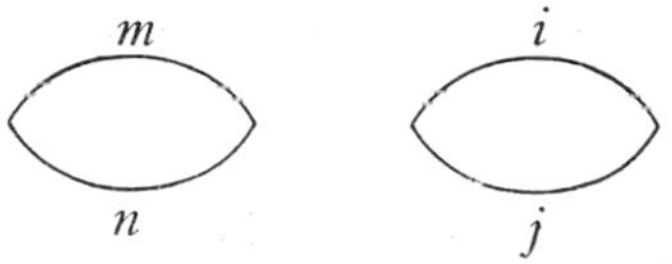

these latter are classed as many-body diagrams and do not contribute significantly to the evolution of dilute systems. Similarly, the four-vertex single diagonal fragment corresponding to overlapping diagrams

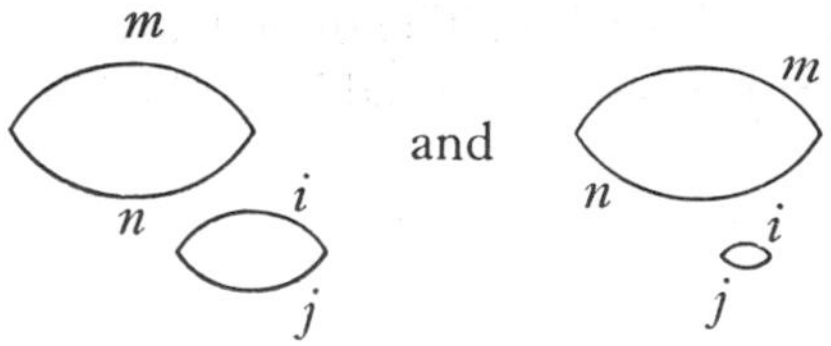

entering to order tc^2 is dropped in the low density limit. We therefore conclude that in the limit of low concentrations the only diagrams retained are the two-body diagrams and disconnected products of such diagrams. Thus (6.14.29) may be supplemented by terms or order λ^3 to give

$$\rho_{\{0\}}(t) = \rho_{\{0\}}(0) + t\{\lambda^2 \bigcirc + \lambda^3 \bigcirc\!\bigcirc + \lambda^4 \bigcirc\!\bigcirc\!\bigcirc + \dots\}\rho_{\{0\}}(0)$$
$$+\frac{t^2}{2!}\{\ \}\{\ \}\rho_{\{0\}}(0) + \frac{t^3}{3!}\{\ \}\{\ \}\{\ \}\rho_{\{0\}}(0) + \dots . \quad (6.14.38)$$

The time derivative of (6.14.38) yields the master equation in the low density strong-coupling limit, from which the classical Boltzmann equation may be deduced. We observe that only a single class of diagrams corresponding to repeated two-body interaction of the same pair of particles need be retained.

We now come to consider explicitly the non-diagonal fragments $\langle\{\boldsymbol{k}\}|\delta\mathscr{L}|\{\boldsymbol{k}'\}\rangle$ which clearly contain creation or destruction processes since $\{\boldsymbol{k}'\} \neq \{\boldsymbol{k}\}$. Again we use a diagram technique to represent the various interactions in which we explicitly depict only the non-zero wave vectors. In constructing a diagram account is taken of the conservation law for wave vectors

$$\boldsymbol{k}_i' + \boldsymbol{k}_j' = \boldsymbol{k}_i + \boldsymbol{k}_j$$

which arises from the fact that the potential energy $\Phi(ij)$ depends only upon the separation $|\boldsymbol{q}_i - \boldsymbol{q}_j|$, and is therefore invariant with respect to translation, since

$$\Phi(|\boldsymbol{q}_i - \boldsymbol{q}_j|) = \Phi(|(\boldsymbol{q}_i + \boldsymbol{a}) - (\boldsymbol{q}_j + \boldsymbol{a})|).$$

This condition ensures the conservation of momentum within the system.

The six types of interaction which we need to consider in the discussion of non-diagonal transitions are listed below.

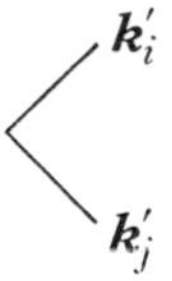

(*a*)

$\boldsymbol{k}_i', \boldsymbol{k}_j' \neq 0, \quad \boldsymbol{k}_i, \boldsymbol{k}_j = 0$

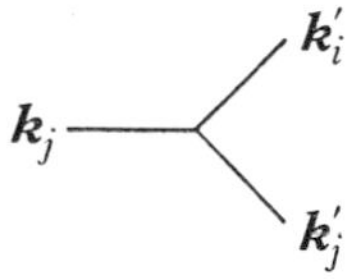

(*b*)

$\boldsymbol{k}_i', \boldsymbol{k}_j' \neq 0, \quad \boldsymbol{k}_j \neq 0, \quad \boldsymbol{k}_i = 0$

The two diagrams (*a*) and (*b*) correspond to the *destruction of correlations*, the destruction process occurring at the vertex. In a destruction diagram the number of wave vectors decreases from right to left. Conversely we may consider the *creation* diagrams (*c*) and (*d*):

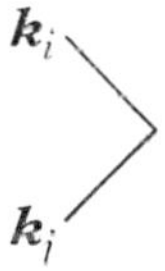

(*c*)

$\boldsymbol{k}_i', \boldsymbol{k}_j' = 0, \quad \boldsymbol{k}_i, \boldsymbol{k}_j \neq 0$

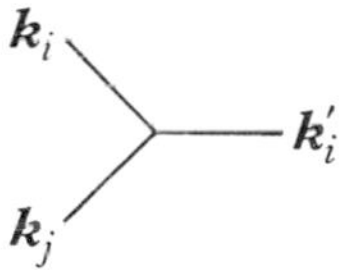

(*d*)

$\boldsymbol{k}_i' \neq 0, \quad \boldsymbol{k}_j' = 0, \quad \boldsymbol{k}_i, \boldsymbol{k}_j \neq 0$

Finally, we have to consider diagrams corresponding to the *propagation of correlations*, (*e*) and (*f*):

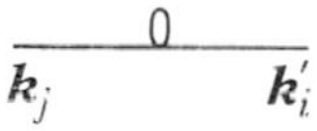

(*e*)

$\boldsymbol{k}_i' \neq 0, \quad \boldsymbol{k}_j' = 0, \quad \boldsymbol{k}_i = 0, \quad \boldsymbol{k}_j \neq 0$

(*f*)

$\boldsymbol{k}_i', \boldsymbol{k}_j' \neq 0, \quad \boldsymbol{k}_i, \boldsymbol{k}_j \neq 0$

where for the process (*e*) we mean, as before

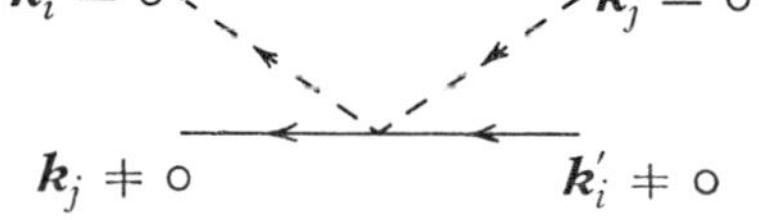

In these diagrams the number of lines is not modified by the vertex. For weakly interacting systems diagrams (c) and (a) combine to form the diagonal element $\bigcirc$ representing a transient excitation. We understand non-diagonal transitions to arise from the inelastic scattering of wave vectors in the field of the particle interaction. To a Fourier component of the transformed potential $\tilde{\Phi}$ proportional to exp $i(\boldsymbol{m}.|\boldsymbol{q}_i-\boldsymbol{q}_j|)$ there corresponds a transition element

$$\langle \boldsymbol{k}\,|\delta\mathscr{L}|\,\boldsymbol{k}\pm\boldsymbol{m}\rangle.$$

We shall now implement these diagrams in a graphical discussion of the processes contributing to the evolution of a given Fourier coefficient. In this diagrammatic representation of the Liouville equation we shall be concerned with a certain number of 'fixed' particles which we shall designate by the indices 1, 2, 3, ... whilst for the itinerant particles which become involved in the interaction we shall use $i, j, k, \ldots$. Summation has to be performed over these particles, of course. We do not show permutation over indices in these diagrams. It is important to observe that conservation of wave vectors is preserved in all these diagrams, (6.14.39), the sum of wave vectors vanishing for a homogeneous system. Of course, the primary

$$\frac{\partial \rho_0^{(1)}}{\partial t} = \Big\langle\ \rho_{\boldsymbol{k},-\boldsymbol{k}}^{ij} \tag{6.14.39a}$$

$$\frac{\partial \rho_{\boldsymbol{k},-\boldsymbol{k}}^{(12)}}{\partial t} = \Big\rangle \rho_0 + \times \rho_{\boldsymbol{k},-\boldsymbol{k}}^{12}$$

$$+ \ \underline{\quad}\ \rho_{\boldsymbol{k},-\boldsymbol{k}}^{i2} + \ \prec\ \rho_{\boldsymbol{k}+\boldsymbol{k}',\,\boldsymbol{k},\,-\boldsymbol{k}'}^{12i} \tag{6.14.39b}$$

$$+ \ \Big\langle\ \rho_{\boldsymbol{k},-\boldsymbol{k},\,\boldsymbol{k}',\,-\boldsymbol{k}'}^{12ij}$$

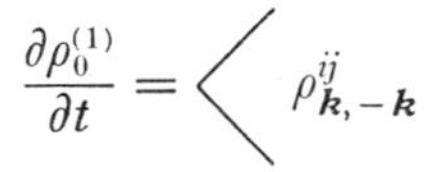

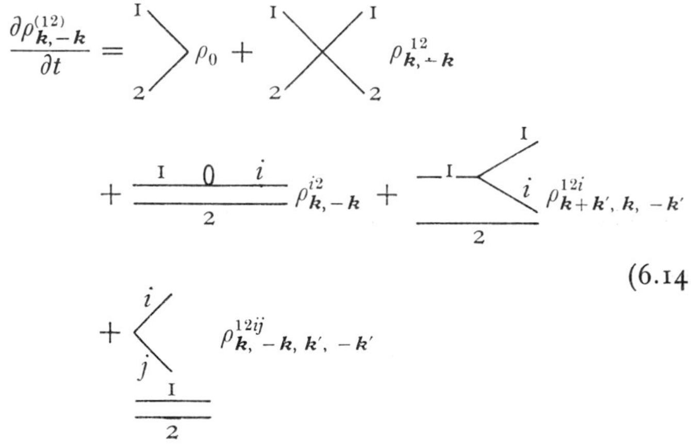

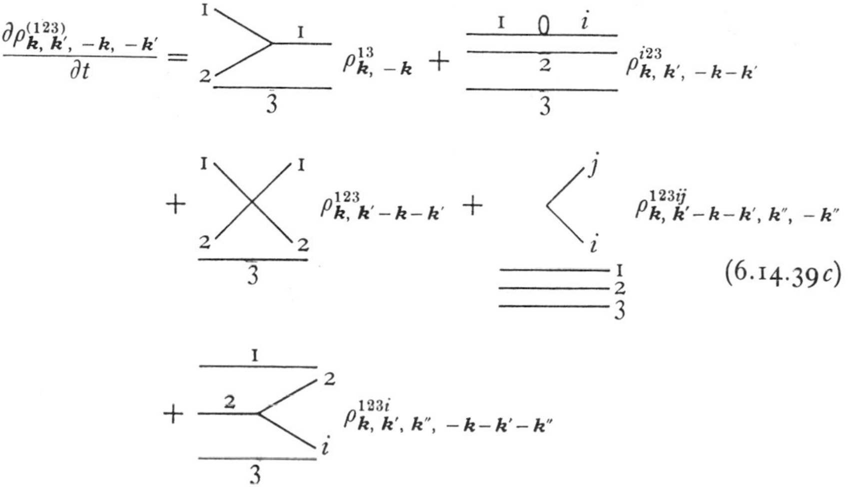

$$\frac{\partial \rho_{k,\,k',\,-k,\,-k'}^{(123)}}{\partial t} = \rho_{k,\,-k}^{13} + \rho_{k,\,k',\,-k-k'}^{i23}$$

$$+ \rho_{k,\,k'-k-k'}^{123} + \rho_{k,\,k'-k-k',\,k'',\,-k''}^{123ij} \qquad (6.14.39c)$$

$$+ \rho_{k,\,k',\,k'',\,-k-k'-k''}^{123i}$$

objective of Prigogine's development is to provide an understanding of the nature of irreversibility. The infinite hierarchy of short-time equations (6.14.39) arising from the Fourier analysis of the Liouville equation are reversible in time. Whereupon the flow of coupling may be displayed in reversible form:

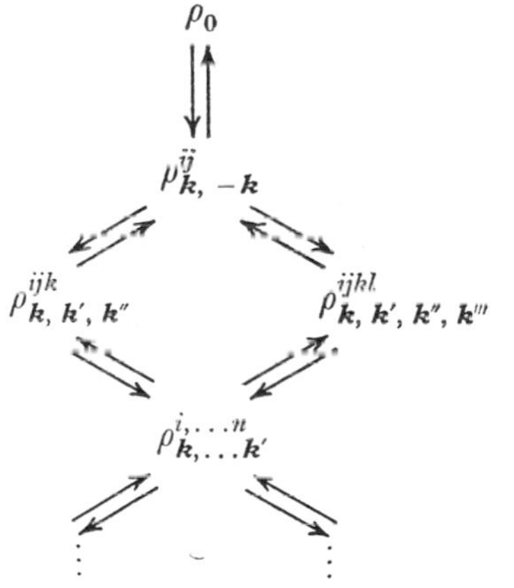

However, in the limit $t \to \infty$, $N \to \infty$, $V \to \infty$, $N/V =$ constant, only the dominant contributions are retained, and the set of equations become partially decoupled. Thus, for example, in the limit of long times $\rho^{ij}_{k,-k}$ is determined only by ρ_0. The asymptotic behaviour then adopts its irreversible form:

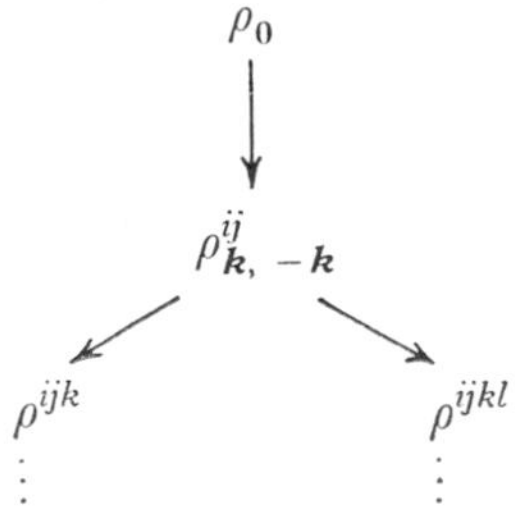

from which we see that the symmetry in time is destroyed[128], thereby defining a direction in time. Instead of the reversible kinetics arising at short times we now have a temporally directed, irreversible cascade of correlations. This macroscopic move towards equilibrium is seen as a spread of correlation throughout the molecules of the system, rather in the form of a diffusion process already familiar in the time perturbation theory of quantum mechanics.

At liquid densities it has been apparent from much of what we have discussed in previous sections that the evolution of a phase variable shows a short-time non-Markovian behaviour, whilst in the limit of long times the evolution tends asymptotically to that of a Markovian, characteristic of a stochastic variable. This arises naturally in our discussion of the memory function $K(t)$ in §6.13: the finite 'memory' of the itinerant particle renders its subsequent evolution independent of its initial dynamical history – a delayed stochastic development. Moreover, the asymptotic form of the velocity autocorrelation is that characteristic of a Markovian, whilst at short times we observe strongly correlated motions of a decidedly non-Markovian nature. This is substantiated by a departure at intermediate frequencies of the spectral density $\tilde{\psi}(\omega)$ (fig. 6.13.2) from the Lorentzian distribution, (6.13.23). Again, as Rahman has explicitly demonstrated[81] (§5.9), at short times there is a pronounced departure from a Markovian evolution which is, nevertheless, regained asymptotically. Rahman investigates the moments of the mean square atomic displacement in liquid argon at 94.4 °K,

1.374 g cm^{-1}, and concludes that the Gaussian form is recovered only after $\sim 10^{-11}$ s. It is therefore important in the development of a general kinetic theory to establish an evolution of the phase density at an arbitrary time, but one which, nevertheless, reduces to Markovian form in the limit of long times. We have already obtained this limiting form for a weakly coupled dilute system of particles, (6.14.31), by a rather qualitative consideration of the cyclic class of diagrams arising in the binary encounters.

The present development, very much that of Résibois[112], develops the formal solution (6.14.8) in resolvent operator form:

$$f_{(N)}(t) = -\frac{1}{2\pi i}\int_{\infty+ic}^{-\infty+ic} (\mathscr{L}-z)^{-1}\exp(-izt)f_{(N)}(0)\,dz.$$

Insertion of (6.14.14) in the above expression yields the spectral amplitude

$$\rho_{\{k\}}(\boldsymbol{p}_N, t) = -\frac{1}{2\pi i}\oint_c \left(\frac{8\pi^3}{V}\right)\exp(-izt)$$
$$\times \sum_{\{k'\}}\langle\{\boldsymbol{k}\}|(\mathscr{L}-z)^{-1}|\{\boldsymbol{k}'\}\rangle\,\rho_{\{k'\}}(\boldsymbol{p}_N, 0)\,dz. \quad (6.14.40)$$

In (6.14.40) there appears the resolvent operator $(\mathscr{L}-z)^{-1}$ corresponding to the Liouville operator $\mathscr{L}$. With $\mathscr{L}$ separated as before

$$\mathscr{L} = \mathscr{L}_0 + \delta\mathscr{L}$$

it follows that

$$(\mathscr{L}-z)^{-1}-(\mathscr{L}_0-z)^{-1} = (\mathscr{L}_0-z)^{-1}\{(\mathscr{L}_0-z)-(\mathscr{L}-z)\}(\mathscr{L}-z)^{-1},$$

i.e. $$(\mathscr{L}-z)^{-1} = (\mathscr{L}_0-z)^{-1}(\mathscr{L}_0-z)^{-1}(\lambda\delta\mathscr{L})(\mathscr{L}-z)^{-1}. \quad (6.14.41)$$

This equation, whilst exact, cannot be rearranged to provide an expression for $(\mathscr{L}-z)^{-1}$ without introducing approximations. If we assume λ to be small, an iterative scheme may be implemented. As a first approximation we set $\lambda = 0$, yielding

$$(\mathscr{L}-z)^{-1} = (\mathscr{L}_0-z)^{-1}$$

which, when iteratively substituted in (6.14.41), leads to the result

$$(\mathscr{L}-z)^{-1} = \sum_{n=0}^{\infty}(-\lambda)^n(\mathscr{L}_0-z)^{-1}\{\delta\mathscr{L}(\mathscr{L}_0-z)^{-1}\}^n$$

whereupon (6.14.40) becomes

$$\rho_{\{k\}}(\boldsymbol{p}_N, t) = -\frac{1}{2\pi i}\oint_c \left(\frac{8\pi^3}{V}\right) \exp(-izt)$$
$$\times \sum_{\{k'\}} \langle\{\boldsymbol{k}\}\,|(\mathscr{L}_0 - z)^{-1} \sum_{n=0}^{\infty} (-\lambda)^n \{\delta\mathscr{L}(\mathscr{L}_0 - z)^{-1}\}^n\,|\{\boldsymbol{k}'\}\rangle$$
$$\times \rho_{\{k'\}}(\boldsymbol{p}_N, 0)\,dz. \quad (6.14.42)$$

Equation (6.14.42) is a purely formal (and therefore exact) result which, in principle, permits the calculation of $f_{(N)}(t)$ if $f_{(N)}(0)$ is known. The formula is convenient because $(\mathscr{L}_0 - z)^{-1}$ is simple in the Fourier representation,

$$\langle\{\boldsymbol{k}\}\,|(\mathscr{L}_0 - z)^{-1}|\,\{\boldsymbol{k}'\}\rangle = \left[\sum_j \left(\boldsymbol{k}_j . \frac{\boldsymbol{p}_j}{m} - z\right)\right]^{-1} \delta^{\mathrm{Kr}}_{\{k\}\{k'\}}, \quad (6.14.43)$$

i.e. the unperturbed resolvent leads only to diagonal terms. This again, as in the case of the orthogonalized eigenfunctions of the unperturbed Liouville operator $\mathscr{L}_0$, effects an analytic separation of the non-interacting 'ground state' from the interacting 'excitations', and herein lies the merit of the approach.

To study the analytic behaviour of the resolvent let us first consider the diagonal transition $\{0\} \to \{0\}$. From (6.14.42) and (6.14.43) we may conveniently write

$$\rho_0(\boldsymbol{p}_N, t) = -\frac{1}{2\pi i}\oint_c \left(\frac{8\pi^3}{V}\right) \exp(-izt) \sum_{n=0}^{\infty}$$
$$\times \left\{-\frac{1}{z}\left[\Psi_{00}(z)\left(-\frac{1}{z}\right)\right]^n\right\} \rho_0(0)\,dz, \quad (6.14.44)$$

where we define the operator

$$\Psi_{\{k\}\{k\}}(z) = \sum_{n=0}^{\infty} \left\langle \{\boldsymbol{k}\} \middle| \delta\mathscr{L}\left[\frac{\delta\mathscr{L}}{(\mathscr{L}_0 - z)}\right]^n \middle| \{\boldsymbol{k}\} \right\rangle. \quad (6.14.45)$$

The first term corresponds to the unperturbed motion, and the second to the effect of cyclic transitions discussed earlier. The sum extends to include all possible diagonal fragments $\{\boldsymbol{k}\} \to \{\boldsymbol{k}\}$. Similarly we define

$$\mathscr{C}_{\{k\}\{k'\}}(z) = \sum_{n=1}^{\infty} (-\lambda)^n \left\langle \{\boldsymbol{k}\} \middle| \frac{1}{(\mathscr{L}_0 - z)} \delta\mathscr{L} \middle| \{\boldsymbol{k}'\} \right\rangle \quad (6.14.46)$$

which is the sum of all possible creation fragments $\{\boldsymbol{k}'\} \to \{\boldsymbol{k}\}$, and

$$\mathscr{D}_{\{k'\}\{k''\}}(z) = \sum_{n=1}^{\infty} (-\lambda)^n \left\langle \{\boldsymbol{k}'\} \left| \delta\mathscr{L} \frac{1}{(\mathscr{L}_0 - z)} \right| \{\boldsymbol{k}''\} \right\rangle \quad (6.14.47)$$

which is the sum of all possible destruction fragments subject to the restriction of course, that the number of non-zero elements $\{\boldsymbol{k}'\} < \{\boldsymbol{k}''\}$. Ψ, $\mathscr{C}$ and $\mathscr{D}$ depend essentially upon the form of the pair interaction through $\delta\mathscr{L}$.

We may now consider the time evolution of the velocity distribution $\rho_0(t)$[113]. The most general contribution to $\rho_0(t)$ must consist of some destruction region followed by an arbitrary number of diagonal fragments, for clearly there can be no creation processes terminating in $\{\boldsymbol{k}\} = 0$. Therefore the initial density of states is

$$\rho_0(0) + \sum_{\{k''\} \neq 0} \mathscr{D}_{0\{k''\}}(z)\, \rho_{\{k''\}}(0),$$

i.e. the initial amplitude $\rho_0(0)$ supplemented by destruction elements terminating in $\{\boldsymbol{k}\} = 0$ at $t = 0$. Then the velocity distribution is given as:

$$\rho_0(t) = -\frac{1}{2\pi i} \oint_c \left(\frac{8\pi^3}{V}\right) e^{-izt} \sum_{n=0}^{\infty} \left\{ -\frac{1}{z} \left[\Psi_{00}(z) \left(-\frac{1}{z} \right)^n \right] \right\}$$
$$\times \left[\rho_0(0) + \sum_{\{k''\} \neq 0} \mathscr{D}_{0\{k''\}}(z)\, \rho_{\{k''\}}(0)\right] dz \quad (6.14.48)$$

which can be taken to represent the Laplace transform of $\rho_0(t)$. Consider now the operator

$$D_{00}(z) = \sum_{n=0}^{\infty} \left(-\frac{1}{z}\right) \left\{ \Psi_{00}(z) \left(-\frac{1}{z}\right) \right\}^n \quad (6.14.49)$$

which has been extracted from (6.14.48). $D_{00}(z)$ may be manipulated into the form

$$D_{00}(z) = -\frac{1}{z} - \frac{1}{z} \Psi_{00}(z)\, D_{00}(z), \quad (6.14.50)$$

whereupon substitution of (6.14.50) and (6.14.49) into (6.14.48), and subsequent differentiation with respect to time, yields

$$\frac{\partial \rho_0}{\partial t} = -\frac{1}{2\pi} \oint_c \left(\frac{8\pi^3}{V}\right) e^{-izt} \sum_{\{k''\} \neq 0} \mathscr{D}_{0\{k''\}}(z)\, \rho_{\{k''\}}(0)$$
$$-\frac{1}{2\pi} \oint_c \left(\frac{8\pi^3}{V}\right) e^{-izt} \Psi_{00}(z)\, D_{00}(z)$$
$$\times \left[\rho_0(0) + \sum_{\{k''\} \neq 0} \mathscr{D}_{0\{k''\}}(z)\, \rho_{\{k''\}}(0)\right] dz. \quad (6.14.51)$$

Finally, the use of the two Laplace transforms

$$\mathfrak{D}_0(t, \rho_{\{k''\}}(0)) = -\frac{1}{2\pi}\oint_c \left(\frac{8\pi^3}{V}\right) e^{-izt} \sum_{\{k''\}\neq 0} \mathscr{D}_{0\{k''\}}(z)\, \rho_{\{k''\}}(0)\, dz,$$

$$G_{00}(\tau) = \frac{1}{2\pi i}\oint_c \left(\frac{8\pi^3}{V}\right) e^{-iz\tau}\, \Psi_{00}(z)\, dz$$

and the convolution theorem on the right hand side of (6.14.51) yields the final master equation[116]

$$\frac{\partial \rho_0}{\partial t} = \mathfrak{D}_0(t, \rho_{\{k''\}}(0)) + \int_0^t G_{00}(t-t')\,\rho_0(t')\, dt'. \qquad (6.14.52)$$

This is the general equation for the evolution of the velocity distribution function, and is remarkable in that it is essentially non-Markovian in character. This may be directly ascribed to the finite duration, t_c, of the collision process: $\partial \rho_0/\partial t$ at time t is seen to depend upon collisions which started at some previous time $t - t_c$, and were completed at t. For this reason an 'integration over the past' is involved, and the velocity distribution evidently has a memory $\sim t_c$. For times $\gg t_c$, the master equation (6.14.52) reduces to its Markovian counterpart, (6.14.34), as we shall see.

The inhomogeneous term, $\mathfrak{D}_0$ gives the contribution at time t of the influence of the preceding destruction region, and contains the effect of correlations which existed at the initial time $t = 0$ on the velocity distribution. Note that information on the initial correlations and conditions are contained within $\mathfrak{D}_0$, whilst G_{00} refers only to the diagonal fragments. The general nondiagonal evolution $\partial \rho_{\{k\}}/\partial t$ is discussed in Prigogine's monograph[107].

As we mentioned above, passage to the limit $t \to \infty$ reduces (6.14.52) to Markovian form: the destruction effects contained in $\mathfrak{D}_0$ are rejected to the infinite past, whilst $\mathfrak{D}_0$ tends to zero with increasing t since correlations of finite extent in the intial state can only interact for a finite time. Moreover, $(t_c/t) \to 0$, and the interaction time has no eventual bearing upon the asymptotic evolution. For large t the quasi-stationary velocity distribution will vary negligibly during the interval t_c and the operator

$$\operatorname*{Lt}_{t \gg t_c} \int_0^\infty G_{00}(t-t')\, dt \to \mathscr{G}_{00}$$

(note the upper limit) where $\mathscr{G}_{00}$ is a constant, whereupon we regain the Markovian form:

$$\frac{\partial \rho_0(t)}{\partial t} = \mathscr{G}_{00}\rho_0(t).$$

Comparison with (6.14.34) reaffirms the essentially diagonal nature of $\mathscr{G}_{00}$. The non-Markovian kernel of (6.14.52) serves to connect the distribution function over a period of the order of t_c, and as such effects a kind of time-smoothing over t_c. The importance of such an event depends critically upon the variation of ρ_0 over such time intervals, and herein lies the essence of Kirkwood's coarse graining hypothesis, and the attainment of irreversibility in terms of the statistical independence of the previous dynamical history. Notice, however, that Kirkwood's time-smoothing procedure, (6.4.2), takes a simple time average, whilst here an average weighted by the function G_{00} is taken. The important point about the kinetic equation (6.14.52) is that it is exact; moreover, it is strictly reversible, even though it exhibits an asymptotic approach to equilibrium as $t \to \infty$[114]. The behaviour of the destruction fragment $\mathfrak{D}_0$ is crucial in this. As we have seen, it has little effect upon the long time evolution of $\rho_0(t)$ – indeed, $\mathfrak{D}_0$ decays to zero – but its inclusion ensures that the initial distribution is properly recovered on reversing the momenta. Without this term the initial distribution could never be regained, and Kirkwood's hypothesis effectively sets this operator to be zero, reducing the evolution to a Markov process. We must therefore conclude that coarse graining has the effect of eliminating all information concerning the initial correlations[118].

We may finally conclude, then, that the master equation is essentially non-Markovian at arbitrary time, but for times long in comparison to the time of an encounter (and this will vary from system to system) the equation tends to a Markovian. Moreover, in this quasi-stationary limit, only the asymptotic form of the diagonal fragment enters the collision operator. In discussing 'collisions' and 'encounters' there is the tacit assumption that we are dealing with a dilute system in the weak-coupling limit. The difficulty is more than one of semantics: van Hove[115] had derived a form of the master equation of non-Markovian nature valid to all orders of the coupling factor. It is, however, derived for the quantum mechanical case and does not have an obvious extension to the classical case.

Very few computed results have as yet been derived from this aspect of the theory. Apart from the deduction of the Boltzmann equation for the description of a dilute gas and the Fokker–Planck equation for the study of crystal dynamics[(107)] little else other than that of a qualitative nature has been achieved. Allen and Cole[(120)] have quoted values of the thermal conductivity of liquid argon which are in an order of magnitude of agreement with experiment, but such work is at this stage no more than exploratory, although Balescu[(110)], and Résibois[(121)] have established a connection between the Prigogine and Kubo formalisms.

6.15 The projection of operator $\mathscr{P}$

An alternative and equivalent approach to the non-Markovian master equation (6.14.52) is provided with great economy by the time-independent linear operator formalism developed by Zwanzig[(119)]. The approach consists essentially in the application of a projection operator $\mathscr{P}$ which serves to divide the full phase distribution $f_{(N)}(\mathbf{p}_N, \mathbf{q}_N, t)$ into two components $f_a(t)$ and $f_b(t)$, the first of which is directly relevant to the non-uniformity, the second of which is not. An evolution equation involving f_a explicitly and independent of f_b is then obtained, f_b entering only to specify the initial conditions.

As we mentioned above, $\mathscr{P}$ is a linear, time-independent operator, and without specifying it further at the moment we may write

$$\left.\begin{aligned} f_a(t) &= \mathscr{P} f_{(N)}(t), \\ f_b(t) &= (1-\mathscr{P}) f_{(N)}(t) = f_{(N)} - f_a(t). \end{aligned}\right\} \qquad (6.15.1)$$

From the operator form of the Liouville equation

$$\frac{\mathrm{i}\partial f_{(N)}}{\partial t} = \mathscr{L} f_{(N)}$$

we may immediately obtain the following equations for $f_a(t)$ and $f_b(t)$ in terms of the complete Liouville operator:

$$\mathscr{P}\mathrm{i}\frac{\partial f_{(N)}}{\partial t} = \mathrm{i}\frac{\partial}{\partial t}(\mathscr{P}f_{(N)}) = \mathrm{i}\frac{\partial f_a}{\partial t} = \mathscr{P}\mathscr{L}(f_a+f_b), \qquad (6.15.2)$$

$$(1-\mathscr{P})\mathrm{i}\frac{\partial f_{(N)}}{\partial t} = \mathrm{i}\frac{\partial f_b}{\partial t} = (1-\mathscr{P})\mathscr{L}(f_a+f_b). \qquad (6.15.3)$$

Equation (6.14.3) is easily integrated with respect to time to yield

$$f_{(b)}(t) = [\exp\{\mathrm{i}t(1-\mathscr{P})\mathscr{L}\}]f_b(0)$$
$$-\mathrm{i}\int_0^t [\exp\{-\mathrm{i}s(1-\mathscr{P})\mathscr{L}\}](1-\mathscr{P})\mathscr{L}f_a(t+s)\,\mathrm{d}s \qquad (6.15.4)$$

whereupon if we substitute (6.15.4) into (6.15.3) for f_b we have an explicit equation for the evolution of the 'relevant' function $f_a(t)$:

$$\frac{\mathrm{i}\,\partial f_a(t)}{\partial t} = \mathscr{P}\mathscr{L}[\exp\{-\mathrm{i}t(1-\mathscr{P})\mathscr{L}\}]f_b(0) + \mathscr{P}\mathscr{L}f_a(t)$$
$$-\mathrm{i}\int_0^t \mathscr{P}\mathscr{L}[\exp\{-\mathrm{i}s(1-\mathscr{P})\mathscr{L}\}](1-\mathscr{P})\mathscr{L}f_a(t-s)\,\mathrm{d}s. \qquad (6.15.5)$$

As we see, f_b enters only to specify the initial conditions. Moreover, comparison of (6.15.4) with (6.14.4) enables us to identify $\exp\{-\mathrm{i}t(1-\mathscr{P})\mathscr{L}\}$ as a *modified propagator*.

The first term on the right of (6.15.5) is the only one containing information about the initial distribution, and is equivalent to the destruction fragment of Prigogine and Résibois[116], (6.14.52) representing as it does the effect of the initial spatial correlations on the evolution of the momentum distribution. The second term has the form of a streaming in phase space, whilst the third term is a non-Markovian collision operator and introduces a memory effect. Equation (6.15.5) is, of course, no more than a formal rearrangement of the propagator and is completely reversible in the sense that $t = -t$, $\boldsymbol{p}_i = -\boldsymbol{p}_i$ $(i = 1 \ldots N)$ and $f_a(-t)$ satisfy it also. We see that (6.15.5) retains a knowledge of the initial distribution at all times, and reversal of the particle momenta at time t will presumably enable the system to regain its initial distribution at $2t$. Nevertheless, in the context of other formalisms, it has been shown that (6.15.5) does tend asymptotically to a stationary state. The exact equation (6.15.5), first given by Zwanzig, may be conveniently written

$$\frac{\partial f_a}{\partial t} = \boldsymbol{D}_1(t) f_b(0) + \boldsymbol{D}_2(t) f_a(t) + \int_0^t \boldsymbol{K}(s) f_a(t-s)\,\mathrm{d}s, \quad (6.15.6)$$

where the operators $\boldsymbol{D}_1$, $\boldsymbol{D}_2$ and $\boldsymbol{K}$ may be determined from (6.15.5) by inspection. In the limit of t being large in comparison with the interaction time, the memory effect disappears and (6.15.6) reduces to a Markov equation in which the rate of change of f_a depends solely upon the current state of the system.

It remains to define the projection operator $\mathscr{P}$, and this varies according to the specific physical application. For example, setting

$$\mathscr{P} = \frac{1}{V^N}\int \mathrm{d}\boldsymbol{q}_N \quad (6.15.7)$$

we obtain directly from (6.15.1)

$$f_a(\boldsymbol{p}_N, \boldsymbol{q}_N, t) = \frac{1}{V^N}\int f_{(N)}(\boldsymbol{p}_N, \boldsymbol{q}_N, t)\,\mathrm{d}\boldsymbol{q}_N = f_a(\boldsymbol{p}_N, t) \quad (6.15.8)$$

which is the momentum distribution function. Moreover, comparison of (6.15.8) and (6.14.14) with $\boldsymbol{k} = 0$ shows that $f_a(\boldsymbol{p}_N, t)$ differs from $\rho_{\{0\}}$ only by a constant factor of normalization. Introduction of the kinetic and interaction contributions to the Liouville operator in the form $\mathscr{L} = \mathscr{L}_0 + \lambda\delta\mathscr{L}$ transforms (6.15.6) to give

$$\frac{\partial f_a}{\partial t}(\boldsymbol{p}_N, t) = \lambda^2 \int_0^\infty \boldsymbol{K}(s) f_a(\boldsymbol{p}_N, s)\,\mathrm{d}s \quad (6.15.9)$$

with the kernel $\quad \boldsymbol{K}(s) = -\mathscr{P}\delta\mathscr{L}\exp(-\mathrm{i}s\mathscr{L}_0)\delta\mathscr{L} \quad (6.15.10)$

in the limit $t \to \infty$. It has been assumed here that since the fluid is initially homogeneous, $f_a(0) = f_b(0) = 0$. Equation (6.15.9) with (6.15.10) is, of course, precisely the master equation (6.14.31) derived by Brout and Prigogine[(117)].

Other applications of the projection operator formalism have been given. Gray[(122)], for example, by suitable choice of the operator $\mathscr{P}$ and an assumption concerning the initial-value conditions has obtained generalizations of the Fokker–Planck equations in which the dissipative terms are non-Markovian. It is further shown that exact equations for the van Hove self- and distinct-correlation function are particular cases of these equations. The skill, of course, lies in making the 'appropriate' choice of operator. Application of this technique has been considered by several authors. Lebowitz and Résibois[(123)], for example, have used this method to derive a kinetic equation for a massive particle in a host fluid of light particles. Lebowitz *et al.*[(124)] have also obtained equations for the self- and distinct-correlation functions, and Muriel and Dresden[(125)] have derived explicit forms of kinetic equations for the limit of a weakly coupled system.

REFERENCES

Chapter 1

(1) J. E. Mayer, *Handbuch der Physik* (ed. S. Flügge), vol. 12, chapter 2, Springer, Berlin (1958).
J. E. Mayer, *Equilibrium Statistical Mechanics*, chapter 4, Pergamon (1968).
J. E. Mayer, *J. Chem. Phys.* **5**, 67 (1937).
(2) B. Kahn and G. E. Uhlenbeck, *Physica*, **5**, 399 (1938).
(3) J. de Boer, 'Contribution to the Study of Compressed Gases', thesis, University of Amsterdam (1940).
(4) G. E. Uhlenbeck and G. W. Ford, *Studies in Statistical Mechanics*, vol. 1, p. 123, North-Holland, Amsterdam (1962).
(5) H. D. Ursell, *Proc. Cambridge Phil. Soc.* **23**, 685 (1927).
(6) F. H. Ree and W. G. Hoover, *J. Chem. Phys.* **40**, 939 (1964).
(7) J. de Boer, *Rep. Prog. Phys.* **12**, 305 (1948–9).
G. H. A. Cole, *Rep. Prog. Phys.* **19**, 1 (1956).
J. S. Rowlinson, *Rep. Prog. Phys.* **28**, 169 (1965).
T. L. Hill, *Statistical Mechanics*, chapter 4, McGraw-Hill, New York (1956).
H. L. Frisch and J. L. Lebowitz (eds.), *The Equilibrium Theory of Classical Fluids*, Benjamin, New York (1964).
E. E. Salpeter, *Annals Phys.* **5**, 183 (1958).
(8) J. Groeneveld, *Physics Letters*, **3**, 50 (1962).
O. Penrose, *J. Math. Phys.* **4**, 1312 (1963).
S. Baer and J. L. Lebowitz, *J. Chem. Phys.* **26**, 2558 (1962).
J. L. Lebowitz and O. Penrose, *J. Math. Phys.* **5**, 841 (1964).
D. Ruelle, *Rev. Mod. Phys.* **36**, 580 (1964).
(9) J. G. Kirkwood, *Phys. Rev.* **44**, 31 (1933).
(10) C. N. Yang and T. D. Lee, *Phys. Rev.* **87**, 404 (1952).
(11) K. Huang, *Statistical Mechanics*, p. 319, Wiley (1963)
(12) T. D. Lee and C. N. Yang, *Phys. Rev.* **87**, 410 (1952).
(13) J. E. Mayer, *Equilibrium Statistical Mechanics*, p. 85, Pergamon (1968).
J. E. Mayer and M. G. Mayer, *Statistical Mechanics*, chapter 13, Wiley (1940).
(14) S. Katsura and M. Fujita, *J. Chem. Phys.* **19**, 795 (1951).
(15) L. Boltzmann, *Verslag. Gewone Vergader.* **7**, 484 (1899).
H. Happel, *Ann. Physik* **21**, 342 (1906).
(16) B. R. A. Nijboer and L. van Hove, *Phys. Rev.* **85**, 777 (1952).
(17) Also obtained by a mixture of analytical, numerical and Monte Carlo techniques by S. Katsura and Y. Abe, *J. Chem. Phys.* **39**, 2068 (1963); J. S. Rowlinson, *Proc. Roy. Soc.* A **279**, 147 (1964).
(18) N. Metropolis, A. W. Rosenbluth, M. N. Rosenbluth, A. H. Teller and E. Teller, *J. Chem. Phys.* **21**, 1087 (1953).
B_3 is calculated by L. Tonks, *Phys. Rev.* **50**, 955 (1936).
(19) B_4 and B_5 have been calculated by Metropolis *et al.*[18] using Monte Carlo integration, and they obtain $B_4/b^3 = 0.5327 \pm 0.0005$.

(20) B. J. Alder and T. E. Wainwright, *J. Chem. Phys.* **33**, 1439 (1960).
B. J. Alder and T. E. Wainwright, *Phys. Rev.* **127**, 359 (1962).
(21) L. D. Landau and E. M. Lifshits, *Statistical Physics*, pp. 230–3, Pergamon (1969).
B. Kahn and G. E. Uhlenbeck, *Physica*, **5**, 399 (1938).
B. Kahn, doctoral dissertation, Utrecht (1938).
(22) N. F. Mott and H. S. Massey, *The Theory of Atomic Collisions*, Oxford University Press (1949).
(23) J. de Boer, doctoral dissertation, Amsterdam (1940).
(24) J. de Boer, J. van Kranendonk and K. Compaan, *Physica*, **16**, 545 (1950); J. de Boer and A. Michels, *Physica*, **6**, 409 (1939). These calculations are based on the Lennard-Jones (6–12) interaction.
H. S. W. Massey and R. A. Buckingham, *Proc. Roy. Soc.* A **168**, 378 (1938), and A **169**, 205 (1939), calculate phase shifts for ^{4}He using the Buckingham–Corner potential.
(25) J. E. Kilpatrick and M. F. Kilpatrick, *J. Chem. Phys.* **19**, 930 (1951).
(26) R. J. Lunbeck, doctoral dissertation, Amsterdam (1950).
(27) J. O. Hirschfelder, C. F. Curtiss and R. B. Bird, *Molecular Theory of Gases and Liquids*, p. 164, Wiley, New York (1967).
(28) *Ibid.*, p. 166.
(29) *Ibid.*, pp. 1110–13.
(30) T. L. Hill, *Introduction to Statistical Mechanics*, p. 484. Addison Wesley, Reading, Mass. (1960).
(31) See J. S. Rowlinson, chapter 3 in *Physics of Simple Liquids* (eds. H. N. V. Temperley, J. S. Rowlinson and G. S. Rushbrooke), North-Holland, Amsterdam (1968).
(32) R. Bergeon, *Compt. Rend.* **234**, 1039 (1952).
(33) J. S. Rowlinson, *Mol. Phys.* **6**, 75, 429 (1963).
(34) S. F. Boys and I. Shavitt, *Proc. Roy. Soc.* A **254**, 487 (1960).
(35) J. A. Barker and J. J. Monaghan, *J. Chem. Phys.* **36**, 2564 (1962).
(36) D. Henderson and L. Oden, *Mol. Phys.* **10**, 405 (1966).
(37) J. A. Barker, P. J. Leonard and A. Pompe, *J. Chem. Phys.* **44**, 4206 (1966).
(38) M. S. Green and J. V. Sengers (eds.), *Critical Phenomena*, National Bureau of Standards Misc. Publ. 273 (1966).
(39) G. E. Uhlenbeck and L. Gropper, *Phys. Rev.* **41**, 79 (1932).
See also K. Huang, *Statistical Mechanics*, pp. 216ff, Wiley, New York (1963).
(40) T. Dunn and A. A. Broyles, *Phys. Rev.* **157**, 156 (1967).
(41) F. Lado, *J. Chem. Phys.* **47**, 5369 (1967).
(42) See, e.g., J. O. Hirschfelder, C. F. Curtiss and R. B. Bird, *Molecular Theory of Gases and Liquids*, chapter 6, Wiley, New York (1954).
(43) J. Ram and Y. Singh, *Mol. Phys.* **26**, 539 (1973).

Chapter 2

(1) J. Groeneveld, *Physics Letters*, **3**, 50 (1962).
O. Penrose, *J. Math. Phys.* **4**, 1312 (1963).
S. Baer and J. L. Lebowitz, *J. Chem. Phys.* **36**, 2558 (1962).

J. L. Lebowitz and O. Penrose, *J. Math. Phys.* **5**, 841 (1964).
D. Ruelle, *Rev. Mod. Phys.* **36**, 580 (1964).
(2) F. Zernike and J. A. Prins, *Z. Physik*, **41**, 184 (1927).
(3) J. de Boer, *Rep. Prog. Phys.* **12**, 305 (1949).
(4) R. Clausius, *Phil. Mag.* **40**, 122 (1870).
(5) J. Yvon, *La Théorie Statistique des Fluides et l'Équation d'État, Actualités scientifiques et Industrielles*, vol. 203, Hermann et Cie., Paris (1935).
(6) N. N. Bogolyubov, *J. Phys. URSS*, **10**, 257 and 265 (1946). (English translation in: *Studies in Statistical Mechanics*, vol. 1 (eds. J. de Boer and G. E. Uhlenbeck), North-Holland Publ. Co., Amsterdam (1962).)
(7) M. Born and H. S. Green, *Proc. Roy. Soc.* A **188**, 10 (1946–7).
(8) H. S. Green, *The Molecular Theory of Fluids*, §6.1, North-Holland Publ. Co., Amsterdam (1952).
(9) J. G. Kirkwood, *J. Chem. Phys.* **3**, 300 (1935).
(10) S. A. Rice and P. Gray, *The Statistical Mechanics of Simple Liquids*, §2.6D p. 79, Wiley, New York (1965).
(11) T. L. Hill, *Statistical Mechanics*, McGraw-Hill (1956).
(12) G. S. Rushbrooke, *Physica*, **26**, 259 (1960).
(13) D. Levesque, *Physica*, **32**, 1985 (1966).
(14) J. G. Kirkwood, E. K. Maun and B. J. Alder, *J. Chem. Phys.* **18**, 1040 (1950).
(15) B. J. Alder and T. Wainwright, *J. Chem. Phys.* **27**, 1209 (1957).
(16) W. W. Wood and J. D. Jacobson, *J. Chem. Phys.* **27**, 1207 (1957).
(17) B. J. Alder, *Phys. Rev. Letters*, **12**, 317 (1964).
(18) A. Rahman, *Phys. Rev. Letters*, **12**, 575 (1964).
P. A. Egelstaff, D. I. Page and C. R. T. Heard, *J. Phys.* C **4**, 1453 (1971).
(19) S. A. Rice and J. Lekner, *J. Chem. Phys.* **42**, 3559 (1965).
(20) J. S. Rowlinson, *Mol. Phys.* **6**, 517 (1963). See also M. J. D. Powell, *Mol. Phys.* **7**, 591 (1964); H. L. Weissberg and S. Prager, *Phys. Fluids*, **5**, 1390 (1921).
(21) E. E. Salpeter, *Ann. Phys.* **5**, 183 (1958).
(22) E. Meeron, *J. Chem. Phys.* **27**, 1238 (1957).
(23) G. A. Baker and J. L. Gammel, *J. Math. Anal. and Appl.* **2**, 21 (1961).
(24) G. E. Uhlenbeck and G. W. Ford in *Studies in Statistical Mechanics*, (eds. J. de Boer and G. E. Uhlenbeck), vol. 1, part B, North-Holland Publishing Co., Amsterdam (1962).
(25) S. A. Rice and D. A. Young, *Disc. Faraday Soc.* **43**, 16 (1967).
(26) J. G. Kirkwood, *J. Chem. Phys.* **3**, 300 (1935).
J. G. Kirkwood and E. Monroe, *J. Chem. Phys.* **9**, 514 (1941).
J. G. Kirkwood and E. M. Boggs, *J. Chem. Phys.* **10**, 394 (1942).
(27) G. H. A. Cole, *Adv. Phys.* **8**, 225 (1959); *J. Chem. Phys.* **28**, 912 (1958).
(28) I. Z. Fisher, *Soviet Phys. Usp.* (English translation), **5**, 239 (1962); *Usp. Fiz. Nauk*, **76**, 499 (1962).
I. Z. Fisher, *Statistical Theory of Liquids*, chapter 5, University of Chicago Press (1964).
(29) F. H. Ree, Y. T. Lee and J. Ree, preprint ACRL-70755, Lawrence Radiation Laboratory.
(30) I. Z. Fisher and B. L. Kopeliovich, *Soviet Phys. Dokl.* **5**, 761 (1960–1) (English translation).

(31) R. Abe, *Prog. Theoret. Phys. (Kyoto)*, **19**, 57 (1958); *ibid*, **19**, 407 (1958).
(32) R. Abe, *Prog. Theoret. Phys. (Kyoto)*, **21**, 421 (1959).
(33) A. Richardson, *J. Chem. Phys.* **23**, 2304 (1956).
(34) L. S. Ornstein and F. Zernike, *Proc. Acad. Sci. Amsterdam*, **17**, 793 (1914). See also: G. S. Rushbrooke, *Statistical Mechanics of Equilibrium and Non-Equilibrium* (ed. J. Meixner), p. 222, North-Holland Publishing Co., Amsterdam (1965); J. S. Rowlinson, *Rep. Prog. Phys.* **28**, 169 (1965).
(35) G. S. Rushbrooke, *Physica*, **26**, 259 (1960).
(36) M. D. Johnson and N. H. March, *Phys. Letters*, **3**, 313 (1963).
M. D. Johnson, P. Hutchinson and N. H. March, *Proc. Roy. Soc.* A **282**, 283 (1964).
(37) P. G. Mikolaj and C. J. Pings, *Phys. Rev. Lett.* **15**, 849 (1965); *ibid.*, **16**, 4 (1966); *J. Chem. Phys.* **46**, 1401 (1967).
C. J. Pings, *Molec. Phys.* **12**, 501 (1967); *Disc. Faraday Soc.* **43**, 89 (1967).
C. J. Pings, chapter 10 in *The Physics of Simple Liquids* (eds. H. N. V. Temperley, J. S. Rowlinson, G. S. Rushbrooke), North-Holland Publ. Co., Amsterdam (1968).
(38) The first steps to generalize the netted-chain approximation were made by:
E. Meeron, *J. Chem. Phys.* **28**, 630 (1958); *Phys. Fluids*, **1**, 139 (1958).
E. Meeron and E. R. Rodemich, *Phys. Fluids*, **1**, 246 (1958).
T. Morita, *Prog. Theor. Phys. (Japan)*, **20**, 920 (1958).
M. S. Green, Hughes Aircraft Report (1959).
R. Abe, *J. Phys. Soc. Japan*, **14**, 10 (1959).
The infinite limit, the hyper-netted chain, is now regarded as the best of these theories. M. Klein, however, (*Phys. Fluids*, **7**, 391 (1964)) maintains that some of the intermediate stages have greater numerical accuracy. The final HNC approximation was the result of the work of:
J. M. J. van Leeuwen, J. Groeneveld and J. de Boer, *Physica*, **25**, 792 (1959).
E. Meeron, *J. Math. Phys.* **1**, 192 (1960).
T. Morita and K. Hiroike, *Prog. Theor. Phys. (Japan)*, **23**, 395 (1960).
G. S. Rushbrooke, *Physica*, **26**, 259 (1960).
L. Verlet, *Nuovo Cim.* **18**, 77 (1960).
L. Verlet and D. Levesque, *Physica*, **28**, 1124 (1962).
(39) J. K. Percus and G. J. Yevick, *Phys. Rev.* **110**, 1 (1958).
J. K. Percus, *Phys. Rev. Letters*, **8**, 462 (1962).
(40) T. Gaskell, *Proc. Phys. Soc.* **89**, 236 (1966).
(41) G. Stell, *Physica*, **29**, 517 (1963).
(42) G. S. Rushbrooke and H. I. Scoins, *Proc. Roy. Soc.* A **216**, 203 (1953).
(43) N. W. Ashcroft and J. Lekner, *Phys. Rev.* **145**, 83 (1966).
(44) J. S. Rowlinson, *Mol. Phys.* **10**, 533 (1966).
(45) L. Verlet, *Physica*, **30**, 95 (1964); *ibid.*, **31**, 959 (1965).
(46) J. S. Rowlinson, *Mol. Phys.* **12**, 513 (1967).
(47) G. S. Rushbrooke and G. Silbert, *Mol. Phys.* **12**, 505 (1967).
(48) E. Thiele, *J. Chem. Phys.* **39**, 474 (1963).
M. S. Wertheim, *Phys. Rev. Letters*, **10**, 321 (1963).
R. J. Baxter, *Phys. Rev.* **154**, 170 (1967).
(49) G. Throop and R. J. Bearman, *J. Chem. Phys.* **42**, 2409 (1965).

(50) H. N. V. Temperley, *Proc. Phys. Soc. (London)*, **84**, 399 (1964); and private communication (February 1973).

(51) P. Hutchinson, *Molec. Phys.* **13**, 495 (1967).

(52) B. Nijboer and L. van Hove, *Phys. Rev.* **85**, 777 (1952).

(53) N. W. Aschroft and N. H. March, *Proc. Roy. Soc.* A **297**, 336 (1967).

(54) H. Reiss, H. L. Frisch and J. L. Lebowitz, *J. Chem. Phys.* **31**, 369 (1959).

(55) M. A. Leontovich, *J. Expt. Theor. Phys.* **5**, 41 (1935).

(56) N. N. Bogoliubov, *J. Phys. URSS*, **10**, 257 and 265 (1946). (English translation in: *Studies in Statistical Mechanics*, vol. 1 (eds. J. de Boer and G. E. Unlenbeck), North-Holland Publ. Co., Amsterdam (1962).)

(57) J. Yvon, *Supp. Nuovo Cim.* **9**, 144 (1958).

(58) J. K. Percus *Phys. Rev. Letters*, **9**, 462 (1962).

(59) G. H. A. Cole, *An Introduction to the Statistical Theory of Simple Dense Fluids*, p. 101, Pergamon Press (1967).

(60) L. Verlet, *Physica*, **30**, 95 (1964).
J. K. Percus, *The Equilibrium Theory of Classical Fluids* (eds. H. L. Frisch and J. L. Lebowitz), Benjamin, New York (1964).

(61) M. S. Wertheim, *J. Math. Phys.* **8**, 927 (1967).

(62) B. M. Axilrod and E. Teller, *J. Chem. Phys.* **11**, 299 (1943); *ibid.*, **19**, 724 (1951).

(63) H. L. Friedman, *Ionic Solution Theory*, chapter 2, Wiley, New York (1962).

(64) R. J. Baxter, *Annls. Phys.* **46**, 509 (1968).

(65) T. Kihara, *Adv. Chem. Phys.* **1**, 267 (1958).

(66) H. W. Graben and R. D. Present, *Phys. Rev. Letters.* **9**, 247 (1962).

(67) A. F. Sherwood and J. M. Prausnitz, *J. Chem. Phys.* **41**, 413 (1964).

(68) A. F. Sherwood, A. G. De Rocco and E. A. Mason, University of Maryland, National Aeronautics and Space Administration Report No. IMP-NASA-52.

(69) P. G. Mikolaj and C. J. Pings, *Phys. Rev. Letters*, **15**, 849 (1965); *ibid.*, **16**, 4 (1966).
P. G. Mikolaj and C. J. Pings, *J. Chem. Phys.* **46**, 1401 and 1412 (1967).
C. J. Pings, *Molec. Phys.* **12**, 501 (1967); *Disc. Faraday Soc.* **43**, 89 (1967).

(70) D. Leveoque and L. Verlet, *Phys. Rev. Letters*, **20**, 905 (1968).
L. Verlet, *Phys. Rev.* **165**, 201 (1968).

(71) R. Jastrow, *Phys. Rev.* **98**, 1479 (1955).

(72) R. Jastrow, in *The Many-Body Problem* (ed. J. Percus), Interscience Publishers, New York (1963).

(73) F. Y. Wu and E. Feenberg, *Phys. Rev.* **122**, 739 (1961).

(74) R. Abe, *Prog. Theoret. Phys. (Kyoto)*, **19**, 57 (1958); *ibid.*, **19**, 407 (1958). See also ref. (87).

(75) R. M. Mazo and J. G. Kirkwood, *J. Chem. Phys.* **41**, 204 (1955); *ibid.*, **28**, 644 (1958).

(76) D. G. Henshaw and D. Hurst, *Can. J. Phys.* **33**, 797 (1955).

(77) B. Widom, *J. Chem. Phys.* **42**, 1128 (1965).

(78) A. Bellemans, *J. Chem. Phys.* **50**, 2784 (1969).

(79) H. J. Raveché and R. D. Mountain, *J. Chem. Phys.* **53**, 3101 (1970).

(80) J. D. Weeks, D. Chandler and H. C. Andersen, *J. Chem. Phys.* **54**, 5237 (1971).

(81) J. S. Rowlinson, *Mol. Phys.* **7**, 349 (1964); *ibid.*, **8**, 107 (1964).
(82) J. A. Barker and D. Henderson *J. Chem. Phys.* **47**, 4714 (1967).
(83) D. Henderson and J. A. Baker, *Phys. Rev.* A **1**, 1266 (1970).
(84) H. C. Anderson, J. D. Weeks and D. Chandler, *Phys. Rev.* A **4**, 1597 (1971).
(85) L. Verlet and J. J. Weis, unpublished.
N. F. Carnahan and K. E. Starling, *J. Chem. Phys.* **51**, 635 (1969).
(86) W. G. Hoover, M. Ross, K. W. Johnson, D. Henderson, J. A. Barker and B. C. Brown, *J. Chem. Phys.* **52**, 4931 (1970).
J. Hansen, *Phys. Rev.* **42**, 221 (1970).
(87) T. B. Davison and D. K. Lee, *Phys. Rev.* A **4**, 1110 (1971).
(88) D. Henderson, *Mol. Phys.* **21**, 841 (1971).
(89) J. D. Weeks, unpublished.
(90) J. A. Barker and D. Henderson, *Mol. Phys.* **21**, 187 (1971).
(91) F. Lado, *J. Chem. Phys.* **49**, 3092 (1968).
(92) This also follows from a comparison of the virial expansion (§1.4) and the compressibility equation of state

$$\frac{1}{kT}\left(\frac{\partial P}{\partial \rho}\right)_T = 1-4\pi\rho\int_0^\infty c(r)\,r^2\mathrm{d}r \quad \text{where } \beta_n = (n-1)^{-1}\int \alpha_n\,\mathrm{d}\mathbf{r}.$$

Chapter 3

(1) S. Kim, D. Henderson and L. Oden, *J. Chem. Phys.* **45**, 4030 (1966).
(2) L. Tonks, *Phys. Rev.* **50**, 955 (1936).
K. F. Hertzfeld and M. G. Mayer, *J. Chem. Phys.* **2**, 38 (1934).
(3) R. J. Riddel and G. E. Uhlenbeck, *J. Chem. Phys.* **21**, 2056 (1953).
(4) L. Tonks, *Phys. Rev.* **50**, 955 (1936).
(5) J. S. Rowlinson, *Mol. Phys.* **7**, 593 (1964).
(6) F. H. Ree and W. G. Hoover, *J. Chem. Phys.* **40**, 939 (1964).
(7) G. S. Rushbrooke and P. Hutchinson, *Physica*, **27**, 647 (1961).
P. Hutchinson and G. S. Rushbrooke, *Physica*, **29**, 675 (1963).
(8) G. S. Rushbrooke, *J. Chem. Phys.* **38**, 1262 (1963).
(9) G. H. A. Cole, *J. Chem. Phys.* **36**, 1680 (1962).
B. R. A. Nijboer and R. Fieschi, *Physica*, **19**, 545 (1953).
(10) G. S. Rushbrooke, in *Statistical Mechanics of Equilibrium and Non-Equilibrium* (ed. J. Meixner), p. 222, North-Holland Publ. Co., Amsterdam, (1965).
(11) G. H. A. Cole, *An Introduction to the Statistical Theory of Classical Simple Dense Fluids*, p. 265, Pergamon Press, Oxford (1967).
(12) J. S. Rowlinson, *Rep. Prog. Phys.* **28**, 169 (1965).
(13) A. E. Rodriguez, *Proc. Roy. Soc.* A **239**, 373 (1957).
(14) C. Hurst, *Proc. Phys. Soc.* **86**, 193 (1965); *ibid.*, **88**, 533 (1966).
(15) D. Henderson, *Proc. Phys. Soc.* **87**, 592 (1966).
(16) G. H. A. Cole, *J. Chem. Phys.* **44**, 338 (1966).
(17) D. McQuarrie, *J. Chem. Phys.* **40**, 3455 (1964).
(18) T. Kihara, *Nippon Sagaku, Buturigakaishi*, **17**, 11 (1943); *Rev. Mod. Phys.* **25**, 831 (1953).

(19) S. Katsura, *Phys. Rev.* **115**, 1417 (1959); *ibid.*, **118**, 1667 (1960).
J. A. Barker and J. J. Monoghan, *J. Chem. Phys.* **36**, 2558 (1962).
(20) D. Henderson and L. Oden, *Mol. Phys.* **10**, 405 (1966).
Less satisfactory are the calculations for arbitrarily high densities by numerical solution of the integral equations for a Lennard-Jones potential:
A. A. Broyles, S. V. Chung and H. L. Sahlin, *J. Chem. Phys.* **37**, 2462 (1962).
M. Klein and M. S. Green, *J. Chem. Phys.* **39**, 1367 (1963).
A. A. Khan, *Phys. Rev.* **134**, A 367 (1964).
L. Verlet, *Physica*, **30**, 95 (1964).
(21) L. Verlet, *Physica*, **31**, 959 (1965); *ibid.*, **32**, 304 (1966).
J. S. Rowlinson, *Mol. Phys.* **10**, 533 (1966).
D. Henderson, S. Kim and L. Oden, *Disc. Faraday Soc.* **43**, 26 (1967).
(22) D. Levesque, *Physica*, **32**, 1985 (1966).
(23) L. Verlet and D. Levesque, *Physica*, **36**, 254 (1967).
L. Verlet, *Phys. Rev.* **159**, 98 (1967).
(24) W. G. Hoover and J. C. Poirier, *J. Chem. Phys.* **37**, 1041 (1962).
(25) D. McQuarrie, *J. Chem. Phys.* **41**, 1197 (1964).
(26) M. S. Wertheim, *Phys. Rev. Letters*, **8**, 321 (1963); *J. Math. Phys.* **5**, 643 (1964).
(27) E. Thiele, *J. Chem. Phys.* **39**, 474 (1963).
(28) H. Reiss, H. L. Frisch and J. L. Lebowitz, *J. Chem. Phys.* **31**, 369 (1959).
(29) H. S. Green, *The Molecular Theory of Fluids*, pp. 75 *et seq.*, North-Holland Publ. Co., Amsterdam (1952).
Analytical modifications of Green's first approach have been given by:
A. E. Rodriguez, *Proc. Roy. Soc.* A **196**, 73 (1948).
A. G. McLellan, *Proc. Roy. Soc.* A **210**, 509 (1952).
(30) J. G. Kirkwood and E. M. Boggs, *J. Chem. Phys.* **10**, 394 (1942).
(31) N. W. Ashcroft and J. Lekner, *Phys. Rev.* **145**, 83 (1966).
R. G. Ross, *Phil. Mag.* **22**, 573 (1970).
(32) J. G. Kirkwood, E. K. Maun and B. J. Alder, *J. Chem. Phys.* **18**, 1040 (1950).
(33) T. L. Hill, *Statistical Mechanics*, McGraw-Hill (1956).
(34) J. G. Kirkwood and E. M. Boggs, *J. Chem. Phys.* **9**, 514 (1941).
(35) M. N. Rosenbluth and A. W. Rosenbluth, *J. Chem. Phys.* **22**, 881 (1954).
T. Wainwright and B. J. Alder, *Nuovo Cimento Suppl.* **9**, 116 (1958).
B. J. Alder and T. Wainwright, *J. Chem. Phys.* **27**, 1209 (1957); *ibid.*, **31**, 459 (1959).
W. W. Wood and J. Jacobson, *J. Chem. Phys.* **27**, 1207 (1957).
W. W. Wood and R. F. Parker, *J. Chem. Phys.* **27**, 720 (1957).
(36) M. Klein, *J. Chem. Phys.* **39**, 1367 (1963); *ibid.*, **39**, 1388 (1963).
(37) A. A. Broyles, S. V. Chung and H. L. Sahlin, *J. Chem. Phys.* **37**, 2462 (1962).
A. A. Broyles, *J. Chem. Phys.* **33**, 456 (1960); *ibid.*, **34**, 359 (1961); *ibid.*, **34**, 1068 (1961).
(38) A. Rahman, *Phys. Rev.* **136**, A405 (1964).
W. W. Wood and R. F. Parker, *J. Chem. Phys.* **27**, 720 (1957).
(39) A. Michels, H. Wijker and H. K. Wijker, *Physica*, **15**, 627 (1949).

A. Michels, J. M. Levelt and W. de Graaff, *Physica*, **24**, 659 (1958).
A. Michels, J. M. Levelt and G. J. Wolkers, *Physica*, **24**, 769 (1958).
A. Michels, W. de Graaff and T. A. Ten Seldam, *Physica*, **26**, 393 (1960).
A. Michels, J. C. Abels, T. A. Ten Seldam and W. de Graaff, *Physica*, **26**, 381 (1960).
(40) A. A. Khan, *Phys. Rev.* **134**, 367 (1963).
(41) L. Verlet and D. Levesque, *Physica*, **28**, 1124 (1962).
(42) J. D. Weeks, S. A. Rice and J. J. Kozak, *J. Chem. Phys.* **52**, 2416 (1970).
(43) E. Meeron, *Phys. Rev. Lett.* **25**, 152 (1970).
(44) J. G. Kirkwood, in *Phase Transformations in Solids* (ed. R. Smoluchowski), Wiley, New York (1951).
(45) R. L. Kerber, *J. Chem. Phys.* **52**, 2436 (1970).
(46) M. D. Johnson, P. Hutchinson and N. H. March, *Proc. Roy. Soc.* A **282**, 283 (1964).
(47) D. G. Henshaw, *Phys. Rev.* **105**, 976 (1957).
P. A. Egelstaff, C. Duffill, V. Rainey, J. E. Enderby and D. M. North, *Phys. Letters*, **21**, 286 (1966).
J. E. Enderby and N. H. March in *Phase Stability in Metals and Alloys* (eds. P. S. Rudman, J. Stringer and R. I. Jaffee), McGraw-Hill, New York (1966).
(48) P. Ascarelli, *Phys. Rev.* **143**, 36 (1966).
J. E. Enderby and N. H. March, *Adv. Phys.* (*Phil. Mag. Suppl.*), **14**, 453.
J. E. Enderby and N. H. March, *Proc. Phys. Soc.* **88**, 717 (1966).
P. G. Mikolaj and C. J. Pings, *J. Chem. Phys.* **46**, 1412 (1967).
(49) T. Gaskell, *Proc. Phys. Soc.* **89**, 231 (1966).
(50) S. A. Rice and D. A. Young, *Disc. Faraday Soc.* **43**, 16 (1967).
(51) S. A. Rice and J. Lekner, *J. Chem. Phys.* **42**, 3559 (1965).
(52) N. H. March, *Liquid Metals*, Pergamon Press, p. 43 (1968).
(53) J. E. Enderby and N. H. March, *Battelle Conference on Phase Stability*, Geneva (1966).
(54) P. Hutchinson, *Disc. Faraday Soc.* **43**, 50 (1967).
(55) A. Rahman, *Phys. Rev. Letters*, **12**, 575 (1964).
(56) J. A. Barker and D. Henderson, *Disc. Faraday Soc.* **43**, 53 (1967).
(57) F. Y. Wu and E. Feenberg, *Phys. Rev.* **122**, 739 (1961).
(58) R. Jastrow, in *The Many-Body Problem* (ed. J. Percus), Interscience Publishers, New York (1963).
(59) R. Abe, *Prog. Theoret. Phys.* (*Kyoto*), **19**, 57 (1958); *ibid.*, **19**, 407 (1958).
(60) W. E. Massey, *Phys. Rev.* **151**, 153 (1966).
(61) L. Goldstein and J. Reekie, *Phys. Rev.* **98**, 857 (1955).
(62) P. C. Gehlen and J. E. Enderby, *J. Chem. Phys.* **51**, 547 (1969).
(63) B. J. Alder and T. E. Wainwright, *Phys. Rev.* **127**, 359 (1962).
(64) W. Kunkin and H. L. Frisch, *J. Chem. Phys.* **50**, 1817 (1969).
(65) L. van Hove, *The Equilibrium Theory of Classical Fluids* (eds. H. L. Frisch, and J. L. Lebowitz), Benjamin, New York (1964).
(66) D. Levesque and J. Vieillard-Baron, *Physica*, **44**, 345 (1969).
(67) F. Lado, *J. Chem. Phys.* **47**, 4828 (1967).
(68) *Ibid.*, **49**, 3092 (1968).
(69) A. A. Kugler, *Physica*, **50**, 155 (1970).

(70) J. A. Barker, P. J. Leonard and A. Pompe, *J. Chem. Phys.* **44**, 4206 (1966).
(71) D. Henderson, *Proc. Phys. Soc. (London)*, **87**, 592 (1966).
(72) M. S. Wertheim, *J. Math. Phys.* **8**, 927 (1967).
(73) C. A. Croxton, *Phys. Letters*, **42**A, 229 (1972).
C. A. Croxton, *Phys. Letters*, (in press, 1973).
C. A. Croxton, *J. Phys.* C., (in press, 1973).
(74) L. E. Ballentine and J. C. Jones, private communication (October, 1972); see also *Proceedings of the Second International Conference of Liquid Metals (Tokyo)*, Taylor and Francis (1973).
(75) F. Y. Wu and E. Feenberg, *Phys. Rev.* **128**, 943 (1962).
(76) D. Schiff, *Nature*, **243**, 130 (1973).
(77) M. Kalos, D. Levesque and L. Verlet, (unpublished).

Chapter 4

(1) R. H. Fowler, *Proc. Roy. Soc.* A **159**, 229 (1937).
(2) J. G. Kirkwood and F. P. Buff, *J. Chem. Phys.* **17**, 338 (1949).
(3) P. D. Shoemaker, G. W. Paul and L. E. Marc de Chazal, *J. Chem. Phys.* **52**, 491 (1970).
(4) T. L. Hill, *J. Chem. Phys.* **20**, 141 (1952).
(5) I. W. Plesner and O. Platz, *J. Chem. Phys.* **48**, 5361 (1968).
(6) L. Tonks, *Phys. Rev.* **50**, 955 (1936).
(7) H. Reiss, H. L. Frisch and J. L. Lebowitz, *J. Chem. Phys.* **31**, 369 (1959).
E. Helfand, H. L. Frisch and J. L. Lebowitz, *J. Chem. Phys.* **34**, 1037 (1961)
(8) C. A. Croxton and R. P. Ferrier, *J. Phys.* C **4**, 1909 (1971).
C. A. Croxton and R. P. Ferrier, *Phil. Mag.* **24**, 489 (1971).
(9) L. D. Landau and E. M. Lifshits, *Mekhanika Sploshnykh Sred'*, izd. 1-oe, Gostekhizdat (1944).
(10) C. A. Croxton, unpublished (1968).
(11) I. Z. Fisher, *Statistical Theory of Liquids*, pp. 156–79, University of Chicago Press (1964).
(12) I. Z. Fisher and B. V. Bokut', *Zh. Fiz. Khim.* **30**, 2547 (1956).
(13) C. A. Croxton and R. P. Ferrier, J. *Phys.* C **4**, 2417 (1971).
C. A. Croxton and R. P. Ferrier, *Phys. Letters*, **35**A, 330 (1971).
(14) R. Tolman, *J. Chem. Phys.* **16**, 758 (1948).
(15) F. B. Sprow and J. M. Prausnitz, *Trans. Faraday Soc.* **62**, 1097 (1966).
(16) C. A. Croxton and R. P. Ferrier, *J. Phys.* C **4**, 1921 (1971).
C. A. Croxton and R. P. Ferrier, *Phil. Mag.* **24**, 493 (1971).
(17) C. A. Croxton and R. P. Ferrier, *J. Phys.* C **4**, 2433 (1971).
C. A. Croxton and R. P. Ferrier, *Phys. Letters*, **36**A, 183 (1971).
(18) D. W. G. White, *Trans. Metall. Soc., A.I.M.E.* **236**, 796 (1966); *Metals Materials, and Metallurgical Reviews*, p. 73 (July 1968).
(19) D. W. G. White, *J. Inst. Metals*, **99**, 287 (1971).
R. T. Southin and G. A. Chadwick, *Scripta Metallurgica*, **3**, 541 (1969).
(20) J. S. Huang and W. W. Webb, *J. Chem. Phys.* **50**, 3677 (1969).
(21) F. P. Buff and R. A. Lovett, in *Simple Dense Fluids* (eds. H. L. Frisch an Z. W. Salsburg), p. 19, Academic Press, New York (1968).

(22) P. Drude, *Theory of Optics*, pp. 287–95, Dover, New York (1959).
(23) F. P. Buff and R. A. Lovett, *1966 Saline Water Conversion Report*, p. 26, U.S. Government Printing Office, Washington, D.C.
(24) R. M. Goodman and G. A. Samorjai, *J. Chem. Phys.* **52**, 6325 (1970); *ibid* **52**, 6331 (1970).
(25) J. Bohdansky, *J. Chem. Phys.* **49**, 2982 (1968).
(26) B. U. Felderhof, *Phys. Rev.* A **1**, 1185 (1970).
(27) J. D. Bernal, *Proc. Roy. Soc.* A **280**, 299 (1964).
(28) J. W. Cahn and J. E. Hilliard, *J. Chem. Phys.* **28**, 258 (1958). See also: J. W. Cahn, *ibid.*, **30**, 1121 (1959).
E. W. Hart, *Phys. Rev.* **113**, 412 (1959); *ibid.*, **114**, 27 (1959); *J. Chem. Phys.* **39**, 3075 (1963); *ibid.*, **34**, 1471 (1961).
(29) J. D. van der Waals, *Z. Physik. Chem.* **13**, 657 (1894).
(30) T. D. Lee and C. N. Yang, *Phys. Rev.* **87**, 410 (1952).
(31) T. L. Hill, *Statistical Mechanics*, chapter 7, McGraw-Hill, New York (1956).
(32) B. Widom, *J. Chem. Phys.* **41**, 1633 (1964); *ibid.*, **43**, 3898 (1965).
(33) B. Widom and O. K. Rice, *J. Chem. Phys.* **23**, 1250 (1955).
(34) J. S. Kouvel and M. E. Fisher, *Phys. Rev.* **136**A, 1626 (1964).
(35) D. S. Gaunt, M. E. Fisher, M. F. Sykes and J. W. Essam, *Phys. Rev. Letters*, **13**, 713 (1964).
(36) M. E. Fisher, *J. Math. Phys.* **5**, 944 (1964).
(37) E. A. Guggenheim, *J. Chem. Phys.* **13**, 253 (1945).
(38) L. Onsager, *Phys. Rev.* **65**, 117 (1944).
(39) B. Widom, *J. Chem. Phys.* **43**, 3892 (1965).
S. Fisk and B. Widom, **50**, 3219 (1969).
(40) D. Stansfield, *Proc. Phys. Soc.* **72**, 854 (1958).
(41) F. B. Sprow and J. M. Prausnitz, *Trans. Faraday Soc.* **62**, 1097 (1966).
(42) A. J. Leadbetter and H. E. Thomas, *Trans. Faraday Soc.* **61**, 10 (1965).
(43) B. L. Smith, P. R. Gardner and E. N. C. Parker, *J. Chem. Phys.* **47**, 1148 (1967).
(44) P. A. Egelstaff and B. Widom, *J. Chem. Phys.* **53**, 2667 (1970).
(45) D. W. G. White, *Trans. Amer. Soc. for Metals*, **55**, 757 (1962).
(46) L. L. Bircumshaw, *Phil. Mag.* **12**, 596 (1931).
(47) T. R. Hogness, *J. Amer. Chem. Soc.* **43**, 1621 (1921).
L. L. Bircumshaw, *Phil. Mag.* **3**, 1286 (1927).
(48) W. Krause, F. Sauerwald and M. Michalke, *Z. Anorg. Allgem. Chem.* **181**, 353 (1929).
(49) P. P. Pugachevich and V. I. Yashkichev, *Akad. Nauk, SSSR, Otd. Khim. Nauk*, 806 (1959).
(50) K. T. Kurochkin, B. A. Baum and Ye. K. Borodulin, *Phys. Metals Metallog. USSR* (English translation), **15**, 118 (1963).
(51) A. Einstein, *Ann. Phys.* **4**, 513 (1901).
(52) D. W. G. White, private communication (1971).
(53) D. W. G. White, *Met. Trans.* **2**, 3067 (1971) (Pb, Sn and Pb–Sn alloys).
A. F. Crawley, H. R. Thresh, D. W. G. White, J. O. Edwards and J. W. Meier, Report PM-R-67-16, Mines Branch, Department of Energy, Mines and Resources, Ottawa, Canada (1967) (Pb and Pb alloys).

H. R. Thresh, D. W. G. White, J. O. Edwards and J. W. Meier, Report PM-R-64-28, Mines Branch, Department of Energy, Mines and Resources, Ottawa, Canada (1964) (Zn and Zn alloys).
D. W. G. White, Research Report R160, Mines Branch, Department of Energy, Mines and Resources, Ottawa, Canada (1965) (Zn and Zn alloys).
D. W. G. White, *Trans. Met. Soc., AIME*, **236**, 796 (1966).
T. R. Hogness, *J. Amer. Chem. Soc.* **43**, 1621 (1921) (Hg, Cd, Zn, Pb, Sn, Bi).
L. L. Bircumshaw, *Phil. Mag.* **2**, 341 (1926) (general).
Y. Matuyama, *Sci. Rep. Tôhoku Imp. Univ.* **16**, 555 (1927) (general).
G. Drath and F. Sauerwald, *Z. Anorg. Chem.* **162**, 301 (1927) (Sn, Pb, Sb, Cu–Sb, Cu–Sn, Fe).
L. L. Bircumshaw, *Phil. Mag.* **17**, 181 (1934) (Pb–Sn).
H. T. Greenaway, *J. Inst. Metals*, **74**, 133 (1948) (Pb–Sb, Cd–Sb).
D. A. Melford and T. P. Hoar, *J. Inst. Metals*, **85**, 197 (1956–7) (Pb, Sn, In).
D. H. Bradhurst and A. S. Buchanan, *J. Phys. Chem.* **63**, 1486 (1959) (Pb in contact with UO_2).
D. V. Atterton and T. P. Hoar, *J. Inst. Metals*, **81**, 541 (1952) (Sn).
S. M. Kaufman and T. J. Whalen, *Acta. Met.* **13**, 797 (1965) (Au, Au–Sn).
W. Gans and H. Parthey, *Z. Metallkunde*, **57**, 19 (1966) (Sn).
N. L. Pokrovski, P. P. Pugachevich and Kh. I. Ibragimov, *Dokl. Akad., Nauk SSSR*, **172**, 829 (1967) (Sn–Au).
V. K. Semenchenko, *Surface Phenomena in Metals and Alloys*, Pergamon Press (1961) (general).
V. N. Eremenko, *Ukran, Kh. Zh.* **28** (4), 427 (1962) (general).
(54) L. D. Landau and E. M. Lifshits, *Statistical Physics*, p. 457, Pergamon Press, Oxford (1969).
S. L. Chan, *Can. J. Phys.* **50**, 1139 (1972) has very recently experimentally confirmed Atkins' limiting law of surface tension–temperature dependence in superfluid ^4He.
(55) J. Frenkel, *Kinetic Theory of Liquids*, chapter 6, Dover, New York (1955).
(56) M. Born and Th. von Kármán, *Phys. Z.* **15**, 361 (1913).
(57) V. K. Semenchenko, *Surface Phenomena in Metals and Alloys*, Pergamon Press (1961).
(58) K. Huang and G. Wyllie, *Proc. Phys. Soc.* A **62**, 180 (1949).
(59) J. Frenkel, *Phil. Mag.* **33**, 297 (1917).
(60) A. Kh. Breger and A. A. Zhukovitskii, *Zh. Fiz. Khim.* **20**, 355 (1946).
(61) R. Stratton, *Phil. Mag.* **44**, 1247 (1953).
(62) A. G. Samoilovich, *Zh. Eksp. Teor. Fiz.* **16**, 135 (1946).
(63) J. Boiani and S. A. Rice, *Phys. Rev.* **185**, 931 (1969).
(64) A. A. Bloch, and S. A. Rice, *Phys. Rev.* **185**, 933 (1969).
(65) W.-C. Lu, M. S. Jhon, T. Ree and H. Eyring, *J. Chem. Phys.* **46**, 1075 (1967).
(66) C. Jouanin, *C.r. Lebd. Séanc. Acad. Sci. Paris*, B **268**, 1597 (1969).
(67) R. Abe, *Prog. Theoret. Phys. (Kyoto)*, **19**, 57 (1958); *ibid.*, **19**, 407 (1958).
(68) W. Gordon, C. Shaw and J. Daunt, *Phys. Rev.* **96**, 1444 (1954).

(69) C. A. Croxton, *Phys. Lett.* A **41**, 413 (1972).
C. A. Croxton, *J. Phys.* C **6**, 411 (1973).

(70) The tacit assumption made here assumes no radial modification of the pair function $g_{(2)}(r)$ in the vicinity of the liquid surface. The extent of the radial distortion has recently been determined (C. A. Croxton, *J. Phys.* C, in press, 1973) as a solution of the Smoluchowski equation (§6.9) within the liquid–vapour transition zone. The modification $\mathscr{R}(r)$ may be expressed in terms of a zero-order spherical Bessel function of the first kind which is radially and progressively collapsed into the origin. It may be shown that the oscillatory aspect of this function is in fact almost entirely collapsed within the collision diameter, $\mathscr{R}(r)$ adopting its asymptotic value of unity beyond, so that the modified distribution $g_{(2)}(r)\,\mathscr{R}(r) \to g_{(2)}(r)$.

(71) D. Crozier, *Canadian J. Phys.* **50**, 1914 (1972).

(72) C. Maze and G. Burnet, *Surf. Sci.* **27**, 411 (1971).

(73) S. Toxvaerd, private communication (September, 1971).
S. Toxvaerd, *J. Chem. Phys.* **55**, 3116 (1971); *Mol. Phys.* **26**, 91 (1973).

(74) G. M. Nazarian, *J. Chem. Phys.* **56**, 1408 (1972).

(75) C. A. Croxton, *Proceedings of the Second International Conference of Liquid Metals* (*Tokyo*), Taylor and Francis (1972).

(76) C. A. Croxton, J. *Phys.* C (in press, 1973).

(77) M. V. Berry, R. F. Durrans and R. Evans, *J. Phys.* A **5**, 166 (1972).

(78) R. W. Zwanzig, *J. Chem. Phys.* **22**, 1420 (1954).

(79) L. Verlet and J. J. Weiss, *Phys. Rev.* A **5**, 939 (1972).

(80) W. R. Smith, D. Henderson and J. A. Barker, *J. Chem. Phys.* **55**, 4027 (1971).

(81) S. Toxvaerd, *J. Chem. Phys.* **57**, 4092 (1972).

(82) S. Toxvaerd, *J. Chem. Phys.* **55**, 3116 (1971).

(83) H. S. Green, *Handb. Phys.* **10**, 79 (1960).

(84) M. V. Berry and S. R. Reznek, *J. Phys.* A **4**, 77 (1971).

(85) C. A. Croxton, *Advances in Physics* **22**, 385 (1973).

(86) A. D. Singh, *Phys. Rev.* **125**, 802 (1962).

(87) K. R. Atkins and Y. Narahara, *Phys. Rev.* A **138**, 437 (1965).

(88) K. S. C. Freeman and I. R. McDonald, *Mol. Phys.* **26**, 529 (1973).
See also the comment by C. A. Croxton, *Mol. Phys.* (in press, 1974).

(89) D. Fitts, *Physica*, **42**, 205 (1969).

Chapter 5

(1) I. E. Farquhar, *Ergodic Theory in Statistical Mechanics*, pp. 37–41, Interscience, New York (1964).

(2) D. Levesque and L. Verlet, *Phys. Rev.* **182**, 307 (1969).

(3) I. R. McDonald and K. Singer, *J. Chem. Phys.* **47**, 4766 (1967).

(4) A. Rahman, *Phys. Rev.* **36**, A405 (1964).

(5) B. J. Alder and T. E. Wainwright, in *Transport Processes in Statistical Mechanics* (ed. I. Prigogine), p. 97, Interscience, New York (1958).

(6) N. A. Metropolis, A. W. Rosenbluth, M. N. Rosenbluth, A. H. Teller and E. Teller, *J. Chem. Phys.* **21**, 1087 (1953).

(7) J. M. Hammersley and D. C. Handscomb, pp. 117–26, *Monte Carlo Methods*, Methuen, London (1964).

I. Z. Fisher, *Usp. Fiz. Nauk*, **69**, 349 (1959); English translation *Soviet Phys.-Usp.* **2**, 783 (1960).

I. Z. Fisher, *Statistical Theory of Liquids*, pp. 206–31, Moscow (English translation), University of Chicago Press (1961).

M. A. Fluendy and F. B. Smith, *Quart. Rev.* **16**, 241 (1962).

J. A. Barker, *Lattice Theories of the Liquid State*, pp. 14–28, Macmillan, New York, (1963).

A. Münster, 'Theory of the liquid state', in *Physics of High Pressures and the Condensed Phase* (ed. A. van Itterbeek), pp. 281–94, North-Holland Publ. Co., Amsterdam (1965).

(8) W. W. Wood, in *Physics of Simple Liquids* (eds. H. N. V. Temperley, J. S. Rowlinson and G. S. Rushbrooke), pp. 115–230, North-Holland Publ. Co., Amsterdam (1968).

(9) W. W. Wood, *J. Chem. Phys.* **48**, 415 (1968).

(10) I. R. McDonald, *Chem. Phys. Letters*, **3**, 241 (1969).

(11) I. R. McDonald, *Proc. Culham Conference on Computational Physics*, vol. 2, UKAEA, Culham Laboratory and IPPS (July, 1969).

(12) W. W. Wood and F. R. Parker, *J. Chem. Phys.* **27**, 720 (1957).

(13) I. R. McDonald and K. Singer, *Disc. Faraday Soc.* **43**, 40 (1967).

(14) W. W. Wood, *Monte Carlo Calculations of the Equation of State of Systems of 12 and 48 Hard Circles*, Los Alamos Scientific Laboratory Report LA-2827, Los Alamos, New Mexico (July 1st, 1963): available from Office of Technical Services, U.S. Dept. of Commerce, Washington 25, D.C.

(15) A. Rotenberg, *Monte Carlo Studies of Systems of Hard Spheres*, Report NYO-1480-3, AEC Computing and Applied Mathematics Centre, Courant Institute of Mathematical Sciences, New York University (1 October 1964).

(16) W. W. Wood, *J. Chem. Phys.* **48**, 415 (1968).

(17) F. H. Ree and W. G. Hoover, *J. Chem. Phys.* **40**, 939 (1964).

(18) W. G. Hoover and B. J. Alder, *J. Chem. Phys.* **46**, 686 (1967).

(19) W. Feller, *An Introduction to Probability Theory and its Applications*, vol. 1, Wiley, New York (1950).

(20) I. Oppenheim and P. Mazur, *Physica*, **23**, 197 (1957).

(21) J. L. Lebowitz, J. K. Percus and L. Verlet, *Phys. Rev.* **153**, 250 (1967).

(22) B. J. Alder and T. E. Wainwright, *Phys. Rev.* **127**, 359 (1962).

(23) M. N. Rosenbluth and A. W. Rosenbluth, *J. Chem. Phys.* **22**, 881 (1964).

(24) W. W. Wood and J. D. Jacobson, *J. Chem. Phys.* **27**, 1207 (1957).

(25) A. Rotenberg, *J. Chem. Phys.* **42**, 1126 (1965).

(26) B. J. Alder and T. E. Wainwright, *J. Chem. Phys.* **33**, 1439 (1960).

(27) P. L. Fehder, *J. Chem. Phys.* **50**, 2617 (1969).

(28) A. Michels, H. Wijker and H. Wijker, *Physica*, **15**, 627 (1949).

(29) L. Verlet and D. Levesque, *Physica*, **36**, 254 (1967).

(30) A. Michels, J. M. Levelt and W. de Graaff, *Physica*, **24**, 659 (1958).

(31) L. Verlet, *Phys. Rev.* **159**, 98 (1967).

(32) J. M. H. Levelt, *Physica*, **26**, 361 (1960).
A. van Itterbeek, O. Verbeke and K. Stacs, *Physica*, **26**, 361 (1960).
W. van Witzenberg and J. C. Stryland, *Canad. J. Phys.* **46**, 811 (1968).
(33) I. R. McDonald and K. Singer, *J. Chem. Phys.* **50**, 2308 (1969).
(34) W. W. Wood, F. R. Parker and J. D. Jacobson, *Nuovo Cimento Suppl.* (series 10) **9**, 133 (1958).
(35) P. Flubacher, A. J. Leadbetter and J. A. Morrison, *Proc. Phys. Soc.* **78**, 1449 (1961).
(36) F. Din, *Thermodynamic Functions of Gases*, vol. 2, pp. 146–201, Butterworths, London (1956).
(37) M. D. Johnson, P. Hutchinson and N. H. March, *Proc. Roy. Soc.* A **282**, 283 (1964).
(38) W. A. Harrison, *Phys. Rev.* **136**, A1107 (1964).
W. Kohn and S. H. Vosko, *Phys. Rev.* **119**, 912 (1960).
N. H. March and A. M. Murray, *Proc. Roy. Soc.* A **256**, 406 (1960).
N. H. March and A. M. Murray, *Proc. Roy. Soc.* A **261**, 119 (1961).
W.-M. Shyu and G. D. Gaspari, *Phys. Rev.* **163**, 667 (1967).
P. S. Ho, *Phys. Rev.* **169**, 523 (1968).
W-M. Shyu and G. D. Gaspari, *Phys. Rev.* **170**, 687 (1968).
(39) T. J. Rowland, *Phys. Rev.* **119**, 900 (1960).
(40) P. G. Mikolaj and C. J. Pings, *Phys. Rev. Letters*, **15**, 849 (1965).
(41) A. Paskin and A. Rahman, *Phys. Rev. Letters*, **16**, 300 (1966).
(42) N. S. Gingrich and L. Heaton, *Phys. Rev.* **34**, 873 (1961).
(43) W. Cochran, *Proc. Roy. Soc.* A **276**, 308 (1968).
(44) R. E. Meyer and N. H. Nachtrieb, *J. Chem. Phys.* **23**, 1851 (1955).
(45) N. W. Ashcroft and J. Lekner, *Phys. Rev.* **145**, 83 (1966).
(46) L. Verlet, *Phys. Rev.* **165**, 201 (1968).
(47) D. Schiff, *Phys. Rev.* **186**, 151 (1969).
(48) R. Pick, *J. Phys. (Paris)*, **28**, 539 (1967).
(49) W. Harrison, *Pseudopotentials in the Theory of Metals*, p. 319, W. A. Benjamin, Inc., New York (1966)
(50) N. W. Ashcroft and D. C. Langreth, *Phys. Rev.* **159**, 500 (1967).
(51) J. M. H. Levelt, *Physica*, **26**, 361 (1960).
(52) P. Fehder, *J. Chem. Phys.* **52**, 791 (1970).
(53) J. Waser and V. Schomaker, *Rev. Mod. Phys.* **25**, 671 (1953).
(54) A. A. Khan, *Phys. Rev.* **136**, A1260 (1964).
P. G. Mikolaj and C. J. Pings, *J. Chem. Phys.* **46**, 1401 (1967).
(55) W. W. Wood, *J. Chem. Phys.* **52**, 729 (1970).
(56) B. J. Alder, W. G. Hoover and D. A. Young *J. Chem. Phys.* **49**, 3688 (1968).
(57) W. G. Hoover and F. H. Ree, *J. Chem. Phys.* **47**, 4873 (1967); *ibid.*, **49**, 3609 (1968).
(58) W. G. Hoover and B. J. Alder, *J. Chem. Phys.* **45**, 2361 (1966).
(59) M. Ross and B. J. Alder, *Phys. Rev. Letters*, **16**, 1077 (1966).
(60) J. G. Kirkwood, *J. Chem. Phys.* **18**, 380 (1950).
(61) J. A. Barker, R. A. Fisher and R. O. Watts, *Mol. Phys.* **21**, 657 (1971).
(62) B. M. Axilrod and E. Teller, *J. Chem. Phys.* **11**, 299 (1943).
B. M. Axilrod, *J. Chem. Phys.* **17**, 1349 (1949); *ibid.*, **19**, 71 (1951).

(63) J. H. Dymond and B. J. Alder, *Chem. Phys. Lett.* **2**, 54 (1968).
J. H. Dymond and B. J. Alder, *J. Chem. Phys.* **51**, 309 (1969).
(64) J. A. Barker and A. Pompe, *Aust. J. Chem.* **21**, 1683 (1968).
(65) J. A. Barker, D. Henderson and W. R. Smith, *Mol. Phys.* **17**, 579 (1969).
(66) M. V. Bobetic and J. A. Barker, *Phys. Rev.* B **2**, 4169 (1970).
(67) J. A. Barker, M. L. Klein and M. V. Bobetic, *Phys. Rev.* B **2**, 4176 (1970).
(68) J. A. Barker, M. V. Bobetic and A. Pompe, *Mol. Phys.* **20**, 347 (1971).
(69) G. C. Chell and I. J. Zucker, *J. Phys.* C **1**, 35 (1968).
(70) W. Goetze and H. Schmidt, *Z. Phys.* **192**, 409 (1966).
(71) A. Hüller, W. Goetze and H. Schmidt, *Z. Phys.* **231**, 173 (1970).
(72) P. J. Leonard, M.Sc. thesis, Melbourne (1968).
(73) H. W. Graben, *Phys. Rev. Lett.* **20**, 529 (1968).
H. W. Graben and R. Fowler, *Phys. Rev.* **177**, 288 (1969).
(74) D. Levesque and J. Vieillard-Baron, *Physica*, **44**, 345 (1969).
(75) J. A. Barker and A. Pompe, *Aust. J. Chem.* **21**, 1683 (1968).
(76) A. Rahman, *Phys. Rev.* **136**, A405 (1964).
(77) The functions G_s and G_d defined here are closely related but not identical with the van Hove functions (L. van Hove, *Phys. Rev.* **95**, 249 (1954)). For a discussion of the relationship see R. Aamodt, R. Case, M. Rosenbaum and P. F. Zweifel, *Phys. Rev.* **126**, 1165 (1962) and A. Rahman, *Phys. Rev.* **130**, 1334 (1963).
(78) G. H. Vineyard, *Phys. Rev.* **110**, 999 (1958).
(79) A. Rahman, *J. Chem. Phys.* **45**, 2585 (1966).
(80) N. Bogolyubov, *J. Phys.* (*USSR*), **11**, 23 (1947).
(81) T. D. Lee, K. Huang and C. N. Yang, *Phys. Rev.* **106**, 1135 (1967).
(82) K. A. Brueckner and K. Sawada, *Phys. Rev.* **106**, 1117 (1957).
(83) S. T. Beliaev, *Zh. Eksp. i Teor. Fiz.* **34**, 417 (1958); *ibid.*, **34**, 433 (1958). (English Translations: *Soviet Phys. JETP*, **7**, 289 (1958); *ibid.*, **7**, 299 (1958).)
(84) N. M. Hugenholtz and D. Pines, *Phys. Rev.* **116**, 489 (1959).
(85) N. F. Mott, *Phil. Mag.* **40**, 61 (1949).
(86) R. B. Dingle, *Phil. Mag.* **40**, 573 (1949).
(87) R. Jastrow, *Phys. Rev.* **98**, 1479 (1955).
(88) J. B. Aviles, Jr, *Ann. Phys.* (*NY*), **5**, 251 (1958).
(89) W. L. McMillan, *Phys. Rev.* **138**, A442 (1965).
(90) J. de Boer and A. Michels, *Physica*, **5**, 945 (1938); *ibid.*, **6**, 97 (1939).
(91) O. Penrose and L. Onsager, *Phys. Rev.* **104**, 576 (1956).
(92) L. D. Fosdick, *J. Math. Phys.* **3**, 1251 (1962).
(93) L. D. Fosdick and H. F. Jordan, *Phys. Rev.* **143**, 58 (1966).
(94) C. A. Croxton and R. P. Ferrier, *J. Phys.* C **4**, 1909 (1971).
(95) C. A. Croxton and R. P. Ferrier, *Phil. Mag.* **24**, 489 (1971).
(96) C. A. Croxton and R. P. Ferrier, *J. Phys.* C **4**, 2447 (1971).
(97) C. A. Croxton and R. P. Ferrier, *Phys. Letters*, **35**A, 330 (1971).
(98) S. J. Cocking, *Inelastic Scattering of Neutrons in Solids and Liquids*, vol. 1, p. 227, International Atomic Energy Agency, Vienna (1961).
(99) P. D. Randolph, *Phys. Rev.* **134**, A1238 (1964).
(100) See, for example, C. A. Coulson, *Valence*, Oxford University Press (1965); W. Kauzmann, *Quantum Chemistry*, Academic Press, New York (1964).

(101) R. M. Glaeser and C. A. Coulson, *Trans. Faraday Soc.* **61**, 389 (1965).

(102) C. A. Coulson and D. Eisenberg, *Proc. Roy. Soc.* **291** A, 445 and 454 (1966).

(103) N. H. Fletcher, *Phil. Mag.* **18**, 1287 (1968).

(104) J. del Bene and J. A. Pople, *J. Chem. Phys.* **52**, 4858 (1970).

(105) P. Barnes, private communication (March 1972).

(106) J. D. Bernal and R. H. Fowler, *J. Chem. Phys.* **1**, 151 (1933).

(107) In order to obtain a simple geometric series for the polarization effects, Coulson and Eisenberg(102) considered only the field statistically averaged over all possible orientations of the water molecules in the ice lattice. The ice I structure is an unusual one in that the oxygen positions are fairly localized, whereas the hydrogen positions (and therefore the orientations of the water molecules themselves) are continually changing in accordance with the Bernal–Fowler rules. Coulson and Eisenberg's statistical picture is therefore correct geometrically, but it is debatable whether the associated calculations are meaningful. Only by considering specific orientations(105) does one obtain non-zero x- and y-fields along the axes of the water molecule.

(108) F. C. Andrews and J. M. Benson *Phys. Letters*, **20**, 16, 1966.

(109) A. Rotenberg, *J. Chem. Phys.* **43**, 1198 (1965).

(110) P. A. Nelson, 'A Molecular Dynamical Study of a Square-Well Fluid', Princeton University thesis (1966); *Dissertation Abstr.* **27**, 2338-B (1967).

(111) B. J. Alder, *J. Chem. Phys.* **31**, 1666 (1959).

(112) H. T. Davis and K. D. Luks, *J. Chem. Phys.* **40**, 2669 (1964).
H. T. Davis, S. A. Rice and J. V. Sengers, *J. Chem. Phys.* **35**, 2210 (1961).

(113) L. van Hove, *Physica*, **15**, 951 (1949).

(114) K. S. C. Freeman and I. R. McDonald, *Mol. Phys.* **26**, 529 (1973).

(115) A. Rahman and F. H. Stillinger, *J. Chem. Phys.* **55**, 3336 (1971); *ibid.* **57**, 1281 (1972).

(116) J. A. Barker and R. O. Watts, *Mol. Phys.* **26**, 789 (1973).

Chapter 6

(1) J. G. Kirkwood, *J. Chem. Phys.* **14**, 180 (1946).

(2) S. A. Rice and J. G. Kirkwood, *J. Chem. Phys.* **31**, 901 (1959).

(3) S. A. Rice, *J. Chem. Phys.* **33**, 1376 (1960).

(4) J. G. Kirkwood, F. P. Buff and M. S. Green, *J. Chem. Phys.* **17**, 988 (1949).
R. W. Zwanzig, J. G. Kirkwood, K. Stripp and I. Oppenheim, *J. Chem. Phys.* **21**, 2050 (1953).

(5) P. Gray, *Mol. Phys.* **7**, 235 (1964).

(6) F. C. Collins and H. Raffel, *J. Chem. Phys.* **29**, 699 (1958).

(7) J. Ross, *J. Chem. Phys.* **24**, 375 (1956).

(8) E. Helfand, *Phys. Rev.* **119**, 1 (1960).

(9) S. A. Rice and A. R. Allnatt, *J. Chem. Phys.* **34**, 2144 (1961).
A. R. Allnatt and S. A. Rice, *J. Chem. Phys.* **34**, 2156 (1961).

(10) P. Gray, *Physics of Simple Liquids* (eds. H. Temperley, J. Rowlinson and G. Rushbrooke), pp. 509–62, North Holland Publ. Co. (1968).

(11) J. Naghizadeh and S. A. Rice, *J. Chem. Phys.* **36**, 2710 (1962).
(12) S. A. Rice and P. Gray *The Statistical Mechanics of Simple Liquids*, § 5.4 B, p. 320, John Wiley and Sons, New York (1965).
(13) S. A. Rice, *Liquids; Structure, Properties, Solid Interactions*, (ed. T. J. Hughel), p. 129, Elsevier, New York (1965).
(14) S. A. Rice and R. A. Harris, *J. Chem. Phys.* **33**, 1055 (1961).
(15) S. A. Rice, *Mol. Phys.* **4**, 305 (1961).
(16) R. Eisenschitz and A. Crowe, *Suppl. Nuovo Cim.* **9**, 262 (1958).
(17) P. Gray and S. A. Rice, *J. Chem. Phys.* **40**, 3669 (1964).
(18) H. Friedmann and W. A. Steele, *J. Chem. Phys.* **40**, 3669 (1964).
(19) G. Boato, G. Casanova and A. Levi, *J. Chem. Phys.* **40**, 2419 (1964). See also the general review of isotope effects in liquid argon by G. Casanova and A. Levi, *Physics of Simple Liquids* (eds. H. N. V. Temperley, J. Rowlinson and G. S. Rushbrooke), chapter 8, North-Holland Publ. Co. (1968).
(20) H. T. Davis and J. A. Palyvos, *J. Chem. Phys.* **46**, 4043 (1967).
J. A. Palyvos and H. T. Davis, *J. Phys. Chem.* **71**, 439 (1967).
(21) H. T. Davis, S. A. Rice and J. V. Sengers, *J. Chem. Phys.* **35**, 2210 (1961).
(22) H. C. Longuet-Higgins and J. P. Valleau, *Mol. Phys.* **1**, 284 (1956).
(23) F. C. Collins and H. Raffel, *J. Chem. Phys.* **29**, 699 (1958).
(24) A. Rahman, *Phys. Rev.* **136**, A405 (1964).
(25) D. C. Douglass, *J. Chem. Phys.* **35**, 81 (1961).
(26) H. T. Davis, S. A. Rice and L. Meyer, *J. Chem. Phys.* **37**, 2470 (1962); *ibid.*, **37**, 947 (1962).
(27) S. A. Rice, *Mol. Phys.* **4**, 305 (1961).
(28) A. Suddaby and P. Gray, *Proc. Phys. Soc. (London)*, **75**, 109 (1960).
(29) H. A. Kramers, *Physica*, **7**, 284 (1940).
(30) A. Suddaby and J. R. N. Miles, *Proc. Phys. Soc. (London)*, **77**, 1170 (1961).
(31) G. H. A. Cole, *An Introduction to the Statistical Theory of Classical Simple Dense Fluids*, p. 204, Pergamon (1967).
(32) J. G. Kirkwood, F. P. Buff and M. S. Green, *J. Chem. Phys.* **17**, 988 (1949).
(33) R. W. Zwanzig, J. G. Kirkwood, K. Stripp and I. Oppenheim, *J. Chem. Phys.* **21**, 2050 (1953).
(34) C. A. Croxton and R. P. Ferrier, *Phil. Mag.* **24**, 493 (1971).
C. A. Croxton and R. P. Ferrier, *Phys. Letters*, **36**A, 183 (1971).
C. A. Croxton and R. P. Ferrier, *J. Phys.* C **4**, 1921 (1971).
C. A. Croxton and R. P. Ferrier, *J. Phys.* C **4**, 2433 (1971).
(35) R. Kubo, *J. Phys. Soc. Japan*, **12**, 570 (1957).
R. Kubo, *Lectures in Theoretical Physics* (eds. W. E. Brittin and L. G. Dunham), vol. 1, Interscience, New York (1959).
Further developments of the Kubo theory have been given by:
H. Mori, *Phys. Rev.* **112**, 1829 (1958).
H. Mori, *J. Phys. Soc. Japan*, **11**, 1029 (1956).
H. S. Green, *J. Math. Phys.* **2**, 344 (1961).
M. S. Green, *Lectures in Theoretical Physics*, vol. 3, p. 195, Interscience, New York (1961).
H. L. Frisch, *J. Chem. Phys.* **36**, 510 (1962).
R. W. Zwanzig, *J. Chem. Phys.* **40**, 2527 (1964).

R. W. Zwanzig, *Ann. Rev. Phys. Chem.* **16**, 67 (1965).
H. C. Longuet-Higgins, *Mol. Phys.* **6**, 65 (1963).
W. A. Steele, *Transport Phenomena in Fluids* (ed. H. J. M. Hanley), chapter 8, Dekker, New York (1969).

(36) A. Suddaby and J. R. N. Miles, *Proc. Phys. Soc.* (*London*), **77**, 1170 (1961).

(37) R. Eisenschitz and A. Suddaby, *Proc. 2nd. Int. Cong. Rheol.*, Butterworth, London (1961).

(38) R. Eisenschitz, *Proc. Phys. Soc.* (*London*), A **62**, 41 (1949).
An initial analysis of transport phenomena in liquids on the basis of the Smoluchowski equation has been given by Eisenschitz (*ibid.*, **59**, 1030 (1957)). This treatment is, however, based on the cell model, and the influence of the rate of shear or gradient of temperature on the distribution was taken into account only at the surface but not in the interior of the cell.

(39) We have assumed an incompressible liquid for simplicity – such an assumption is by no means essential. For a compressible fluid div $\mathbf{c} \neq 0$, and we are led a consideration of the bulk viscosity, ζ. The Smoluchowski equation, (6.10.3), in this case becomes

$$\nabla^2 w(\boldsymbol{r}) - \langle \boldsymbol{F} \rangle . \operatorname{grad} w(\boldsymbol{r}) = -\left\{ \beta_{(2)} \frac{m}{kT} \mathbf{c} . \langle \boldsymbol{F} \rangle + \operatorname{div} \mathbf{c} \right\}.$$

(40) J. G. Kirkwood, F. P. Buff and M. S. Green, *J. Chem. Phys.* **17**, 988 (1949).
See also J. G. Kirkwood, *J. Chem. Phys.* **14**, 180 (1946).

(41) R. Eisenschitz and G. H. A. Cole, *Phil. Mag.* **45**, 394 (1954).
Experimental data contained in E. N. da C. Andrade and C. Dodd, *Proc. Roy. Soc.* A **204**, 449 (1951).

(42) J. Champion, *Proc. Phys. Soc.* (*London*), **72**, 711 (1958); *ibid.*, **75**, 421 and 799 (1960).

(43) R. Orton, M.Sc. thesis, University of London (1954).

(44) R. W. Zwanzig, J. G. Kirkwood, K. P. Stripp and I. Oppenheim, *J. Chem. Phys.* **21**, 2050 (1953).
R. W. Zwanzig, J. G. Kirkwood, I. Oppenheim and B. J. Alder, *J. Chem. Phys.* **22**, 783 (1954).

(45) R. Eisenschitz and M. J. Wilford, *Proc. Phys. Soc.* (*London*), **80**, 1078 (1962).

(46) B. A. Lowry, S. A. Rice and P. Gray, *J. Chem. Phys.* **40**, 3673 (1964).

(47) N. F. Zhdanova, *Zh. Eksperim. i. Teor. Fiz.* **31**, 724 (1956).
Soviet Phys. JETP (English Translation), **4**, 749 (1957).

(48) R. Scott, Ph.D. thesis, Queen Mary College, University of London; *Program. Kam. Onnes Conf. Ion Temp. Phys.* Leiden (1958).

(49) E. N. da C. Andrade, *Phil. Mag.* **17**, 497 and 698 (1934); *Nature* (*London*), **170**, 794 (1952); *Proc. Roy. Soc.* A **215**, 36 (1952).
Assuming a Brownian form for the friction constant in the Einstein relation, Frenkel (*Z. Phys.* **35**, 652 (1926)) relates the viscosity to the diffusion coefficient.
P. A. Egelstaff, (*An Introduction to the Liquid State*, Academic Press, New York (1967)) gives a useful discussion of the phenomenological relations between the viscosity and diffusion coefficients.

Earlier elementary treatments have been given by
G. Jager, *Handbuch der Physik*, vol. 9, p. 456, Springer, Berlin (1926).
R. O. Herzog and H. C. Kudar, *Z. Phys.* **80**, 217 (1933).
R. Furth, *Proc. Camb. Phil. Soc.* **37**, 281 (1941).
F. C. Auluck and D. S. Kothari, *Proc. Nat. Inst. Sci. India*, **10**, 397 (1944).
M. Born and H. S. Green, *Proc. Roy. Soc.* A **190**, 455 (1947).

(50) C. J. Pings, in *Physics of Simple Liquids* (eds. H. N. V. Temperley, J. S. Rowlinson and G. S. Rushbrooke), chapter 10, North-Holland Publ. Co. (1968), reports extensive investigations of the radial distribution function and direct correlation in liquid argon at various densities at a fixed temperature. It is found that the first peak in the RDF remains virtually independent of density (or pressure). On the other hand the distribution damps out at large radial distances, the location of the first peak remaining more or less fixed, although the co-ordination does of course drop with decreasing density.

(51) S. A. Rice and P. Gray, *The Statistical Mechanics of Simple Liquids*, §6.4E, John Wiley and Sons, New York (1965).

(52) A. Uhlir, *J. Chem. Phys.* **20**, 463 (1962).

(53) H. Ziebland and J. T. A. Burton, *Brit. J. Appl. Phys.* **9**, 52 (1958).

(54) F. Keyes, *Trans. ASME*, **77**, 1395 (1955).

(55) L. D. Ikenberry and S. A. Rice, *J. Chem. Phys.* **39**, 1561 (1963); see also Errata, *J. Chem. Phys.* **41**, 4002 (1964).

(56) S. A. Rice, *Liquids – Structure, Properties, Solid Interactions*, (ed. T. J. Hughel), p. 120, Elsevier, New York (1965).

(57) G. H. A. Cole, *An Introduction to the Statistical Theory of Classical Simple Dense Fluids*, p. 221, Pergamon (1967).

(58) M. S. Green, *J. Chem. Phys.* **20**, 1281 (1952); *ibid.*, **22**, 398 (1954).

(59) R. Kubo, *J. Phys. Soc. Japan*, **12**, 570, 1203 (1957).
R. Kubo, *Lectures in Theoretical Physics*, vol. 1, p. 120, Boulder (1958).

(60) K. G. Denbigh, *The Thermodynamics of the Steady State*, Methuen, London (1951).
D. D. Fitts, *Non-equilibrium Thermodynamics*, McGraw-Hill, New York (1962).
H. J. M. Hanley, 'Non-equilibrium Thermodynamics' in *Transport Phenomena in Fluids*, pp. 37–48, Dekker, New York (1969).

(61) R. Kubo, *J. Phys. Soc. Japan*, **12**, 570 and 1203 (1957).
See also the general review of the theory of irreversible processes: G. V. Chester, *Rep. Prog. Phys.* **26**, 411 (1963).
R. Kubo, *Lectures in Theoretical Physics* (eds. W. E. Brittin and L. S. Dunham, vol. 1, p. 120, Interscience, New York (1959).
H. Mori, *J. Phys. Soc. Japan*, **11**, 1029 (1956).
R. Kubo, M. Yokota and S. Nakajima, *J. Phys. Soc. Japan*, **12**, 1203 (1957).
R. W. Zwanzig, *Phys. Rev.* **124**, 983 (1961).
R. W. Zwanzig, *Lectures in Theoretical Physics* (eds. W. E. Brittin, W. B. Downs and J. Downs), vol. 3, p. 106, Interscience, New York (1962).

(62) L. Onsager, *Phys. Rev.* **37**, 405 (1931); *ibid.*, **38**, 2265 (1931).

(63) M. Fixman, *J. Chem. Phys.* **26**, 1421 (1957).
H. C. Longuet-Higgins, *Mol. Phys.* **6**, 65 (1963).
(64) This development follows that of S. A. Rice and P. Gray, *The Statistical Mechanics of Simple Liquids*, §7.13 B, John Wiley and Sons, New York (1965).
(65) N. Hashitsume, *Prog. Theoret. Phys.* (*Kyoto*), **8**, 461 (1952); *ibid.*, **15**, 369 (1956).
(66) E. W. Montroll, in *Lectures in Theoretical Physics* (eds. W. E. Britten, W. B. Downs and J. Downs), vol. 3, p. 221, Wiley (Interscience), New York (1961).
(67) G. V. Chester, *Rep. Prog. Phys.* **26**, 411 (1963). See also S. A. Rice and P. Gray, *The Statistical Mechanics of Simple Liquids*, §7.3 C, John Wiley and Sons, New York (1965).
(68) S. R. de Groot and P. Mazur, *Non-Equilibrium Thermodynamics*, North-Holland Publ. Co., Amsterdam (1962).
(69) H. Nakano, *Prog. Theory. Phys. Japan*, **22**, 453 (1959); *ibid.*, **23**, 180, 182, 526 and 527 (1960).
(70) M. Kohler, *Z. Phys.* **124**, 772 (1948).
(71) P. Mazur, *Statistical Mechanics of Equilibrium and Non-Equilibrium* (ed. J. Meixner, North-Holland, Amsterdam, p. 69 (1965)).
(72) E. Helfand, *Phys. Rev.* **119**, 1 (1960).
(73) J. H. Irving and J. G. Kirkwood, *J. Chem. Phys.* **22**, 817 (1950).
(74) P. Schofield, *Phys. Rev. Letters*, **4**, 239 (1960).
(75) R. Kubo, *Lectures in Theoretical Physics* (eds. W. E. Brittin and L. S. Dunham), vol. 1, Interscience, New York (1959).
(76) P. Schofield, *Proc. Phys. Soc.* **88**, 149 (1966).
(77) G. V. Chester and A. Thellung, *Proc. Phys. Soc.* **73**, 765 (1959).
L. van Hove, *The Theory of Neutral and Ionized Gases*, p. 149, Wiley, New York (1960).
(78) P. A. Egelstaff, *Rep. Prog. Phys.* **89**, 747 (1966).
(79) V. F. Sears, *Proc. Phys. Soc.* **86**, 953 (1965).
(80) P. Gray, *Mol. Phys.* **7**, 235 (1964).
(81) A. Rahman, *Phys. Rev.* **136**, A405 (1964).
(82) R. Kubo, *Statistical Mechanics of Equilibrium and Non-Equilibrium* (ed. J. Meixner), p. 81, North-Holland Publ. Co., Amsterdam (1965). See also ref. (87).
(83) A summary of the neutron scattering experiments of S. J. Cocking and P. D. Randolph will be found in Egelstaff's review[(78)].
(84) See a general discussion in P. A. Egelstaff, *An Introduction to the Liquid State*, p. 141, Academic Press, London (1967).
(85) M. C. Wang and G. E. Uhlenbeck, *Rev. Mod. Phys.* **17**, 323 (1945).
(86) J. L. Doob, *Ann. Math.* **43**, 351 (1942).
(87) R. Kubo, *Tokyo Summer Lectures in Theoretical Physics*, part 1: *Many Body Theory* (ed. R. Kubo), Benjamin, New York (1966).
(88) Y. Nakahara and H. Takahashi, *Proc. Phys. Soc.* (*London*), **89**, 747 (1966).
(89) P. S. Damle, A. Sjölander and K. S. Singwi, *Phys. Rev.* **165**, 277 (1968).
(90) R. Kubo, *Many Body Theory* (ed. R. Kubo), part 1, p. 1, Syokabo Publishing Co., Tokyo (1966).

(91) See, e.g.,
P. A. Egelstaff, *Rept. Prog. Phys.* **29**, 333 (1966).
K. Skold and K. E. Larsson, *Phys. Rev.* **161**, 102 (1967).
K. E. Larsson, *Thermal Neutron Scattering* (ed. P. A. Egelstaff), p. 347, Academic Press Inc., New York (1965).
(92) B. R. A. Nijboer and A. Rahman, *Physica*, **32**, 415 (1966).
(93) K. S. Singwi and M. P. Tosi, *Phys. Rev.* **157**, 153 (1966).
(94) J. A. Tjon, *Phys. Rev.* **143**, 259 (1966).
(95) R. Zwanzig, *Lectures in Theoretical Physics* (ed. W. E. Brittin), p. 106, Interscience, New York (1961).
(96) H. Mori, *Prog. Theor. Phys. (Kyoto)*, **33**, 423 (1965).
(97) B. J. Berne, J.-P. Boon and S. A. Rice, *J. Chem. Phys.* **45**, 1086 (1966).
(98) See K. S. Singwi and A. Sjölander, *Phys. Rev.* **147**, 152 (1968).
(99) Results at 85.5 °K: *J. Chem. Phys.* **45**, 2585 (1966).
Results at 94.4 °K: *Phys. Rev.* **136**, A405 (1964).
(100) See, for example:
J.-P. Boon and S. A. Rice, *J. Chem. Phys.* **47**, 2480 (1968).
B. J. Berne, J.-P. Boon and S. A. Rice, *J. Chem. Phys.* **45**, 1086 (1966).
K. S. Singwi and M. P. Tosi, *Phys. Rev.* **157**, 153 (1966).
K. S. Singwi and A. Sjölander, *Phys. Rev.* **147**, 152 (1968).
(101) T. Gaskell and N. H. March, *Phys. Letters*, **33**A, 460 (1970).
(102) M. H. Ernst, E. H. Hauge and J. M. J. Van Leeuwen, *Phys. Rev. Letters*, **25**, 1254 (1970).
(103) P. A. Egelstaff, *An Introduction to the Liquid State*, p. 142, Academic Press, New York (1967).
See also S. J. Cocking, *Inelastic Scattering of Neutrons in Solids and Liquids*, vol. 1, p. 227, International Atomic Energy Agency, Vienna (1961).
(104) P. D. Randolph, *Phys. Rev.* **134**, 1483 (1964).
(105) T. Gaskell, *J. Phys.* C **4**, 1466 (1971).
T. Gaskell and M. I. Barker, *J. Phys.* C **5**, 353 (1972).
(106) A. Rahman, *J. Chem. Phys.* **45**, 2585 (1966).
(107) I. Prigogine, *Non-Equilibrium Statistical Mechanics*, Interscience, New York (1962).
(108) R. C. Tolman, *Principles of Statistical Mechanics*, Oxford University Press (1938).
(109) J. G. Kirkwood, *J. Chem. Phys.* **14**, 180 (1946).
(110) R. Balescu, *Statistical Mechanics of Charged Particles*, Wiley, New York (1964).
P. Résibois, *Kinetic Theory of Classical Gases*, Gordon and Breach, New York (1967).
(111) J. Mathews and R. L. Walker, *Mathematical Methods of Physics*, p. 108, Benjamin, New York (1970).
M. Abramowitz and I. A. Segun (eds.), *Handbook of Mathematical Functions*, p. 1020, Dover, New York (1968).
Spiegel, M. R. *Laplace Transforms*, Schaum, New York (1965).
(112) P. Résibois, *Physica*, **27**, 541 (1961).

(113) I. Prigogine, *Non-Equilibrium Statistical Mechanics*, chapters 8 and 11, Interscience, New York (1962).
S. A. Rice and P. Gray, *The Statistical Mechanics of Simple Liquids*, §7.2C, John Wiley and Sons, New York (1965).
G. H. A. Cole, *Rep. Prog. Phys.* **33**, 737 (1970).

(114) P. Résibois, *Diagrammatic Analysis of Classical Gases* (ed. E. Meeron), Gordon and Breach, New York (1966).
See also G. R. Dowling and H. T. Davis, *Can. J. Phys.* **50**, 317 (1972), who have recently made a numerical study of non-equilibrium diagram theory.

(115) L. van Hove, *Physica*, **23**, 441 (1957).

(116) I. Prigogine and P. Résibois, *Physica*, **27**, 629 (1961).

(117) R. Brout and I. Prigogine, *Physica*, **22**, 621 (1956).

(118) A discussion of the reversibility paradox has been given for electrolytes by H. T. Davis and P. Résibois, *J. Chem. Phys.* **40**, 3276 (1964); for dilute gases by P. Résibois, *J. Math. Phys.* **4**, 166 (1963) and *Phys. Letters*, **9**, 139 (1964); and approximate solutions for dense gases and liquids by P. Allen and G. H. A. Cole, *Mol. Phys.* **14**, 413 (1968) and **15**, 549, 557 (1968). I. Prigogine, G. Nicolis and J. Misguich, *J. Chem. Phys.* **43**, 4516 (1965); J. Misguich, *J. Phys. Paris*, **30**, 221 (1969); J. Misguich, G. Nicolis, J. A. Palyvos and H. T. Davis, *J. Chem. Phys.* **48**, 951 (1968); J. A. Palyvos, H. T. Davis, J. Misguich and G. Nicolis, *J. Chem. Phys.* **49**, 4088 (1968).

(119) R. Zwanzig, *Phys. Rev.* **124**, 983 (1961).

(120) P. M. Allen and G. H. A. Cole, *Mol. Phys.* **15**, 557 (1968).
See also, P. M. Allen, *Mol. Phys.* **18**, 349 (1970).

(121) P. Résibois, *J. Chem. Phys.* **41**, 2979 (1964).

(122) P. Gray, *Mol. Phys.* **21**, 675 (1971).

(123) J. L. Lebowitz and P. Résibois, *Phys. Rev.* **139**, A1101 (1965).

(124) J. L. Lebowitz, J. K. Percus and J. Sykes, *Phys. Rev.* **188**, 487 (1969).

(125) A. Muriel and M. Dresden, *Physica*, **43**, 424 and 449 (1969).

(126) P. A. Nelson, 'A Molecular Dynamical Study of a Square-Well Fluid', Princeton University thesis (1966); *Dissertation Abstr.* **27**, 2338-B (1967).

(127) D. J. Isbister and D. A. McQuarrie, *J. Chem. Phys.* **56**, 736 (1972).
A. Suddaby and J. Hales, *Proc. Phys. Soc.* **89**, 1003 (1966).

(128) See a short discussion of irreversibility regarded as a symmetry-breaking process, I. Prigogine, *Nature*, **246**, 67 (1973).

INDEX

Page references in italics are to tables or figures in which experimental data or the results of theoretical calculations are presented. Chemical elements are listed under their conventional symbols, only if they are the subject of special comment in the text.